Logic, Language, and the World

Volume 2

Time and Space in Formal Logic

Richard L. Epstein

Advanced Reasoning Forum

COPYRIGHT © 2022 by Richard L. Epstein.

ALL RIGHTS RESERVED. No part of this work covered by the copyright hereon may be reproduced or used in any form or by any means—graphic, electronic, or mechanical, including photocopying, recording, taping, Web distribution, information storage and retrieval systems, or in any other manner—without the written permission of the author.

The moral rights of the author have been asserted.

Names, characters, and incidents relating to any of the characters in this text are used fictitiously, and any resemblance to actual persons, living or dead, is entirely coincidental. *Honi soit qui mal y pense.*

For more information visit our website:
    www.AdvancedReasoningForum.org
Or contact us:
    Advanced Reasoning Forum
    P. O. Box 635
    Socorro, NM 87801 USA
    rle@AdvancedReasoningForum.org

ISBN paperback 978-1-938421-68-6

ISBN e-book 978-1-938421-69-3

CONTENTS

PREFACE

LOGIC and LANGUAGE
    1 Propositions and Inferences 1
    2 Logic and Language 5

RELATIVE TIMES
TEMPORAL PROPOSITIONAL CONNECTIVES
    3 Times and Propositions 10
    4 Temporal Connectives 14
    5 Time as an Ordering 17
    6 Classical Propositional Logic with Temporal Connectives 20
    7 Reduced Models 24
    8 An Axiom System for Classical Propositional Logic with
        Temporal Connectives in Linear Time 26
    9 Examples of Formalizing 35
   10 Other Orderings of Time 46
   11 Past, Present, and Future 51
        Summary 54

QUANTIFYING over RELATIVE TIMES
   12 Quantifying over Relative Times via Indices 56
   13 Examples of Formalizing 63
        Summary 66

TIMES and LOCATIONS as THINGS
TIME in PREDICATE LOGIC
   14 The Timelessness of Classical Predicate Logic 70
   15 Time and Reference in Predicate Logic 73
   16 Times as Things 78
   17 Basic Assumptions about Times 81
   18 Times and Predications 85
   19 Identity and the Equality Predicate 89
   20 When Things Exist 93
   21 The Logic **QT** for Quantifying over Times in Classical Predicate Logic 98
   22 Examples of Formalizing: The Nature of Time 109
   23 The Internal Structure of Atomic Predicates 114

FORMALIZING with QUANTIFYING over TIMES
   24 Quasi-Linear Time 122
   25 Formalizing English Tenses 126
   26 Examples of Formalizing: Context of Utterance and Tenses 130

27	Examples of Formalizing: Existence	133
28	Examples of Formalizing: Attributes	138
29	Relativizing Quantifiers	142
30	Examples of Formalizing: Meaning Axioms	146
31	Examples of Formalizing: How Many Times	147
32	Examples of Formalizing: The Internal Structure of Atomic Predicates	150
33	Measuring Time?	156
34	Other Tenses	160
35	The Habitual	163
	Summary of Quantifying over Times	165

SPACE in PREDICATE LOGIC

36	Things in Time and Space	168
37	Locations	171
38	References to Things, Times, and Locations Are Independent	174
39	Assumptions about Locations	176
40	True in a Location	179
41	Locational Predicates	183
42	Existence in Space and Time	186
43	Where Things Are	188
44	A Formal Language for Reasoning about Things in Time and Space	190
45	Semantics for a Formal Theory for Reasoning about Things in Time and Space	198
46	An Axiom System for **QTS**	207
47	Examples of Formalizing: The Nature of Space	212
48	Informal Examples	216
49	Examples of Formalizing	218
50	Location-Orienting Predicates	221
51	Parts of Physical Things	227

METAPHYSICAL BASES of LOGICS of TIME and SPACE   a summary and introduction   231

APPENDICES

A. Tenses as Propositional Operators	242
B. The Tapestry of Time	247
C. Events	249
D. Intentions	250
E. Are Things Needed to Pick Out Locations?	266
F. Descriptive Names in the Logic of Quantifying over Times	267

Bibliography	270
Index of Examples	275
Index of Notation	280
Index	282

Logic, Language, and the World

Volume 0   An Introduction To Formal Logic

The Basics of Logic
Reasoning with Compound Propositions
Classical Propositional Logic: Form
Classical Propositional Logic: Meaning
Using Classical Propositional Logic
Proofs
Reasoning about Things
The Grammar of Things
A Formal Language for Predicate Logic
A Predicate Applies to an Object or Objects
Models for Classical Predicate Logic
Substitution of Variables and Distribution of Quantifiers
An Axiom System for Classical Predicate Logic
Formalizing in Classical Predicate Logic
Identity
Formalizing with the Equality Predicate
Possibilities
Appendices

Volume 1 The Internal Structure of Predicates and Names

Classical Predicate Logic
The Internal Structure of Predicates
    Restrictors of Unary Predicates
    Other Modifiers
    Internal Conjunctions and Disjunctions
    Relations
    A Formal Theory
    Predicates Used as Restrictors
The Internal Structure of Names
    Functions and Descriptive Functions
    Non-Referring Simple Names
Appendices

Volume 3 Reasoning about the World as the Flow of All

In preparation; to appear 2023.

Preface

What is time? What is space? I will not try to answer these questions, if they even make sense. What we can do is try to understand what role time and space play in our talk and reasoning, hoping to come to some better understanding of what it is we believe and perhaps a better understanding of the world.

In this book I present two very different approaches. Both start with the basics of propositions and inferences set out in Chapter 1 and the relation of formal logic to ordinary language set out in Chapter 2. Each part can then be read independently.

Relative Times
We say "Spot barked before Dick yelled"; we say "Tom met Suzy after Tom broke his foot". We talk of before and after. But before and after what? We can and do pick out times with true descriptions: "Spot barked", "Dick yelled", "Tom met Suzy", "Tom broke his foot". We order them as describing before and after. This is all we need to take account of time in our reasoning: a minimal metaphysics of before and after, codified with temporal propositional connectives.

Times and Locations as Things
A different approach assumes instead that we can talk about times and locations as if they were things, quantifying over them in an extension of classical predicate logic. This is closer to what has been done by others in trying to include talk of time and space in the scope of formal logic and mathematics. But paying attention to the metaphysical assumptions on which to proceed, there are many hard questions to investigate, even before setting up formal systems, and then many more arise in formalizing ordinary language propositions and inferences.

Throughout, I have tried to be clear about what metaphysical assumptions are assumed in establishing the formal systems. Those assumptions and the mechanics of the formal systems are tested by many examples of formalizing ordinary language propositions and inferences. To facilitate comparisons of those there is an Index of Examples. The last chapter, "Metaphysical Bases of Logics of Time and Space" reviews all of the book, setting out what metaphysical assumptions were introduced, why they were needed, and how they were used. That chapter could be read as an extended introduction.

There is so much more we could do, so much begun but not done. And surely there will be more questions, doubts, and interests that will arise. I hope that you can go farther.

* * * * * * * * *

This is the third volume in the series *Logic, Language, and the World*. Volume 0, *An Introduction to Formal Logic*, gives the background for all the work here and can be referred to as needed. The material in Volume 1, *The Internal Structure of Predicates and Names*, is discussed in separate sections and used in examples of formalizing, but those can be skipped without loss of continuity.

Works cited in the text not attributed to anyone are by me.

Supplements to this text which examine extensions of the logics here can be found at <www.AdvancedReasoningForum.org> .

Acknowledgements
The work on temporal propositional connectives is based on earlier work I did with Esperanza Buitrago-Díaz which appeared as "A Propositional Logic of Temporal Connectives" in *Logic and Logical Philosophy*, vol. 24, no. 2, pp. 155–200, 2014.

Chapter 14 on the timelessness of classical predicate logic first appeared as an essay in my book *Reasoning and Formal Logic*.

Appendix A, "Tenses as Propositional Operators", comes from my paper "Reflections on Temporal and Modal Logic" that appeared in *Logic and Logical Philosophy*, vol. 24, no. 1, pp. 111–139, 2014, and which was reprinted in my *Reasoning and Formal Logic*, pp. 98–126.

Appendix D, "Intentions", was developed in concert with members of the Advanced Reasoning Forum.

I have been helped by many people who talked with me about this project or commented on parts of earlier drafts. I am indebted to Melissa Axelrod, Walter Carnielli, William S. Robinson, Fred Kroon, Achille Varzi, Sherman Wilcox, Charlie Silver, Ian Grant, Peter Adams, and Paulo Santos. This last year Eduardo Ribeiro and Juan Francisco Rizzo worked through a draft with me, while Chad Hansen and Chris Sinha led me to relate this work better to reasoning in Chinese and other languages. Even when I disagreed or poorly understood these people, their comments forced me to clarify what I had written. To these and any others I may have forgotten, thank you.

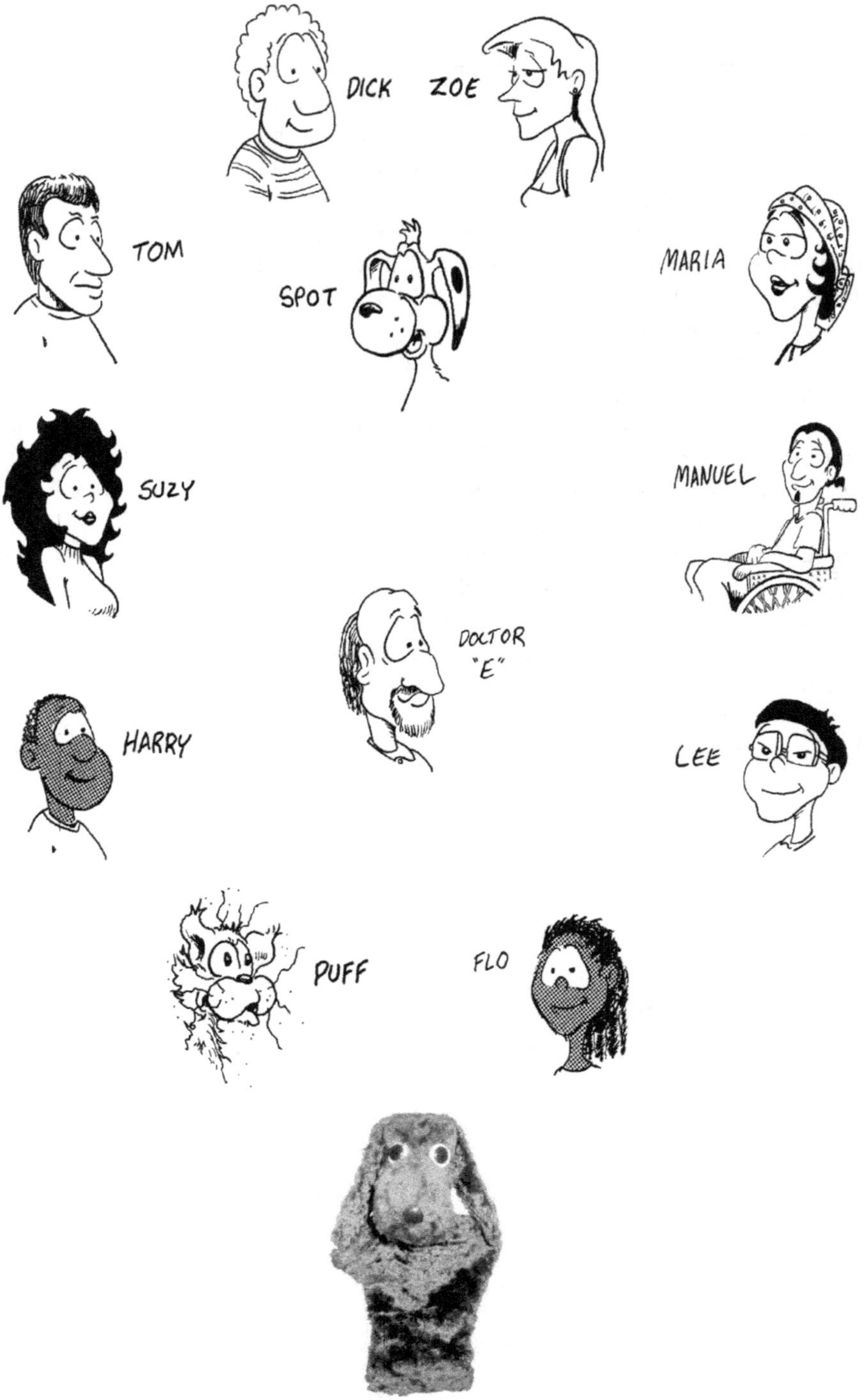
DICK
ZOE
TOM
SPOT
MARIA
SUZY
MANUEL
DOCTOR "E"
HARRY
LEE
PUFF
FLO
Ralph

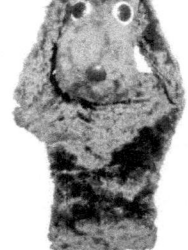

LOGIC and LANGUAGE

1   Propositions and Inferences

Propositions
We would like to know how to reason well. We would like rules to guide us in finding truths and when that is not straightforward to determine what are the consequences of what we believe. To do so, we must first agree about what it is that is true or false.

Proposition   A *proposition* is a written or uttered part of speech used in such a way that it is true or false, but not both.

By "uttered" I include silent uttering to oneself, what we might call thinking of the speech. The propositions we'll focus on in this text are sentences.

Some say that propositions are abstract objects or thoughts shared by all people and that what I have defined are physical linguistic representatives of propositions. But those who hold such views reason using linguistic propositions, which can serve as a common basis on which to begin our work.

Others say that there are not two truth-values but many or that some propositions have no truth-value. But all who reason and all who devise formal models for reasoning divide propositions into those we can proceed on to arrive at what we are justified in believing or acting on and those we cannot. We are right to label any such dichotomy a true-false division, as I explain in "Truth and Reasoning".

A proposition is meant as a description of the world, or at least some part of it. For a proposition to be true it must in some way be an accurate description of the world, relative to whatever you reckon the world is.[1]

Propositions are types
When I say that "dog" is a word, you understand that I do not mean just that particular inscription but any inscription like it or any utterance that we deem the same, for example "$\mathcal{DOG}$", or "DoG" or "**dog**". To treat a word or sentence as a *type* in this way is to draw equivalences: the inscriptions and utterances are not the same but only the same for the uses we plan to make of them. We say that they are *equiform*. We identify equiform words for our reasoning. We might or might not say that the English "dog" and the French "chien" are the same word. Similarly, we identify equiform propositions, though for those it is more difficult to be clear about what is or is not important for reasoning. We can say that our

---

[1] Or an accurate description of how we do or should use words, as I discuss in *Reasoning about the World as the Flow of All*. The coherence theory of truth invokes consistency rather than description as the basis of truth, using only a syntactic analysis. But unexamined metaphysical assumptions about how language connects to the world are needed to justify whatever notion of consistency is used.

identifications are provisional, allowing that we may later uncover differences that are important for reasoning. Explicitly, we make the following assumption.

Words and propositions are types Throughout any particular discussion, equiform words can be treated as the same for our reasoning. We identify them and treat them as if they were the same word. Equiform propositions, too, will be identified and treated as the same for our reasoning. Briefly, *a word or a proposition is a type*.[2]

Inferences
What does it mean to say that a proposition follows from one or more propositions?

Inference An *inference* is a collection of two or more propositions—one of which is designated the *conclusion* and the others the *premises*—that is intended by the person who sets it out either to show that the conclusion follows from the premises or to investigate whether that is the case.[3]

When does an inference show that the conclusion follows from the premises? That depends in part on what kind of reasoning we are analyzing. Different conditions apply depending on whether we are concerned with evaluating arguments, explanations, reasoning in mathematics, reasoning about cause and effect, reasoning with prescriptive claims, or conditionals.[4] However, for all kinds of reasoning a fundamental criterion for whether the conclusion follows is that the inference is valid or strong.

Valid and strong inferences An inference is *valid* means that there is no way the premises could be true and conclusion false in the same way and at the same time.
    An inference is *strong* means that there is a way for the premises to be true and conclusion false, but any such way is unlikely.

For example, the following is valid:

---

[2] Some say that types are abstract objects in accord with their belief that propositions are abstract. In that case the assumption that words and propositions are types concerns which inscriptions and utterances (which are what we use in reasoning together) represent or express or point to the same abstract thing.

[3] Note that intent is crucial in determining whether what has been uttered is an inference, as can be seen in hundreds of examples in *Critical Thinking*. The examples of inferences in this book should be understood as prefaced by "imagine that someone has put forward the following inference".
    Some say that inferences, too, are abstract things. But all who reason use linguistic inferences, and it is those we can study, whether or not we consider them to be representatives of abstract entities.

[4] See my *Reasoning in Science and Mathematics*, *The Fundamentals of Argument Analysis*, *Prescriptive Reasoning*, and *Cause and Effect, Conditionals, Explanations*.

(1) Ralph is a dog.
    All dogs bark.
    Therefore, Ralph barks.

I can't prove that to you. At best I can rephrase it in other words. If you understand English, it's clear that it's valid. That's not to say it's a good inference. More is required for an inference to be good, but what that more is depends on what kind of reasoning we are analyzing.

Similar inferences are also valid:

Dick is a student.
All students study hard.
Therefore, Dick studies hard.

Suzy is a cheerleader.
All cheerleaders have a liver.
Therefore, Suzy has a liver.

It would facilitate our reasoning if we could clarify in what way these inferences are similar, for it seems that we don't need to know anything about cheerleaders, or students, or dogs to see that they are valid. Somehow, it is the forms of the propositions in these inferences that matter.

Formal logic *Formal logic* is the analysis of inferences for validity in terms of the structure of the propositions appearing in the inference as well as the analysis of propositions for truth in terms of their structure.

Formal logic does not in itself give us rules for how to reason well. It is a tool we can use in the analysis of inferences and propositions in our reasoning.

Propositions and schemes of propositions
Consider another inference that looks a lot like (1):

(2) Ralph barks.
    All dogs bark.
    Therefore, Ralph is a dog.

Is it valid? We consider ways the premises could be true and conclusion false. One way is that Ralph could be a fox or seal that barks and that all dogs bark; then the premises would be true and conclusion false. So the inference is not valid.

If (2) is an inference, then the premises and conclusion are propositions, true or false. But in analyzing whether the conclusion follows from the premises, we take each of those sentences to be a scheme of propositions, needing a specification of how the world is in order for it to be true or false.

In formal logic, we establish rules for what sentences we'll consider to be propositions in terms of form relative to some assumptions about the nature of the world. Then given a collection of propositions, a way the world could be is a description in which each of those propositions is true or is false, what we call *a model*. We use schemes of propositions to establish relations among propositions in terms of not whether they are true or false but how they could be true or false.

Given a sentence such as "Ralph barks", for us to agree that it has a truth-value we need to know what thing "Ralph" is meant to name and what "barks" means. Often it's obvious, and we don't bother to make the agreements explicit. Sometimes it's not. But agreements there are that turn the sentence into a proposition; those are what determine that the sentence is used in such a way that it is true or false.

Thus "Spot is a dog" is a proposition relative to the choice of what "Spot" refers to and what "is a dog" means. If the choice is our usual understanding of "is a dog" and "Spot" refers to my burro, then it's false. If by "Spot" we mean Dick's dog, and "is a dog" means it's a purebred show dog, then it's false. If by "Spot" we mean Dick's dog and have our usual understanding of "is a dog", then it's true. These are three different propositions. They are not one proposition that has different truth-values, but a scheme that is made into three different propositions by choosing how we understand the parts of the scheme, that is, relative to a way the world could be. We reason with propositions. We reason about propositions using schemes of propositions.

Schemes of propositions A (type of) utterance is a *scheme of propositions* means that for each choice relative to a specified range of choices of semantic values for its parts it is true or false, that is, it is a proposition.

Classical logic
In Volume 0 (*An Introduction to Formal Logic*), we saw how to formulate and use classical propositional logic and classical predicate logic. In Volume 1 (*The Internal Structure of Predicates and Names*) we saw how to extend the scope of that work by looking at the internal structure of what we had taken to be atomic in classical predicate logic. Now consider:

> I never hit an old man.
> My brother is an old man.
> Therefore, I never hit my brother.

Even in that richer system, the formalization of this is valid. But it is clearly invalid: I hit my brother more than a few times when he and I were children. To evaluate this inference, we need to take account of time in our reasoning. How shall we proceed?

2 Logic and Language

Developing logic from the point of view of formalizing reasoning in ordinary language is essential for us to see why we should accept its norms and how we can apply it.

In *An Introduction to Formal Logic* (Volume 0) and in *The Internal Structure of Predicates and Names* (Volume 1), I have motivated formal logics based on formalizing reasoning using hundreds of examples from English. This was suitable because the metaphysical assumptions used in establishing classical propositional logic and classical predicate logic are compatible with the usually unstated metaphysical assumptions in speaking English. In particular, classical propositional logic is based on the assumptions of the previous chapter; for classical predicate logic we further assume that the world is made up at least in part of individual things.

In order to take account of time and space in our formalizations of ordinary reasoning using the tools of formal logic, we will have to assume more. But time is treated differently in different languages. In English we treat time in two ways. One is with the use of tenses along with temporal connectives such as "before" to indicate time relationships. The other is by talking of times, using names such as "May 31, 2009" and informal quantifiers like "sometimes" and "always". Modalities, such as duration, are treated mainly through the use of specific words. In contrast, in American Sign Language modalities such as duration are treated as forms of verbs, and in Chinese time indications are established with specific words.

How, then, shall we proceed? Shall we continue to formalize reasoning in English, knowing that to do so will be to build into our formal models particular conceptions of time? Or shall we look for what is common to the treatment of time in all languages, some more universal assumptions?

If we do not base our formal systems on abstracting from reasoning in some ordinary language we will not be able to test our work with examples. We can do no experiments to see if what we have done is apt even for the metaphysics we have assumed. We must begin somewhere. But how we proceed will depend on how we understand our project in terms of the relation of formal logic to reasoning in ordinary language. Consider what Benjamin Lee Whorf says in "Grammatical Categories":

> English adjectives form two main cryptotypes with subclasses. A group referring to "inherent" qualities—including color, material, physical state (solid, liquid, porous, hard, etc.), provenience, breed, nationality, function, use—has the reactance of being placed nearer the noun than the other group, which we may call one of noninherent qualities, though it is rather the residuum outside the first group—including adjectives of size, shape, position, evaluation (ethical, esthetic, or economic). These come before the inherent group, e.g. "large red house" (not

"red large house"), "steep rocky hill, nice smooth floor." The order may be reversed to make a balanced contrast, but only by changing the normal stress pattern, and the form is at once sensed as being reversed and peculiar. p. 179

Whorf is describing the grammar of English: what is correct and incorrect to say in our language. From that perspective, "Anubis is a wild big dog" is just a bad or odd way to say "Anubis is a big wild dog". But those two sentences can have distinct truth-values and distinct consequences: wild dogs are generally small so that a big wild dog may not be a big dog. Since our formalizations must respect consequences, we formalize the two differently (Chapter 14 of Volume 1). To follow grammar strictly as a guide would leave us unable to say much less to note the logical role of sentences that use an "odd" order of adjectives. Grammar can suggest but cannot be a strict guide to our logical investigations. As Whorf says in "Languages and Logics":

> I can sympathize with those who say, "Put it into plain, simple English," especially when they protest against the empty formalism of loading discourse with pseudolearned words. But to restrict thinking to the patterns merely of English, and especially to those patterns which represent the acme of plainness in English, is to lose a power of thought which, once lost, can never be regained. It is the "plainest" English which contains the greatest number of unconscious assumptions about nature. . . . Western culture has made, through language, a provisional analysis of reality and, without correctives, holds resolutely to that analysis as final. The only correctives lie in all those other tongues which by aeons of independent evolution have arrived at different, but equally logical, provisional analyses. pp. 235–236

I will start with English as the source of the reasoning we'll formalize. But at each point I'll try to make clear what metaphysical assumptions we are building into our formal system so we can go back and adjust or abandon those as needed in developing models of how to reason well for other languages or simply to investigate other assumptions. We shall see as we begin with English that many issues of how to deal with time and later space will lead us to consider ways of conceiving of experience and reasoning that are far from what we usually accept as speakers of English. Being aware of those assumptions, we can begin to have a more general view of formalizing reasoning about time and space, and a more ample understanding of different ways of encountering the world.

RELATIVE TIMES

TEMPORAL PROPOSITIONAL CONNECTIVES

3   Times and Propositions

We cannot point to a time as we point to a cow or a table. We cannot point to a time as some say we point through our intellects to a number. We pick out a part of time we are paying attention to with a description: "the time when Spot barked", "the time when George Washington was inaugurated as the first President of the U.S.A.", "the time when dinosaurs lived". Perhaps you or I can pick out times by some sense, some feeling or intellectual apprehension. But reasoning together, we have only linguistic descriptions. And those descriptions are propositions. Loosely speaking, we might say that we pick out a time by what happened at it. But that's too loose. It assumes that the part of time we are discussing existed as an individual unit before we made the description, which is no more reasonable than to say that the water that fills the glass I submerged in the bathtub was a separate and distinct thing before I put the glass in the tub. It is also too loose in assuming that "what happened" is not just a description but some doing or configuration. Yet that doing or configuration would have to be all that happened at that time. So June 6, 1970 is the time picked out by all that "occurred" in the universe at that moment. But that is impossible for us to use in our talk and reasoning. We have no idea of "all that happened". We don't even have a single clock we can use to pick out "what happened" at distant places, physicists tell us. Simply, we pick out or describe times using linguistic propositions. "Spot barked" picks out a part of time which, if we know enough more about the world, like who Spot is, is sufficiently clear that we can treat it as an individual unit of time.

Often we need and can use several temporal propositions together to pick out a time: "Spot barked", "Dick yelled", "Suzy closed the gate". For example, Suzy and Tom were talking last week:

> Suzy: Spot barked.
> Tom: When?
> Suzy: Before Dick yelled.

If what Suzy said is true, Spot barked. When did he bark? He barked when he barked. That's very unsatisfying. Tom wants to know more. We orient ourselves in time by relating one time to another, each established with a proposition. We relate the "when" of one proposition by comparing:

> Spot barked before Dick yelled.

If that's true, then "Spot barked" picks out a time that is before the time that "Dick yelled" picks out. Suzy has given Tom a better idea of the "when" of "Spot barked".

If "Spot barked" is true, we can use it along with "Dick yelled" and other true propositions to establish time, relative time. The "when" of this, the time it describes, is before the "when" of that.

What if "Spot barked" is false? Then it does not describe the world, it establishes no time, just as "Spot jumped three meters high" picks out no time. A false proposition is not false of a time; it marks no time relative to any other time.

Yet Zoe said:

Dick yelled at 3 p.m.

Surely if this is true, it's true about 3 p.m.; and if it's false, it's false about 3 p.m.

We have a standard measure for time, one no less arbitrary than saying that the distance on a rod in Paris is one meter. We have a standard clock. We think that this measures time quite independently of us. Perhaps it does. But the times of the standard measure are no less picked out with propositions:

Dick yelled *at the same time as* the standard clock read 3 p.m.

or

Dick yelled *at the same time as* Zoe looked at her watch and saw "3 p.m.".

We do not locate times. We locate ourselves in time by comparing the "when" of true propositions. We use the words and phrases "before", "after", "at the same time as", "during", "while", and others to connect propositions in order to make assertions about relative times.

Suzy: Spot barked.
Tom: When?
Suzy: Before Dick yelled.

Zoe: Suzy is going to the grocery store.
Dick: When?
Zoe: After I go to her apartment to take care of Puff.

Tom: Puff scratched Zoe.
Dick: When?
Tom: While Suzy was at the grocery store.

It's not that one proposition becomes true before another. No, if "Spot barked" is true, it was true when first uttered, it is true now, and it is, was, and will be true whenever those words are meant to make the same description of the world.

We can make further comparisons:

(1) Spot barked. Then Dick yelled. And then Spot barked.

If these are true, then there are three times that are established here. But the first and the last are both described as when "Spot barked" is true. They are distinguished from each other by one being before when Dick yelled and one being after when Dick yelled. There might be no other words we can use to pick out those times: we might not have fuller descriptions of "what happened" then. In (1), there are two inscriptions that we classify as the same linguistic type, a sentence-type, yet we cannot identify them for our reasoning. They are (or represent) distinct propositions.

We need to distinguish them, as in:

(2)   (Spot barked)$_1$. Then Dick yelled. Then (Spot barked)$_2$.

"(Spot barked)$_1$" is a linguistic type that is also a type of a proposition: that is, we can identify distinct inscriptions and utterances of this linguistic type as being (or representing) the same proposition. Generally, any sentence-type that we previously treated as an atomic proposition we will now want to index. Let's use "1", "2", . . . for that . To make the discussion easier to read when there's only one occurrence of a sentence-type in an example I'll suppress the index, understanding it to be "1". So in (2), "Dick yelled" is shorthand for "(Dick yelled)$_1$".

A true proposition establishes a time but only in relation to other true propositions. A false proposition is not about any time. Perhaps "Dick yelled" picks out a time that Suzy remembers; perhaps it picks out a "real time". But together in our reasoning we have only relative times marked by propositions. So when I use the phrase "the time of" I mean it as shorthand for something like "the time that this proposition establishes relative to the times established by other true propositions we are considering". I'm not assuming—though you may if you wish—a reality of time "out there", parts of which are picked out by propositions. Perhaps all we have are propositions we classify as true that together in their temporal relations create our shared web of time. Whether that ordering is objective or intersubjective doesn't matter here—if it even makes sense to ask which it is.

That "Dick yelled" establishes a time relative to other true propositions is primitive in our work here. We leave open to many interpretations what we mean by the truth of a proposition in time, just as we leave open to many interpretations what we mean by the truth of an atomic proposition in classical propositional logic. If we say that "Spot barked" is true of some time (that we establish with the help of other propositions), does that mean he never stopped barking for even a second to take a breath or look around during that time? Did sound have to keep coming out of his mouth the entire time? These are not questions we need to resolve if we want to make our work open to many conceptions of truth in time.

To summarize, here's what we've assumed in this chapter.

Propositions and times

- A *temporal proposition* is a proposition meant as a description of the world in time. Being a proposition, it is true or false; it is not a scheme of propositions, true of some times and false of others.

- A true temporal proposition establishes a time of which it is a correct description of "the world", however you conceive that, relative to the times established by other true temporal propositions.

- A false temporal proposition establishes no time. There is no "when" of a false proposition.

- Times as established by propositions are ordered with some notion of before and after.
- We cannot take sentence types as (representing) temporal propositions because distinct instances of a single type can be used for distinct propositions. We use instead indexed versions of sentence types as propositions.
- The truth-value of a proposition in time is taken as primitive.

4 Temporal Connectives

True temporal propositions establish times. To relate those times, we relate the propositions. Relating propositions is what we do with propositional connectives. We already have the four standard connectives of classical propositional logic: $\neg, \rightarrow, \wedge, \vee$. They do not take account of time. What temporal propositional connectives should we add?

The most fundamental of our assumptions about time is that there is a before and after of times. So let's start by formalizing "before". Consider:

(1) Spot barked before Dick yelled.

This is ambiguous. It could mean any of:

(2) Spot began barking before Dick began yelling.

Spot ended barking before Dick began yelling.

Spot began barking before Dick ended yelling.

Spot ended barking before Dick ended yelling.

To reason with (1) we have to be clear and mark our use of "before" to show that choice:

before$_{bb}$ began before began

before$_{eb}$ ended before began

before$_{be}$ began before ended

before$_{ee}$ ended before ended

These four readings of "before" are basic in that we need to distinguish them, and none can be defined in terms of the others. And using them we can formalize a great deal of reasoning that depends on taking account of relative times.

But where in the propositions at (2) are "Spot barked" and "Dick yelled"? In English we modify the verb with auxiliaries such as "began" and "ended" or "started" and "stopped" so that we take the first one at (2) rather than "Spot barked beginning before the ending of when Dick yelled" or some other clumsy locution. Moreover, because all sentences in English are tensed, we modify those auxiliaries to reflect the tense of the propositions that are meant to be connected. Thus, joining "Spot will bark" and "Dick will yell" we might say, "Spot will begin barking before Dick will end yelling". We can regiment these uses of "before" independently of the vagaries of English grammar by adopting the following four connectives:

$p \wedge_{bb} q$ is true iff both p and q are true and the time that p establishes begins before the beginning of the time that q establishes

$p \wedge_{eb} q$ is true iff both p and q are true and the time that p establishes ends before the beginning of the time that q establishes

$p \wedge_{be} q$ is true iff both p and q are true and the time that p establishes begins before the ending of the time that q establishes

$p \wedge_{ee} q$ is true iff both p and q are true and the time that p establishes ends before the ending of the time that q establishes

These are temporal versions of "and", which is why I've used the symbol for conjunction in them.

To use these connectives we need to think of the time a proposition establishes as an interval with a beginning and an ending. What are these beginnings and endings?

We point and say, "That is a caterpillar". Then we start a video camera to film it. Day after day, hour after hour, minute after minute, until we notice and say, "That is a butterfly". Running the video slowly, ever so slowly, we look to see a "point in time", an "instant" at which it stopped being a caterpillar and started being a butterfly. At the instant that divides those periods, was it a caterpillar or a butterfly? Or was it both? Or perhaps there was no instant dividing those periods, and hence there are gaps in time. It seems we need to resolve these questions if the interval of time of a true proposition is to have a beginning and an ending.

To assume that there is a sharp dividing line between when something is a caterpillar and when it is a butterfly is to assume a greater precision in our use of the words "caterpillar" and "butterfly" than we have or need. Our language is vague, and this shows that it needs to be vague. It is entirely a matter of choice for us to say at that "instant" whether the thing was a caterpillar or butterfly. We can impose the precision, but that isn't and needn't be a precision that exists "in the world". In the flow of time, the flow of the world, it is we who delimit beginnings and endings. They are our way of encapsulating a portion of our experience. There is no sharp end of a person's life, the brain functions a little while, the blood pumps a bit more, a scream lingers, and decay is already there. We make limits, we stipulate—a stipulation we forget is a stipulation and then look for in the "reality" of that which was "out there."

But I will not be dogmatic. Perhaps there are objective beginnings and endings to when this was a caterpillar and when this was a butterfly. Perhaps there is an objective first instant of the time of "Dick yelled". We need not resolve the metaphysical issue. All we need is that the "when" of any true proposition we consider is an interval that has a beginning and ending—though that interval might be (considered to be) a single "instant" whose beginning and ending are the same.[5] To be explicit, we assume the following.

---

[5] The issues that need to be resolved if beginnings and endings are taken as real in the world beyond our stipulations are complex and were investigated deeply by the medievals, as you can read in "Continuity, Contrariety, Contradiction, and Change" by Norman Kretzmann.

Beginnings and endings The time of a true atomic proposition is an interval with a beginning and an ending, though the interval may be (conceived of as) a single instant so that the beginning is the same as the ending.

So if "(Spot barked)$_1$" is true, then it describes a time in which Spot was barking. Immediately outside that interval, Spot wasn't barking because the beginning and the ending of the interval is meant to mark the beginning and ending of when Spot barked. So if "(Spot barked)$_4$" is true, the time it describes either is the same as the time of "(Spot barked)$_1$" or has no overlap with that.

Times of indexed sentence types are equal or have no overlap Two true indexed versions of a sentence type either pick out the same time or pick out times that have no overlap.

We find the beginning and ending of "Dick yelled" in the same way we find other times: by comparing that proposition to other true ones. If we know that "Spot barked $\wedge_{eb}$ Dick yelled" is true, we have located the beginning of the time of "Dick yelled" a bit—not completely, but a bit. Comparing "Dick yelled" to other true propositions we know more. We pick out the beginning and ending of "Dick yelled" as we pick out the time of "Dick yelled" relative to other true propositions.

Do compound propositions pick out times? We can describe the time when Spot barked and Suzy closed the gate by making comparisons to both "Spot barked" and "Suzy closed the gate" rather than saying that the compound "Spot barked $\wedge$ Suzy closed the gate" picks out a time. There's no need to assume that a conjunction establishes a time. Nor is there any need to assume that a disjunction establishes a time because we can make comparisons to one or the other of the disjuncts.

What about negations? Puff is a cat, so "Puff barked" is false. What time does it describe? No time at all. So it seems that the true proposition "¬ (Puff barked)" is true of all time. If so, that's a worthless piece of information, for the negation of any false atomic proposition picks out the same time as the negation of any other false atomic proposition: all time.

What about conditionals? "If Spot barked, then Spot is a dog" is hypothetical, not meant as a description of the time of either the antecedent or the consequent or any other time. In any case, using the classical evaluation of $\rightarrow$, there is no more reason to assume that "Spot barked $\rightarrow$ Spot is a dog" picks out a time than there is to accept that "¬ (Spot barked $\wedge$ ¬ Spot is a dog)" does.

A compound proposition is about a time only derivatively by way of the times of its parts. We need not and indeed do not want to allow our temporal connectives to join compound propositions. So from now on, when I speak of a *temporal proposition* I'll mean an atomic temporal proposition.

5 Time as an Ordering

Our conception of time has a before and after. What the before and after are of, what it is that is before and after, may or may not be "really out there". What we pay attention to in our talk and reasoning is the before and after that we pick out with true atomic propositions. And the before and after that those determine are the before and after of the beginnings and endings of the times of those propositions. Those beginnings and endings, and only those, are the "instants" of time that are ordered in our talk and reasoning here.

If we are to investigate time as an ordering, we will need to have instants and orderings of instants in our models. It seems the wrong direction to start by allowing in a model of our formal logic any ordered collection as our representation of time, since the order comes from the true propositions in the model. But it is much easier to start that way and then, as we'll see, reduce any model so that time is represented by just the "instants" that are the beginnings and endings of the true propositions of the model.

To begin, we need some definitions.

Orderings An *ordering* is a non-empty collection T and a relation < on T such that for all a, b, c in T:

Not a < a .   the ordering is *anti-reflexive*.

If a < b and b < c, then a < c .   the ordering is *transitive*.

We define:

a ≤ b ≡$_{Def}$ a < b or a = b

From this definition we can deduce that every ordering satisfies:

If a < b, then not-(b < a) .   the ordering is *anti-symmetric*

That's because if a < b and b < a, then by transitivity a < a, which would contradict that the ordering is anti-reflexive.

We have now ruled out giving a formal model of circular time, since in that conception every instant is both before and after itself.

In order to simplify the presentation here, I am going to make an assumption that is often implicit in our talk and reasoning: all times can be compared as before or after, as if lined up in a row. That is, to begin I'll assume that the ordering of times is linear.

Linear orderings An ordering ⟨T, <⟩ is *linear* if for all a and b in T:

a < b or b < a or b = a   *trichotomy*

This is only a pedagogical choice that will make it easier to follow the discussions. In Chapter 10 we'll see how to modify what follows to allow for non-linear orderings of times.

In what follows, I'll sometimes use the notation of set theory as a convenient shorthand. It's not meant to suppose anything about collections understood as sets.

We've made the assumption that each time is an interval with a beginning point and ending point. Here's a formal definition for that.

Intervals An *interval* in an ordering $\langle T, < \rangle$ is a non-empty subset I of T such that for any a, b, c in T, if a and b are in I and $a < c < b$, then c is in I.

An interval I has *endpoints* if there are c and d in T such that one of the following holds:

$I = \{x : c < x < d\}$ an *open interval*

$I = \{x : c \leq x \leq d\}$ a *closed interval*

$I = \{x : c \leq x < d\}$ a *half-open, half-closed interval*

$I = \{x : c < x \leq d\}$ a *half-open, half-closed interval*

An interval can be of more than one type. For example, suppose $\langle T, < \rangle$ is $c_0 < c_1 < c_2 < c_3$. Then the following are the same:

$\{x : c_0 < x < c_3\}$

$\{x : c_1 \leq x < c_3\}$

$\{x : c_0 < x \leq c_2\}$

$\{x : c_1 \leq x \leq c_2\}$

We can take the intervals that true propositions establish to be open, closed, or half-open and half-closed. But if we allow a mix of those, there's a problem defining a connective that formalizes "at the same time". The beginnings and endings of two intervals might be the same, yet the intervals are different: in the example above $\{x : c_0 < x < c_3\}$ is not the same as $\{x : c_0 \leq x < c_3\}$. We can avoid this problem by allowing only open or only closed intervals. It is easier to use only closed intervals. After the development using only those, I'll show that we get the same formal logic if we use only open intervals.

"Electrons have spin" is temporal: it is meant to describe the world of all time. So we need to take the entire collection of times to count as an interval. Since we've said that a proposition picks out an interval with a beginning and ending point, we need to assume that the entire collection T has a beginning and ending point. Doing so need not commit us to the view that that there is a beginning and ending of all time, any more than talking of points at infinity need commit us to the reality of a beginning and an ending of the real numbers. We can treat the beginning and ending points of the entire timeline as just labels that simplify our discussions.

Orderings with endpoints   An ordering $\langle T, < \rangle$ has *endpoints* if there are b_T and e_T in T such that for all σ in T, $b_T \leq \sigma \leq e_T$.

Now we can define a formal logic.

Aside: Interval Temporal Logics

Valentin Goranko, Angelo Montanari, and Guido Sciavicco in *A Road Map of Interval Temporal Logics and Duration Calculi* discuss how intervals can be taken as primitive without invoking points or endpoints. However, temporal relations in the propositional logics they consider are formalized syntactically with what they call "modal operators", which are propositional operators. These have the same problems as logics in the tradition of Arthur Prior that make them unsuitable for giving formal models for reasoning, as I explain in Appendix A.

Much work on intervals of time derives from the work of James F. Allen in "Maintaining Knowledge about Temporal Intervals". His paper seems to be directed to reasoning about time—what is true of time—rather than taking account of time in reasoning, as we are concerned with here.

A much earlier and interesting logic based on taking times as intervals was presented by C. L. Hamblin in "Instants and Intervals".

6   Classical Propositional Logic with Temporal Connectives

The formal language $L(\neg, \to, \wedge, \vee, \wedge_{bb}, \wedge_{ee}, \wedge_{be}, \wedge_{eb}, p_0, p_1, \ldots)$

The symbols $p_0, p_1, \ldots$ are called *sentence-type symbols*.

Well-formed formulas (wffs)
  i. If p is a sentence-type symbol, each *indexed* version of it is a wff of length 1:
     $((p)_1)$ $((p)_2)$ $((p)_3)$ $\ldots$.
     The subscript is the *index*.
  ii. If p and q are wffs of length 1, then each of the following is a wff of length 2:
     $(p \wedge_{bb} q)$
     $(p \wedge_{ee} q)$
     $(p \wedge_{be} q)$
     $(p \wedge_{eb} q)$
  iii. If A and B are wffs and the maximum of the lengths of A and B is n, then each of the following is a wff of length $n + 1$:
     $(\neg A)$
     $(A \wedge B)$
     $(A \to B)$
     $(A \vee B)$
  iv. A concatenation of symbols is a *wff* iff it is a wff of length n for some $n \geq 1$.

Wffs of length 1 are *atomic*; all others are *compound*.

$\neg, \to, \wedge, \vee$ are the *classical connectives*.

$\wedge_{bb}, \wedge_{ee}, \wedge_{be}, \wedge_{eb}$ are the *temporal connectives* or *temporal conjunctions*.

I'll leave to you to show there is a unique reading of every wff of this language (compare the proof for classical propositional logic in Volume 0).

We can use the following metavariables:

p, q, r, s, $q_1, q_2, \ldots, r_1, r_2, \ldots, s_1, s_2, \ldots$ stand for atomic wffs
  except when it's said they stand for sentence-type symbols.

A, B, C, D, and subscripted versions stand for wffs of L.

Γ, Δ, Σ, and subscripted versions stand for collections of wffs.

Informally, we delete parentheses by agreeing that $\wedge_{bb}, \wedge_{ee}, \wedge_{be}, \wedge_{eb}$ bind equally and more strongly than $\neg$, which binds more strongly than $\wedge$ and $\vee$, which bind more strongly than $\to$. The outermost parentheses may be deleted.

Realizations and semi-formal languages

We assign to some of the sentence-type symbols a sentence-type such that:

- In classical propositional logic the sentence-type could serve as an atomic wff.
- No word or phrase in the sentence-type could be formalized as a temporal connective.
- The sentence-type can be construed as describing in time.

Indexed versions of these are the *temporal atomic wffs* of the realization. The wffs of the formal language whose sentence-type symbols are replaced by their realizations comprise a *semi-formal language*, though not all indexed versions of a sentence-type need be realized. For convenience, let's assume that the indexed versions of a sentence-type begin with index "1" and continue consecutively. We can use the same metavariables as above for parts of the semi-formal language.

Semantics

A model M is a realization and:

- A *valuation* v that assigns to each atomic wff p the value true or the value false: $v(p) = T$ or $v(p) = F$.
- A linear ordering $\langle T, < \rangle$ with endpoints, called the *timeline* of the model. The elements of T are called *instants*, with b_T the *beginning point* of T and e_T the *ending point* of T.
- A *time assignment* t such that:
  i. For every atomic wff p such that $v(p) = T$, $t(p)$ is a closed interval of $\langle T, < \rangle$ with beginning point b_p and ending point e_p.
  ii. If q realizes a sentence-type symbol, then for any indices i and j, either $t((q)_i) = t((q)_j)$ or $t((q)_i) \cap t((q)_j) = \emptyset$.

Since intervals are not empty; if $t(p)$ is defined, it is not empty.

The valuation v is extended to all wffs of the realization by:

$v(\neg A) = T$ iff $v(A) = F$

$v(A \rightarrow B) = T$ iff $v(A) = F$ or $v(B) = T$

$v(A \wedge B) = T$ iff $v(A) = T$ and $v(B) = T$

$v(A \vee B) = T$ iff $v(A) = T$ or $v(B) = T$

$v(p \wedge_{bb} q) = T$ iff $v(p) = T$ and $v(q) = T$ and $b_p < b_q$

$v(p \wedge_{ee} q) = T$ iff $v(p) = T$ and $v(q) = T$ and $e_p < e_q$

$v(p \wedge_{be} q) = T$ iff $v(p) = T$ and $v(q) = T$ and $b_p < e_q$

$v(p \wedge_{eb} q) = T$ iff $v(p) = T$ and $v(q) = T$ and $e_p < b_q$

Here are pictures for the temporal parts of true temporal conjunctions:

$p \wedge_{bb} q$ — with $t(p)$ starting at b_p and $t(q)$ starting at b_q, b_p before b_q.

$p \wedge_{ee} q$ — with $t(p)$ ending at e_p and $t(q)$ ending at e_q, e_p before e_q.

$p \wedge_{be} q$ — with $t(p)$ beginning at b_p and $t(q)$ ending at e_q.

$p \wedge_{eb} q$ — with $t(p)$ ending at e_p and $t(q)$ beginning at b_q.

A model is designated $M = \langle \upsilon, \langle T, \langle \rangle, t \rangle$.

A wff B is *true in* M means that $\upsilon(B) = T$, which we write as $M \vDash B$; it is false means that $\upsilon(B) = F$, which we write as $M \nvDash B$. If Γ is a collection of wffs, we write $M \vDash \Gamma$ to mean that $M \vDash B$ for every B in Γ.

A formal wff A is *valid* or a *tautology* iff in every model its realization is true; in that case we write $\vDash A$. The formal inference Γ therefore A is *valid*, written $\Gamma \vDash A$, means that there is no model in which the realizations of all the wffs in Γ are true and the realization of A is false; this is the *semantic consequence relation*.

Sufficiency of the collection of models
Any realization, any assignment of truth-values to the atomic propositions, any linear ordering with endpoints, and any assignment of closed intervals from that ordering as times of the true atomic propositions defines a model.

The logic TPC-linear The logic of *temporal propositional connectives in linear time*, **TPC-linear**, is the formal language and the collection of these models with their tautologies and semantic consequence relation.

We'll need some defined connectives:

$p \approx_{bb} q \equiv_{Def} (p \wedge q) \wedge \neg (p \wedge_{bb} q) \wedge \neg (q \wedge_{bb} p)$

$p \approx_{ee} q \equiv_{Def} (p \wedge q) \wedge \neg (p \wedge_{ee} q) \wedge \neg (q \wedge_{ee} p)$

$p \approx_{be} q \equiv_{Def} (p \wedge q) \wedge \neg (p \wedge_{be} q) \wedge \neg (q \wedge_{eb} p)$

$p \approx_{eb} q \equiv_{Def} q \approx_{be} p$

$p \approx_T q \equiv_{Def} (p \approx_{bb} q) \wedge (p \approx_{ee} q)$

The following lemma, whose proof I'll leave to you, depends on the ordering being linear.

Lemma $\ \ v(p \approx_{bb} q) = T$ iff $v(p) = T$ and $v(q) = T$ and $b_p = b_q$

$v(p \approx_{ee} q) = T$ iff $v(p) = T$ and $v(q) = T$ and $e_p = e_q$

$v(p \approx_{be} q) = T$ iff $v(p) = T$ and $v(q) = T$ and $b_p = e_q$

$v(p \approx_{eb} q) = T$ iff $v(p) = T$ and $v(q) = T$ and $e_p = b_q$

$v(p \approx_T q) = T$ iff $v(p) = T$ and $v(q) = T$ and $t(p) = t(q)$

7 Reduced Models

In a model the only instants that enter into evaluations of wffs are the beginning and ending points of the intervals assigned to the atomic propositions. We can show this by constructing reduced models that include only those points along with the beginning and ending points of the entire timeline.

Reduced models Given a model $M = \langle v, \langle T, \langle \rangle, t \rangle$, its *reduced model* is $M_R = \langle v_R, \langle T_R, \langle_R \rangle, t_R \rangle$ where:

$v_R(p) = v(p)$

$T_R = \{b_p, e_p : v_R(p) = T\} \cup \{b_T, e_T\}$

$z <_R w$ iff $z < w$

If $v_R(p) = T$, then $t_R(p) = t(p) \cap T_R$

Note that if $v_R(p) = T$, then

$t_R(p) = \{b_q : b_q \in t(p)\} \cup \{e_q : e_q \in t(p)\}$
$\cup \{b_T \text{ if } b_T \in t(p)\} \cup \{e_T \text{ if } e_T \in t(p)\}$

So if $v_R(p) = T$, then $t_R(p)$ is not empty since both b_p and e_p are in $t(p)$. And if for all p, $v(p) = F$, then $T_R = \{b_T, e_T\}$.

The reduced model of a wff Given a model M and a wff A, the *reduced model for A* is $M_A = \langle v_A, \langle T_A, \langle_A \rangle, t_A \rangle$ where

$v_A(p) = v(p)$

$T_A = \{b_p, e_p : p \text{ appears in A and } v_A(p) = T\} \cup \{b_T, e_T\}$

$z <_A w$ iff $z < w$

If p is in A and $v_A(p) = T$, then $t_A(p) = t(p) \cap T_A$.

If p is not in A and $v_A(p) = T$, then $t_A(p) = T_A$.

Theorem 1   Models and reduced models
  a. Given any model M and wff A, $M_R \models A$ iff $M \models A$.
  b. Given any model M and wff A, $M_A \models A$ iff $M \models A$.

Proof a. I'll let you show that for every p, $t_R(p)$ is a closed interval and that in M_R, if $v_R(p) = T$, then the beginning point of $t_R(p)$ is b_p and the ending point is e_p.

The proof is then by induction on the length of wffs. It's true for atomic wffs. I'll do just one case for compound wffs and leave the rest and part (b) to you.

$v(p \wedge_{bb} q) = T$ iff $v(p) = T$ and $v(q) = T$ and $b_p < b_q$
iff $v_R(p) = T$ and $v_R(q) = T$ and $b_p <_R b_q$
iff $v_R(p \wedge_{bb} q) = T$ ∎

A model is *finite* if its timeline T is finite; otherwise it is *infinite*. Note that for every wff A and for every model M, M_A is finite.

Corollary 2 For every A, if for some M, $M \nvDash A$, then there is a finite M' such that $M' \nvDash A$.

Corollary 3 $\vDash A$ iff for every finite M, $M \vDash A$.

Corollary 4 **TPC-linear** is decidable.

Proof Suppose A has exactly *n* propositional variables. If $\nvDash A$, then A fails in some model in which there are at most $2n + 2$ instants. There are only a finite number of such models, and in each the truth-value of A can be calculated. So A is valid iff it is true in each of those. ∎

Another consequence of our theorem on reduced models is that **TPC-linear**, like classical propositional logic, is *local*: all that matters to an evaluation of a wff are the semantic values assigned to the atomic propositions appearing in it.

Corollary 5 **TPC-linear** *is local*

Suppose that:

- The atomic wffs that appear in A are $q_1, \ldots, q_n$.
- $\langle v, \langle T, <\rangle, t\rangle$ and $\langle v', \langle T', <'\rangle, t'\rangle$ are models.
- For every i, $v'(q_i) = v(q_i)$.
- For every p, q among $q_1, \ldots, q_n$,

   $b'_p <' b'_q$ iff $b_p < b_q$
   $b'_p <' e'_q$ iff $b_p < e_q$
   $e'_p <' b'_q$ iff $e_p < b_q$
   $e'_p <' e'_q$ iff $e_p < e_q$

Then $\langle v, \langle T, <\rangle, t\rangle \vDash A$ iff $\langle v', \langle T', <'\rangle, t'\rangle \vDash A$.

8 An Axiom System for Classical Propositional Logic with Temporal Connectives in Linear Time

I'll present an axiom system for **TPC-linear** and a proof that it is strongly complete. That proof is reminiscent of work in linear algebra where keeping track of indices is complicated but nonetheless a straightforward working out of the underlying ideas. Marking each axiom as it is needed in the completeness proof will make that clear.

An axiom system for TPC-linear
Language
TPC-linear in $L(\neg, \rightarrow, \wedge, \vee, \wedge_{bb}, \wedge_{ee}, \wedge_{be}, \wedge_{eb}, p_0, p_1, \ldots)$

We define:

$p \approx_{bb} q \equiv_{Def} (p \wedge q) \wedge \neg (p \wedge_{bb} q) \wedge \neg (q \wedge_{bb} p)$

$p \approx_{ee} q \equiv_{Def} (p \wedge q) \wedge \neg (p \wedge_{ee} q) \wedge \neg (q \wedge_{ee} p)$

$p \approx_{be} q \equiv_{Def} (p \wedge q) \wedge \neg (p \wedge_{be} q) \wedge \neg (q \wedge_{eb} p)$

$p \approx_T q \equiv_{Def} p \approx_{bb} q \wedge p \approx_{ee} q$

We use the usual notion of proof (Chapter 6 of Volume 0), so that $\vdash A$ means that A is a theorem of the system, and $\Gamma \vdash A$ means that A is a syntactic consequence of the wffs in Γ; similarly, $\Gamma \nvdash A$ means that A is not a syntactic consequence.

Axioms of classical propositional logic, **PC**
The axiom schemes of classical propositional logic (Chapter 6 of Volume 0) where A, B, and C stand for any wffs of the formal language.

$\neg A \rightarrow (A \rightarrow B)$

$B \rightarrow (A \rightarrow B)$

$(A \rightarrow B) \rightarrow ((\neg A \rightarrow B) \rightarrow B)$

$(A \rightarrow (B \rightarrow C)) \rightarrow ((A \rightarrow B) \rightarrow (A \rightarrow C))$

$A \rightarrow (B \rightarrow (A \wedge B))$

$(A \wedge B) \rightarrow A$

$(A \wedge B) \rightarrow B$

$A \rightarrow (A \vee B)$

$B \rightarrow (A \vee B)$

$(A \rightarrow C) \rightarrow ((B \rightarrow C) \rightarrow ((A \vee B) \rightarrow C))$

An Axiom System

Temporal axioms

The following axiom schemes, where p, q, r, s stand for any atomic wff:

1. $\neg (p \wedge_{bb} p)$
2. $\neg (p \wedge_{ee} p)$
3. $(p \approx_{bb} q) \wedge (q \approx_{bb} r) \rightarrow (p \approx_{bb} r)$
4. $(p \approx_{bb} q) \wedge (q \approx_{be} r) \rightarrow (p \approx_{be} r)$
5. $(p \approx_{be} q) \wedge (r \approx_{be} q) \rightarrow (p \approx_{bb} r)$
6. $(p \approx_{be} q) \wedge (q \approx_{ee} r) \rightarrow (p \approx_{be} r)$
7. $(p \approx_{ee} q) \wedge (q \approx_{ee} r) \rightarrow (p \approx_{ee} r)$
8. $(p \approx_{ee} q) \wedge (q \approx_{eb} r) \rightarrow (p \approx_{eb} r)$
9. $(q \approx_{be} p) \wedge (q \approx_{be} r) \rightarrow (p \approx_{ee} r)$
10. $(q \approx_{be} p) \wedge (q \approx_{bb} r) \rightarrow (r \approx_{be} p)$
11. $[(p \wedge_{bb} q) \wedge (p \approx_{bb} r) \wedge (q \approx_{bb} s)] \rightarrow (r \wedge_{bb} s)$
12. $[(p \wedge_{bb} q) \wedge (p \approx_{bb} r) \wedge (q \approx_{be} s)] \rightarrow (r \wedge_{be} s)$
13. $[(p \wedge_{bb} q) \wedge (p \approx_{be} r) \wedge (q \approx_{bb} s)] \rightarrow (r \wedge_{eb} s)$
14. $[(p \wedge_{bb} q) \wedge (p \approx_{be} r) \wedge (q \approx_{be} s)] \rightarrow (r \wedge_{ee} s)$
15. $[(p \wedge_{be} q) \wedge (p \approx_{bb} r) \wedge (q \approx_{eb} s)] \rightarrow (r \wedge_{bb} s)$
16. $[(p \wedge_{be} q) \wedge (p \approx_{be} r) \wedge (q \approx_{eb} s)] \rightarrow (r \wedge_{eb} s)$
17. $[(p \wedge_{be} q) \wedge (p \approx_{be} r) \wedge (q \approx_{ee} s)] \rightarrow (r \wedge_{ee} s)$
18. $[(p \wedge_{be} q) \wedge (p \approx_{bb} r) \wedge (q \approx_{ee} s)] \rightarrow (r \wedge_{be} s)$
19. $[(p \wedge_{eb} q) \wedge (p \approx_{eb} r) \wedge (q \approx_{bb} s)] \rightarrow (r \wedge_{bb} s)$
20. $[(p \wedge_{eb} q) \wedge (p \approx_{ee} r) \wedge (q \approx_{bb} s)] \rightarrow (r \wedge_{eb} s)$
21. $[(p \wedge_{eb} q) \wedge (p \approx_{eb} r) \wedge (q \approx_{be} s)] \rightarrow (r \wedge_{be} s)$
22. $[(p \wedge_{eb} q) \wedge (p \approx_{ee} r) \wedge (q \approx_{be} s)] \rightarrow (r \wedge_{ee} s)$
23. $[(p \wedge_{ee} q) \wedge (p \approx_{eb} r) \wedge (q \approx_{eb} s)] \rightarrow (r \wedge_{bb} s)$
24. $[(p \wedge_{ee} q) \wedge (p \approx_{eb} r) \wedge (q \approx_{ee} s)] \rightarrow (r \wedge_{be} s)$
25. $[(p \wedge_{ee} q) \wedge (p \approx_{ee} r) \wedge (q \approx_{eb} s)] \rightarrow (r \wedge_{eb} s)$

26. $[(p \wedge_{ee} q) \wedge (p \approx_{ee} r) \wedge (q \approx_{ee} s)] \rightarrow (r \wedge_{ee} s)$

27. $(p \wedge_{bb} q) \wedge (q \wedge_{bb} r) \rightarrow (p \wedge_{bb} r)$

28. $(p \wedge_{bb} q) \wedge (q \wedge_{be} r) \rightarrow (p \wedge_{be} r)$

29. $(p \wedge_{be} q) \wedge (q \wedge_{eb} r) \rightarrow (p \wedge_{bb} r)$

30. $(p \wedge_{be} q) \wedge (q \wedge_{ee} r) \rightarrow (p \wedge_{be} r)$

31. $(p \wedge_{ee} q) \wedge (q \wedge_{ee} r) \rightarrow (p \wedge_{ee} r)$

32. $(p \wedge_{ee} q) \wedge (q \wedge_{eb} r) \rightarrow (p \wedge_{eb} r)$

33. $(p \wedge_{eb} q) \wedge (q \wedge_{be} r) \rightarrow (p \wedge_{ee} r)$

34. $(p \wedge_{eb} q) \wedge (q \wedge_{bb} r) \rightarrow (p \wedge_{eb} r)$

35. $(p \wedge_{bb} q) \rightarrow p$

36. $(p \wedge_{bb} q) \rightarrow q$

37. $(p \wedge_{ee} q) \rightarrow p$

38. $(p \wedge_{ee} q) \rightarrow q$

39. $(p \wedge_{be} q) \rightarrow p$

40. $(p \wedge_{be} q) \rightarrow q$

41. $(p \wedge_{eb} q) \rightarrow p$

42. $(p \wedge_{eb} q) \rightarrow q$

43. $((p)_i \wedge (p)_j) \rightarrow [((p)_i \wedge_{eb} (p)_j) \vee ((p)_i \wedge_{be} (p)_j) \vee ((p)_i \approx_T (p)_j)]$
    where i and j are any indices.

rule $\quad \dfrac{A, A \rightarrow B}{B} \quad$ *modus ponens*

We adopt the usual definitions of completeness, consistency, and theory as for classical propositional logic:

Γ is *consistent* iff for no A does $\Gamma \vdash A$ and $\Gamma \vdash \neg A$.

Γ is *complete* iff for every A, $\Gamma \vdash A$ or $\Gamma \vdash \neg A$.

Γ is a *theory* iff for every A, if $\Gamma \vdash A$ then A is in Γ.

The proof of the following is as for classical propositional logic, **PC** (Appendix 4 of Volume 0).

Theorem 1

a. *The Deduction Theorem* $\Gamma \cup \{A\} \vdash B$ iff $\Gamma \vdash A \rightarrow B$.

b. If Γ is complete and consistent, then Γ is a theory.

c. If $\Gamma \nvdash A$, then there is some complete and consistent $\Sigma \supseteq \Gamma$ such that $A \notin \Sigma$.

d. If A has the form of a valid wff in **PC**, then $\vdash A$.

Theorem 2 If Γ is a complete and consistent collection of wffs of L, then Γ has a model.

Proof I'll note the first use of each axiom scheme in bold face. I'll write "by PC" to mean that the wff is PC-valid and so by Theorem 1 is a theorem of this system.

Given Γ we'll construct a model $\langle \mathsf{v}, \langle \mathsf{T}, \langle \rangle, \mathsf{t} \rangle$ such that $\mathsf{v}(A) = \mathsf{T}$ iff $A \in \Gamma$. First note that by Theorem 1, Γ is a theory, and $\Gamma \supseteq$ **PC**.

We begin by setting for each atomic wff p, $\mathsf{v}(p) = \mathsf{T}$ iff $p \in \Gamma$.

In order to construct a timeline, we first define:

$$\mathsf{S} = \{\mathsf{x}_p, \mathsf{y}_p : p \text{ is an atomic wff and } p \in \Gamma\} \cup \{\mathsf{b}, \mathsf{e}\}$$

The letters x_p, y_p, and b, e do not stand for anything; they are not symbols but simply letters.

We define a relation $<$ on **S**. For all atomic wffs p, q such that $p \in \Gamma$ and $q \in \Gamma$:

$\mathsf{b} < \mathsf{x}_p$

$\mathsf{b} < \mathsf{y}_p$

$\mathsf{x}_p < \mathsf{e}$

$\mathsf{y}_p < \mathsf{e}$

$\mathsf{b} < \mathsf{e}$

$\mathsf{x}_p < \mathsf{x}_q$ iff $(p \wedge_{bb} q) \in \Gamma$

$\mathsf{y}_p < \mathsf{y}_q$ iff $(p \wedge_{ee} q) \in \Gamma$

$\mathsf{x}_p < \mathsf{y}_q$ iff $(p \wedge_{be} q) \in \Gamma$

$\mathsf{y}_p < \mathsf{x}_q$ iff $(p \wedge_{eb} q) \in \Gamma$

The letters x_p, y_p are meant to be the beginning and ending points of the interval assigned to the atomic wff p. However, there may be distinct p, q with $(p \approx_{bb} q) \in \Gamma$, yet $\mathsf{x}_p \neq \mathsf{y}_p$, since those are different letters. So we define an equivalence relation on **S** and take the equivalence classes to be the instants of our models.

$\mathsf{b} \sim \mathsf{b}$ $\quad \mathsf{e} \sim \mathsf{e}$

$\mathsf{x}_p \sim \mathsf{x}_q$ iff $(p \approx_{bb} q) \in \Gamma$

$\mathsf{y}_p \sim \mathsf{y}_q$ iff $(p \approx_{ee} q) \in \Gamma$

$\mathsf{x}_p \sim \mathsf{y}_q$ iff $(p \approx_{be} q) \in \Gamma$

$\mathsf{y}_p \sim \mathsf{x}_q$ iff $(p \approx_{eb} q) \in \Gamma$

Lemma A The relation ~ is an equivalence relation. That is,
~ is reflexive, symmetric, and transitive.

Proof

reflexive For all z, $z \sim z$:
$x_p \sim x_p$ by **Axiom 1**
$y_p \sim y_p$ by **Axiom 2**
$b \sim b$ and $e \sim e$ by definition

symmetric For all z, w, if $z \sim w$, then $w \sim z$:

If $z = w$, this follows by reflexivity.

If z is b, then $w = b$, and we are done by reflexivity.

If z is e, then $w = e$, and we are done by reflexivity.

$x_p \sim x_q$ iff $(p \approx_{bb} q) \in \Gamma$ by definition
   iff $(p \wedge q) \wedge \neg (p \wedge_{bb} q) \wedge \neg (q \wedge_{bb} p) \in \Gamma$ by definition/PC
   iff $(q \wedge p) \wedge \neg (q \wedge_{bb} p) \wedge \neg (p \wedge_{bb} q) \in \Gamma$ by definition/PC
   iff $(q \approx_{bb} p) \in \Gamma$ by definition
   iff $x_q \sim x_p$

Similarly, if $y_p \sim y_q$, then $y_p \sim y_q$.

$y_p \sim x_q$ iff $(p \approx_{be} q) \in \Gamma$ by definition
   iff $x_q \sim y_p$ by definition

transitive For all z, w, if $z \sim w$ and $w \sim u$, then $z \sim u$:

If $z = w$ or $w = u$, we are done.

If z, w, or u is b, then all three are b, and we are done.

If z, w, or u is e, then all three are e, and we are done.

If $x_p \sim x_q$ and $x_q \sim x_r$, then $x_p \sim x_r$.   by **Axiom 3**
If $x_p \sim x_q$ and $x_q \sim y_r$, then $x_p \sim y_r$.   by **Axiom 4**
If $x_p \sim y_q$ and $y_q \sim x_r$, then $x_p \sim x_r$.   by **Axiom 5**
If $x_p \sim y_q$ and $y_q \sim y_r$, then $x_p \sim y_r$.   by **Axiom 6**
If $y_p \sim y_q$ and $y_q \sim y_r$, then $y_p \sim y_r$.   by **Axiom 7**
If $y_p \sim y_q$ and $y_q \sim x_r$, then $y_p \sim x_r$.   by **Axiom 8**
If $y_p \sim x_q$ and $x_q \sim y_r$, then $y_p \sim y_r$.   by **Axiom 9**
If $y_p \sim x_q$ and $x_q \sim x_r$, then $y_p \sim x_r$.   by **Axiom 10**

This completes the proof of Lemma A. ■

Lemma B The relation ~ respects <:

If $z < w$ and $z \sim u$ and $w \sim v$, then $u < v$.

Proof: If $b < w$ and $b \sim u$ and $w \sim v$, then u is b, and $w \neq b$. Hence $v \neq b$, so $b \sim v$. I'll leave to you the other cases when any of z, w, u, or v is either b or e.

If $x_p < x_q$ and $x_p \sim x_r$ and $x_q \sim x_s$, then $x_r < x_s$.   by **Axiom 11**
If $x_p < x_q$ and $x_p \sim x_r$ and $x_q \sim y_s$, then $x_r < y_s$.   by **Axiom 12**

An Axiom System 31

If $x_p < x_q$ and $x_p \sim y_r$ and $x_q \sim x_s$, then $y_r < x_s$.   by **Axiom 13**

If $x_p < x_q$ and $x_p \sim y_r$ and $x_q \sim y_s$, then $y_r < y_s$.   by **Axiom 14**

If $x_p < y_q$ and $x_p \sim x_r$ and $y_q \sim x_s$, then $x_r < x_s$.   by **Axiom 15**

If $x_p < y_q$ and $x_p \sim y_r$ and $y_q \sim x_s$, then $y_r < x_s$.   by **Axiom 16**

If $x_p < y_q$ and $x_p \sim y_r$ and $y_q \sim y_s$, then $y_r < y_s$.   by **Axiom 17**

If $x_p < y_q$ and $x_p \sim x_r$ and $y_q \sim y_s$, then $x_r < y_s$.   by **Axiom 18**

If $y_p < x_q$ and $y_p \sim x_r$ and $x_q \sim x_s$, then $x_r < x_s$.   by **Axiom 19**

If $y_p < x_q$ and $y_p \sim y_r$ and $x_q \sim x_s$, then $y_r < x_s$.   by **Axiom 20**

If $y_p < x_q$ and $y_p \sim x_r$ and $x_q \sim y_s$, then $x_r < y_s$.   by **Axiom 21**

If $y_p < x_q$ and $y_p \sim y_r$ and $x_q \sim y_s$, then $y_r < y_s$.   by **Axiom 22**

If $y_p < y_q$ and $y_p \sim x_r$ and $y_q \sim x_s$, then $x_r < x_s$.   by **Axiom 23**

If $y_p < y_q$ and $y_p \sim x_r$ and $y_q \sim y_s$, then $x_r < y_s$.   by **Axiom 24**

If $y_p < y_q$ and $y_p \sim y_r$ and $y_q \sim x_s$, then $y_r < x_s$.   by **Axiom 25**

If $y_p < y_q$ and $y_p \sim y_r$ and $y_q \sim y_s$, then $y_r < y_s$.   by **Axiom 26**

This completes the proof of Lemma B. ∎

For $z \in S$, denote the equivalence class of z by $[z]$. Note that $[b] = \{b\}$ and $[e] = \{e\}$.

Define: $T = \{ [z] : z \in S \}$

$[z] < [w]$ iff $z < w$.

By Lemma B this is well-defined: the definition does not depend on the choice of representative of the equivalence class.

Lemma C $\langle T, < \rangle$ is a linear ordering with endpoints.

Proof We first show that it has endpoints.

For every $z \neq b$, $b < z$ and not-($b < b$). Hence, for every $[z] \neq [b]$, $[b] < [z]$, and also not-($[b] < [b]$). Similarly, for every $[z] \neq [e]$, $[z] < [e]$, and not-($[e] < [e]$).

We now show that $<$ is anti-reflexive and transitive.

anti-reflexive All $[z]$, not $[z] < [z]$.

If $[z] = [b]$ or $[z] = [e]$, we are done by what we just noted above.

Otherwise, for some p, $[z] = [x_p]$ or $[z] = [y_p]$. We have not-($x_p < x_p$) by **Axiom 1**, and we have not-($y_p < y_p$) by **Axiom 2**.

transitive If $[z] < [w]$ and $[w] < [v]$, then $[z] < [v]$:

I'll leave to you the cases when any one of $[z]$, $[w]$, or $[v]$ is $[b]$ or $[e]$.

If $[x_p] < [x_q]$ and $[x_q] < [x_r]$, then $[x_p] < [x_r]$.   by **Axiom 27**

If $[x_p] < [x_q]$ and $[x_q] < [y_r]$, then $[x_p] < [y_r]$.   by **Axiom 28**

32   Chapter 8

    If $[x_p] < [y_q]$ and $[y_q] < [x_r]$, then $[x_p] < [x_r]$.   by **Axiom 29**

    If $[x_p] < [y_q]$ and $[y_q] < [y_r]$, then $[x_p] < [y_r]$.   by **Axiom 30**

    If $[y_p] < [y_q]$ and $[y_q] < [y_r]$, then $[y_p] < [y_r]$.   by **Axiom 31**

    If $[y_p] < [y_q]$ and $[y_q] < [x_r]$, then $[y_p] < [x_r]$.   by **Axiom 32**

    If $[y_p] < [x_q]$ and $[x_q] < [y_r]$, then $[y_p] < [y_r]$.   by **Axiom 33**

    If $[y_p] < [x_q]$ and $[x_q] < [x_r]$, then $[y_p] < [x_r]$.   by **Axiom 34**

To show trichotomy, recall that x_p and y_p are defined only if $p \in \Gamma$. Then:

    $[x_p] < [x_q]$ iff $x_p < x_q$ iff $(p \wedge_{bb} q) \in \Gamma$

    $[x_p] < [x_q]$ iff $x_p < x_q$ iff $(q \wedge_{bb} p) \in \Gamma$

    $[x_p] = [x_q]$ iff $x_p \sim x_q$ iff $(p \approx_{bb} q) \in \Gamma$ iff $\neg(p \wedge_{bb} q) \wedge \neg(q \wedge_{bb} p) \in \Gamma$

By PC one of these must hold. The other cases follow similarly.

    This completes the proof of Lemma C. ■

    Now we can define the model:

        $\langle T, < \rangle$ is the timeline.

        $v(p) = T$ iff $p \in \Gamma$

        If $v(p) = T$, then $t(p) = \{ \sigma \in T : [x_p] \leq \sigma \leq [y_p] \}$.

By definition, for each p, $t(p)$ is an interval with endpoints, for which b_p is $[x_p]$ and e_p is $[y_p]$.

    The valuation v is extended to all wffs by the standard evaluations in a model.

Lemma D $v(A) = T$ iff $A \in \Gamma$.

Proof If A has length 1, A is p and the lemma follows by definition.

    We consider next wffs of length 2 where the connective is a temporal one. We first show that if $v(p \wedge_{bb} q) = T$, then $(p \wedge_{bb} q) \in \Gamma$:

        If $v(p \wedge_{bb} q) = T$ then $v(p) = T$ and $v(q) = T$, and $[x_p] < [x_q]$,

                          then $p \in \Gamma$ and $q \in \Gamma$, and $x_p < x_q$ by Lemma B,

                          then $(p \wedge_{bb} q) \in \Gamma$.

Now we show that if $(p \wedge_{bb} q) \in \Gamma$, then $v(p \wedge_{bb} q) = T$:

        If $(p \wedge_{bb} q) \in \Gamma$ then $x_p < x_q$ by definition,

                        then $[x_p] < [x_q]$ by Lemma B.

        If $(p \wedge_{bb} q) \in \Gamma$ then $p \in \Gamma$ and $q \in \Gamma$ by **Axiom 35** and **Axiom 36**,

                        then $v(p) = T$ and $v(q) = T$.

The cases for the other temporal connectives follow similarly using **Axiom 37**, **Axiom 38**, **Axiom 39**, **Axiom 40**, **Axiom 41**, and **Axiom 42**.

    For all other wffs of length ≥ 2, the proof follows as for **PC** (see Appendix 4 of Volume 0). This completes the proof of Lemma D. ■

Lemma E If p is any sentence-type symbol and *i* and *j* are any indices,
$$t((p_i)) = t((p_j)) \text{ or } t((p_i)) \cap t((p_j)) = \varnothing.$$

Proof This follows from **Axiom 43**. ∎

This completes the proof of Theorem 2. ∎

Theorem 3 *Strong Completeness of the Axiomatization*
For all wffs A, $\Gamma \vdash A$ iff $\Gamma \vDash A$.

Proof I'll leave to you to check that each of the axioms is true in every model and that the rule preserves truth in a model. Hence, if $\Gamma \vdash A$ then $\Gamma \vDash A$.

In the other direction, suppose $\Gamma \nvdash A$. Then by Theorem 1.c there is a complete and consistent Σ such that $\Sigma \supseteq \Gamma$ and $A \notin \Sigma$. By Theorem 2 there is a model M such that for all wffs B, $M \vDash B$ iff $B \in \Sigma$. Hence, $M \nvDash A$, and so $\Sigma \nvDash A$, and hence $\Gamma \nvDash A$. ∎

If in a model $p \approx_T q$, then p and q have the same semantic values. Hence, the following is a scheme of valid wffs:

$$(p \approx_T q) \to (A(p) \to A(q))$$

where q replaces some but not necessarily all occurrences of p in A

By the completeness theorem, this is a scheme of theorems, too.

Open Intervals in the Semantics

Theorem 4 *Strong Completeness of the Axiomatization for Open Intervals*
The same axiom system is strongly complete for the class of models defined as before except that the time of each true atomic proposition is an open interval.

Proof After the proof of Lemma C above, we have to add another step. We would like to take $\langle T, < \rangle$ as the timeline for our model, with the assignments:

$$t(p) = \{ [z] : [x_p] < [z] < [y_p] \}$$

But that could be empty if $x_p \sim y_p$. So we add extra points to T. Set:

$$U = \{ [z] : z \in S \} \cup \{ m_p : x_p \sim y_p \}$$

$[z] < [w]$ iff $z < w$

$[z] < m_p$ iff $[z] < [x_p]$ or $[z] = [x_p]$

$m_p < [w]$ iff $[y_p] < [w]$ or $[w] = [y_p]$

I'll let you show that $\langle U, < \rangle$ is a linear ordering with endpoints.

Now we can define the model:

$\langle U, < \rangle$ is the timeline.

$v(p) = T$ iff $p \in \Gamma$

If $v(p) = T$, then $t(p) = \{ a \in T : [x_p] < a < [y_p] \}$

For each p, $t(p)$ is an interval with endpoints, for which b_p is $[x_p]$ and e_p is $[y_p]$.

The valuation ∪ is extended to all wffs by the standard evaluations in a model. The proof then follows as before reading U for T. ∎

Using the same methods as in this last proof, you can modify the proofs in Chapter 7 about reduced and finite models to apply to the semantics when only open intervals are allowed.

It seems, then, that it does not matter whether we choose to use closed or open intervals for our time assignments so long as we use only one of those kinds. But there is a difference. With closed intervals the only "instants" that matter are those that are endpoints of the intervals assigned to true propositions, as shown in the completeness proof here and in Theorem 1 of Chapter 7. But with open intervals we must ensure that the intervals assigned to true propositions are not empty, and hence we must "insert" or accept that there can be instants in reduced models and in the completeness proof that are not picked out by being the endpoints of any interval.

9 Examples of Formalizing

Example 1 Spot barked before Dick yelled.

Analysis To formalize this, we must choose an index for "Spot barked" and for "Dick yelled". Since this is the first example, let's use "1" for both. Then we have four connectives that we could use to formalize "before". Here are pictures for those possibilities (where I've left off the indices to make them easier to read). As usual the timeline runs from left to right.

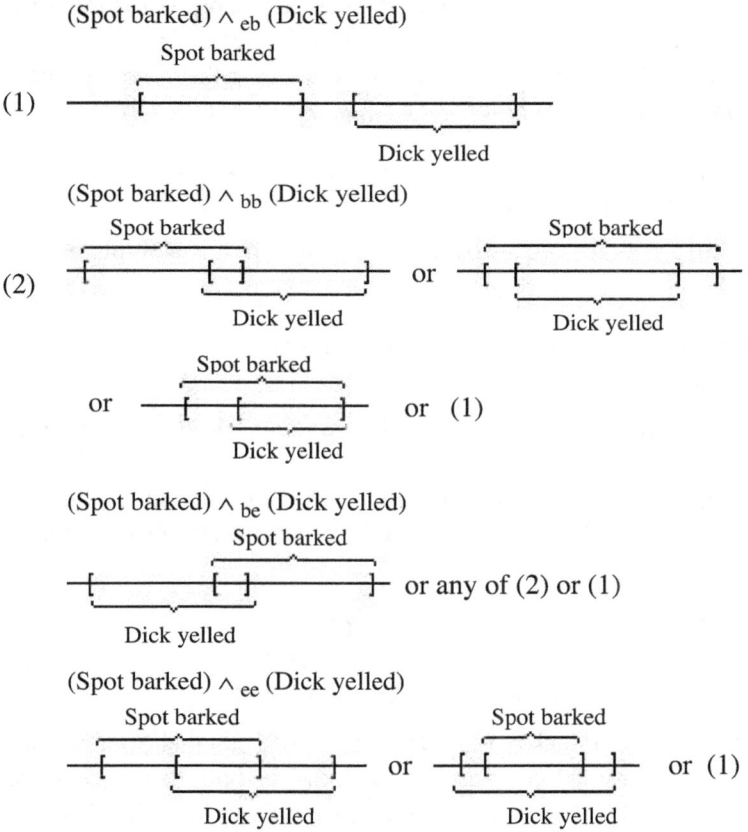

All but the first have interpretations that would not be suitable for formalizing "before" in the example. So we formalize the example as:

(Spot barked)$_1 \wedge_{eb}$ (Dick yelled)$_1$

Example 2 Dick yelled after Spot barked.

   Formalization as for Example 1.

Analysis The order of the propositions is reversed in the formalization, taking as a convention to read "after" as the reverse of "before".

36   Chapter 9

Example 3   Dick yelled, then Spot began barking, and then Spot finished barking before Dick stopped yelling.

((Dick yelled)$_8$ ∧$_{bb}$ (Spot barked)$_{12}$) ∧ ((Spot barked)$_{12}$ ∧$_{ee}$ (Dick yelled)$_8$)

Analysis   It seems we have four propositions here:

Dick yelled.

Spot began barking.

Spot finished barking.

Dick stopped yelling.

But we're treating "began", "finished", and "stopped" as incorporated into the connectives rather than within the atomic propositions, since their role is to relate times of propositions. Again, we have to choose indices for the atomic propositions. There's no reason to think they should be the same as in the last example.

Example 4   Sometime after Dick stopped eating, Spot began to bark.

(Dick ate)$_1$ ∧$_{eb}$ (Spot barked)$_3$

Analysis   Despite the apparent quantification over times in the example, we can formalize this.

Example 5   Dick yelled at the same time as Spot barked.

(Dick yelled)$_1$ ≈$_T$ (Spot barked)$_1$

Analysis   Assuming that someone is contradicting Example 1, we should use the same indices as there.

Here "≈$_T$" is the defined connective from p. 22. This is to understand "at the same time" to mean that the times these propositions establish are the same, not merely overlapping times.

Example 6   Spot barked at 7:12 am May 5th, 2005.
          Spot barked around 7:12 am May 5th, 2005.
          Spot barked at exactly 7:12 am May 5th, 2005.

Analysis   It is not clear what these mean. Does the last mean that Spot began barking at 7:12 am May 5th, 2005, or does it mean that he both began and ended barking at 7:12 am May 5th, 2005? If the latter, then we can formalize it as:

(Spot barked)$_1$ ≈$_T$ (the standard clock read "7:12 am May 5th, 2005")$_1$

This is to take the person who asserts this to be talking about the times as in the first example. In our models we need only one indexed version of "the standard clock read '7:12 am May 5th, 2005' ". The first two are less clear.

Example 7   Spot barked. Then Dick yelled. And then Spot barked.

((Spot barked)$_1$ ∧$_{eb}$ (Dick yelled)$_1$) ∧ ((Dick yelled)$_1$ ∧$_{eb}$ (Spot barked)$_2$)

Examples of Formalizing 37

Analysis I understand "then" to mean that the time of the first is completely before the time of the second. I've used the same indices as in the first example, imagining someone to be clarifying when Dick yelled.

Note that the indices for "Spot barked" by themselves do not mark specific times. We use them only to distinguish two occurrences of the same sentence-type as distinct propositions. We could as well have marked them in different colors.

Example 8 *Suzy took off her clothes and went to bed.*

(Suzy took off her clothes)$_5$ $\wedge_{eb}$ (Suzy went to bed)$_1$

Analysis I understand "and" here as "and then".

Example 9 *If Suzy went to bed, then she took off her clothes.*

(Suzy went to bed)$_1$ $\to$ (Suzy took off her clothes)$_5$

Analysis The "then" in this example is not meant temporally but as part of the conditional. Both this and the previous example could be true. Or perhaps just this one is true if Suzy climbed into bed and then took off her clothes.

Example 10 *If Suzy went to bed, then she took off her clothes.*
   Suzy went to bed.
   Therefore, *Suzy took off her clothes.*

(Suzy went to bed)$_1$ $\to$ (Suzy took off her clothes)$_5$
(Suzy went to bed)$_1$
_____

(Suzy took off her clothes)$_5$

Analysis The inference is valid.

Example 11 *Dick yelled within the time that Spot was barking.*

Analysis We have four possibilities for the times of "Spot barked" and "Dick yelled" that could make this true:

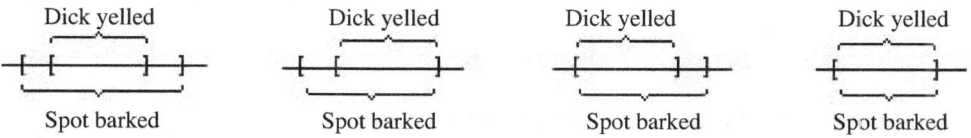

We can define a connective to formalize that the time of one atomic proposition is *within* the time of another:

$p \wedge_w q \equiv_{Def} (p \wedge q) \wedge \neg(p \wedge_{bb} q) \wedge \neg(q \wedge_{ee} p)$

This has truth-conditions:

$v(p \wedge_w q) = T$ iff $v(p) = T$ and $v(q) = T$ and $t(p) \subseteq t(q)$

This evaluation depends on the timeline being linear. Assuming that it is meant to contradict Example 1, we use the same indices as there to formalize it as:

(Dick yelled)$_1$ $\wedge_w$ (Spot barked)$_1$

Example 12 Spot barked yesterday.

Analysis To understand "Spot barked" as a proposition that describes yesterday would be a very poor description of a day, since we don't think that Spot barked all of yesterday. The example is true iff there is some part of yesterday which "Spot barked" picks out. We can formalize this because "yesterday" picks out a time by virtue of a proposition about the standard clock; suppose in this example it is April 27, 2015. Then we can formalize the example as:

$(\text{Spot barked})_{83} \wedge_w (\text{the standard clock read April 27, 2015})_1$

Example 13 Dick yelled during the time that Spot barked.

Analysis This is ambiguous. We have (at least) two choices:

Dick yelled within the time that Spot barked.

(a) $(\text{Dick yelled})_1 \wedge_w (\text{Spot barked})_1$

Dick yelled through all the time that Spot yelled,
beginning before and ending after.

(b) $((\text{Dick yelled})_1 \wedge_{bb} (\text{Spot barked})_1) \wedge ((\text{Spot barked})_1 \wedge_{ee} (\text{Dick yelled})_1)$

We could retain the ambiguity of the original by formalizing the example as the disjunction of (a) and (b).

Example 14 There was a time when both Dick yelled and Spot barked.

Analysis Though the example involves a quantification over time, we need no quantification to formalize overlapping times. We can define:

$$p \wedge_O q \equiv_{Def} ((p \wedge_{bb} q) \wedge (q \wedge_{be} p)) \vee [(q \wedge_{bb} p) \wedge (p \wedge_{be} q))$$
$$\vee (p \approx_{bb} q) \vee (p \approx_{ee} q) \vee (p \approx_{be} q) \vee (p \approx_{eb} q)$$

This has truth-conditions:

$\upsilon(p \wedge_O q) = T$ iff $\upsilon(p) = T$ and $\upsilon(q) = T$ and $t(p) \cap t(q) \neq \emptyset$

Note that this evaluation depends on the timeline being linear. Then we can formalize the example as:

$(\text{Dick yelled})_4 \wedge_O (\text{Spot barked})_5$

Example 15 Dick yelled while Spot was barking.

Analysis The use of "while" as a connective in English is ambiguous. We have (at least) the following four choices:

There was a time when both Dick yelled and Spot barked.
    —formalization as in Example 14 (with appropriate indices).
Dick yelled at the same time as Spot barked.
    —formalization as in Example 5 (with appropriate indices).
Dick yelled within the time that Spot was barking.
    —formalization as in Example 11 (with appropriate indices).

Dick yelled through all the time that Spot barked.

(Spot barked)$_1 \wedge_w$ (Dick yelled)$_1$

Example 16 *Dick likes hamburger, while Wanda is a vegetarian.*

(Dick likes hamburger)$_1 \wedge$ (Wanda is a vegetarian)$_1$

Analysis Sometimes we use "while" as "although", which we formalize as $\wedge$.

Example 17 *Dick yelled when Spot barked.*
Analysis This is not clear enough to formalize.

Example 18 *Spot has been barking since Suzy opened the gate.*
Analysis We often use "since" to introduce the premises of an inference, as in "Since dogs bark, Spot barks". But we also use it as a temporal connective. We have two possibilities for the times of "Spot has been barking" and "Suzy opened the gate" that could make this true:

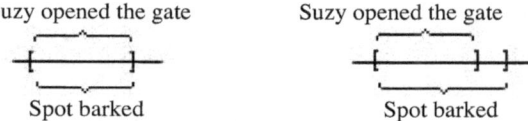

We can define a connective to formalize *since*:

$$p \wedge_{since} q \equiv_{Def} (q \approx_T p) \vee ((q \approx_{bb} p) \wedge (q \wedge_{eb} p))$$

Then we can formalize the example as:

(Spot barked)$_{91} \wedge_{since}$ (Suzy opened the gate)$_{32}$

I've used my informal understanding of the tenses of the atomic propositions, replacing "Spot has been barking" by "Spot barked" with index "91" (Spot barks a lot). In Chapter 11 we'll see how we can take account of tenses more directly.

Example 19 *Dick will cook dinner when Zoe arrives.*

(Zoe will arrive)$_1 \wedge_{eb}$ (Dick will cook dinner)$_1$

Analysis We use the present tense in "Zoe arrives" to talk about some future time, formalizing the example to mean that Dick will start cooking only after Zoe arrives.

If by "when" we mean to assert that Dick will start cooking dinner the moment when Zoe arrives, we should formalize the example instead as:

(Dick will cook dinner)$_1 \approx_{be}$ (Zoe will arrive)$_1$

But for that I'd probably say, "Dick will begin cooking just as soon as Zoe arrives".

Example 20 *Dick will begin cooking no later than when Zoe arrives.*

((Dick will cook)$_1 \wedge_{bb}$ Zoe will arrive)$_1$) $\vee$
((Zoe will arrive)$_1 \approx_{eb}$ Dick will cook)$_1$))

Chapter 9

Analysis I've used the same indices as in the previous example, seeing this one as a different prediction.

Example 21 Sometime between when Zoe arrived and Dick started cooking, Spot ran away.

((Zoe arrived)$_1$ ∧$_{eb}$ (Spot ran away)$_7$) ∧ ((Spot ran away)$_7$ ∧$_{eb}$ (Dick cooked)$_1$)

Analysis We can formalize the time of one atomic proposition being strictly between the times of two others by defining a connective:

$$B(p, q, r) \equiv_{Def} (p \wedge_{eb} q) \wedge (q \wedge_{eb} r)$$

The truth-conditions of this are:

$v(B(p, q, r)) = T$ iff $v(p) = T$ and $v(q) = T$ and $v(r) = T$
and $t(p)$ is completely before $t(q)$,
and $t(q)$ is completely before $t(r)$

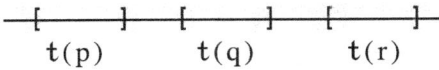
$\quad\quad t(p) \quad\quad\quad t(q) \quad\quad\quad t(r)$

A more general connective for "between" doesn't specify the direction of the ordering:

$$\underline{B}(p, q, r) \equiv_{Def} [(p \wedge_{eb} q) \wedge (q \wedge_{eb} r)] \vee [(r \wedge_{eb} q) \wedge (q \wedge_{eb} p)]$$

In order to make the new methods easier to see, I'm going to let you supply indices in the formalizations in the following examples.

Example 22 Spot barked, then Dick yelled and Zoe got upset.

Analysis Understanding "then" to mean completely before, if we follow the grammar here it seems we should formalize the example:

Spot barked ∧$_{eb}$ (Dick yelled ∧ Zoe got upset)

But we do not allow temporal connectives to join compound wffs. Yet if we understand "then" in the example as meaning that the time when Spot barked ended before the beginning of when Dick yelled and when Zoe got upset, then choosing indices, we can formalize the example as:

(Spot barked ∧$_{eb}$ Dick yelled) ∧ (Spot barked ∧$_{eb}$ Zoe got upset)

We can make temporal comparisons to a conjunction by making temporal comparisons to its parts. We can introduce a defined connective:

If A is a conjunction of atomic propositions,

$p \wedge_{eb} A \equiv_{Def} \bigwedge_{q \text{ in } A} (p \wedge_{eb} q)$

Here ⋀ means the conjunction of the wffs, associating to the left. The truth-conditions for this are:

$v(p \wedge_{eb} A) = T$ iff $v(p) = v(A) = T$ and for every q in A, $e_p < b_q$.

Example 23 *Spot barked and Dick yelled before Dick fell and Zoe laughed.*

(Spot barked $\wedge_{eb}$ Dick fell) $\wedge$ (Dick yelled $\wedge_{eb}$ Dick fell)
$\wedge$ (Spot barked $\wedge_{eb}$ Zoe laughed) $\wedge$ (Dick yelled $\wedge_{eb}$ Zoe laughed)

Analysis If A and B are conjunctions of atomic propositions,

$A \wedge_{eb} B \equiv_{Def} \bigwedge_{p \text{ in } A, q \text{ in } B} (p \wedge_{eb} q)$

The truth-conditions for this are:

$v(A \wedge_{eb} B) = T$ iff $v(A) = v(B) = T$ and for every p in A, q in B, $e_p < b_q$.

I'll leave to you to define $\wedge_{bb}$, $\wedge_{be}$, and $\wedge_{ee}$ for comparing conjunctions.

Example 24 *Spot barked and Dick yelled before Dick fell, and then Zoe laughed.*

Analysis If we follow the grammar of this and take "before" to mean completely before, it seems we should formalize the example (awaiting indices) as:

(Spot barked $\wedge$ Dick yelled) $\wedge_{eb}$ (Dick fell $\wedge_{eb}$ Zoe laughed)

Yet if we understand the example as meaning that the time of when Spot barked came before the time of when Dick fell and the time of when Zoe laughed, and the time of when Dick yelled also came before the time of when Dick fell and when Zoe laughed, and the time of when Dick fell came before the time when Zoe laughed, —and I can find no other good reading of it—then we can formalize the example as:

(Spot barked $\wedge_{eb}$ Dick fell) $\wedge$ (Dick yelled $\wedge_{eb}$ Dick fell)
$\wedge$ (Dick fell $\wedge_{eb}$ Zoe laughed)

Example 25 *Dick yelled or Zoe laughed, and then Spot barked.*

Analysis If we follow the grammar of the example and take "then" to mean completely before, it seems we should formalize the example as:

(Dick yelled $\vee$ Zoe laughed) $\wedge_{eb}$ Spot barked

Yet if we understand "then" in the example as meaning that either when Dick yelled or when Zoe laughed ended before the beginning of when Spot barked—and I can see no other good reading—we can formalize the example:

(Dick yelled $\wedge_{eb}$ Spot barked) $\vee$ (Zoe laughed $\wedge_{eb}$ Spot barked)

We can make temporal comparisons to a disjunction by making temporal comparisons to its parts. I'll leave to you to define connectives $\vee_{eb}$, $\vee_{bb}$, $\vee_{be}$, and $\vee_{ee}$.

Example 26 *Spot barked when either Dick was jumping or Zoe was laughing.*

Analysis There are two ways we can understand this. One we can formalize as:

(Spot barked $\wedge_w$ Dick was jumping) $\vee$ (Spot barked $\wedge_w$ Zoe was laughing)

The other is to assume that the time of "Dick was jumping" overlaps the time of "Zoe was laughing", so that the union of those intervals is an interval, and "Spot barked" picks out a time within that. To say that the time picked out by a proposition r is within the interval of the times picked out by p and q, we can define:

$W(r, p, q) \equiv_{Def} (p \wedge_O q) \wedge [(p \approx_{bb} r) \vee (p \approx_{bb} r) \vee (q \wedge_{bb} r) \vee (q \approx_{bb} r)]$
$\wedge [(r \wedge_{ee} p) \vee (r \approx_{ee} p) \vee (r \wedge_{ee} q) \vee (r \approx_{ee} q)]$

I'll let you show that:

$\nu(W(r, p, q)) = T$ iff $\nu(p) = T$ and $\nu(q) = T$ and $\nu(r) = T$
and $t(p) \cup t(q)$ is an interval
and $t(r) \subseteq (t(p) \cup t(q))$

Then we can formalize the example:

W(Spot barked, Dick was jumping, Zoe was laughing)

Example 27 Spot barked and Dick yelled before Dick fell or Zoe laughed.

[(Spot barked $\wedge_{eb}$ Dick fell) $\wedge$ [(Dick yelled $\wedge_{eb}$ Dick fell)]
$\vee$ [((Spot barked $\wedge_{eb}$ Zoe laughed) $\wedge$ (Dick yelled $\wedge_{eb}$ Zoe laughed)]

Analysis I understand the "or" here to mean "or before". We can define:

If A is a conjunction of atomic propositions and
B is a disjunction of atomic proposition:

$A \wedge_{eb} B \equiv_{Def} W_{p\,in\,A} [\bigwedge_{q\,in\,B} (p \wedge_{eb} q)]$

Here W means the conjunction of the wffs, associating to the left. The truth conditions for this are:

$\nu(A \wedge_{eb} B) = T$ iff $\nu(A) = \nu(B) = T$ and
for every p in A, for some q in B, $e_p < b_q$

Example 28 Dick is sleeping $\leftrightarrow \neg$ (Spot is barking)

Analysis If "Dick is sleeping" is true, doesn't this show that a false proposition, in this case a negation of a true proposition, can pick out a time?

This is an example where not showing the indices leads us astray, as if the example were a timeless truth or true of all times. With indices, we could have:

(Dick is sleeping)$_1 \leftrightarrow \neg$ ((Spot is barking)$_6$)

To say that "$\neg$ ((Spot is barking)$_6$)" establishes a time we'd have to identify that with the time that "(Dick is sleeping)$_1$", and there's no good reason to do that just because they have the same truth-value. There can be no temporal equivalence of any atomic proposition with a compound proposition.

Example 29 Spot didn't bark between the time that Suzy opened the gate and Dick yelled.

Not formalizable.

Analysis If a true negation doesn't pick out a time, how can we identify the time between when Suzy opened the gate and when Dick yelled? All we can use is:

Suzy opened the gate $\wedge_{eb}$ Dick yelled

Example 30 Spot barked twice.

Analysis We can formalize this as:

(Spot barked)$_1$ ∧ (Spot barked)$_2$ ∧ ¬ ((Spot barked)$_1$ ≈$_T$ (Spot barked)$_2$)

Example 31 Spot barked just twice.

Analysis To formalize this, we have to add to the previous formalization that there is no other indexed version of "Spot barked" that is not time-equivalent to one of "(Spot barked)$_1$" or "(Spot barked)$_2$". If there are only a (small) finite number of those in our semi-formal language, we can make all those assertions. But if there are (potentially) infinitely many, we cannot formalize the example because we have no way to quantify over indexed versions of "Spot barked".

Example 32 Electrons have spin.

Analysis This is meant to be a true description of all time: at any time whatsoever, every electron has spin. Let's say that a proposition meant to be true of all times is *omnitemporal*.

Suppose we have a realization where p is a true description of all time. Then we'll have $t(p) = T$. To assert in a semi-formal language that p is about all times we need a unary connective Omni, where:

ᴠ(Omni p) = T   iff   ᴠ(p) = T and $t(p) = T$

This can't be defined in **TPC-linear** because we would need to quantify over atomic propositions to say that t(Omni p) contains the times of all true atomic propositions. But we could add it to **TPC-linear** in order to formalize more.

Example 33 2 + 2 = 4

Analysis Some say that numbers are abstract objects existing outside space and time. In that case, "2 + 2 = 4" does not establish any time. It is timelessly true, not true of all times. Let's say a proposition meant as timelessly true is *atemporal*.

We could extend our temporal propositional logic to accommodate reasoning based on the assumption that there are true atomic propositions about no time at all. We would add a unary connective Atemp to the language, where Atemp p is meant as an assertion that p is true yet establishes no time. We would modify our models to allow the empty set as an interval, so that $t(p) = \emptyset$ is allowed. Then we'd adopt the evaluation:

ᴠ(Atemp p) = T   iff   ᴠ(p) = T and $t(p) = \emptyset$

With this we could formalize some talk of abstract things such as numbers and beauty along with talk about dogs and cats. I'll leave to those who have an intuition for the nature of abstract things to provide more motivation and examples.

Example 34 When Zoe cooks, Dick washes up.

Not formalizable.

Analysis We can't formalize this as "(Zoe cooks)$_1$ ∧ $_{bb}$ (Dick washes up)$_1$" because "(Zoe cooks)$_1$" is meant to establish one particular time, yet the example is about all times that Zoe cooks. That's clear by rewriting the example as "Whenever Zoe cooks, Dick washes up".

Example 35 *There was some time between when Zoe arrived and Dick started cooking.*

Analysis Whoever is asserting this must have in mind a specific instance of Zoe arriving and of Dick cooking, which means some particular indices, say "(Zoe arrived)$_3$" and "(Dick cooked)$_8$". We might try to formalize the example:

(Zoe arrived)$_3$ ∧ $_{eb}$ (Dick cooked)$_8$

To take this as a formalization is to assume that there are always times that "fill in the gaps" between the times of true atomic propositions. But the only times we have are those picked out by true atomic propositions. So for the example to be true in a model, there has to be some true atomic proposition that establishes a time in that gap.

Example 36 *Dick talked only when Spot wasn't barking.*
    Not formalizable.

Analysis If this is meant as true always, then we can't formalize it because we'd have to quantify over propositions of the schemes "Dick talked" and "Spot barked".

Example 37 *Caesar was assassinated.*
Analysis We can formalize this as:

(Caesar was assassinated)$_1$

This leaves open the possibility that there is another indexed version of "Caesar was assassinated" that's true and describes another time. But we all believe that Caesar was assassinated only once, so that any other indexed version of "Caesar was assassinated" that's true should describe the same time. We can formalize that only by adding for every index n for which "(Caesar was assassinated)$_n$" is in the semi-formal language the following:

(Caesar was assassinated)$_1$ ≈$_T$ (Caesar was assassinated)$_n$

Example 38 *Spot didn't bark when Caesar was assassinated.*
    Not formalizable.

Analysis We can assert:

(Caesar was assassinated)$_1$ ∧
¬ ((Caesar was assassinated)$_1$ ∧ $_O$ (Spot barked)$_1$)

But this won't do because it might be that the following is true:

(Caesar was assassinated)$_1$ ∧ $_O$ (Spot barked)$_2$

To formalize this example, we'd have to quantify over all indexed versions of "Spot barked".

Example 39 *Spot barked before Dick yelled.*

Analysis We formalized this in Example 1 as:

(Spot barked)$_1 \wedge_{eb}$ (Dick yelled)$_1$

This would be true if Spot barked on Monday and Dick yelled the following Friday. It would also be true if Spot barked eighty-three times times between Monday and Friday when Dick finally yelled. That seems wrong. I think we understand the example to mean:

(a)   Spot barked shortly before Dick yelled.

To formalize this we need a way to measure time.

We might choose two propositions outside the system and evaluate "shortly" relative to those: no longer than the time between the times of those. But then we'd have to introduce into our system what we mean by "no longer than".

In this chapter we investigated how to use our temporal propositional logic to formalize ordinary language propositions and inferences on the assumption that time is linear. We saw how to formalize:

- Uses of some ordinary language temporal propositional connectives such as "after" and "while".
- Sequences of propositions ordered temporally.
- Some propositions that seem to involve a quantification over time.
- Temporal comparisons that involve compound propositions by comparing their parts.
- One proposition being within the time of two or more others when those other propositions together pick out an interval of time.
- Two propositions together picking out an interval of time.

We saw that we can't formalize:

- Only two indexed versions of a particular sentence-type are distinct.
- Omnitemporal propositions, unless we extend the system.
- Atemporal propositions, unless we extend the system.
- Measures of time.

10 Other Orderings of Time

Whether time is flow or time is composed of indivisible instants is not an issue we will deal with here. All we have, or at least all we need in reasoning together, are times we can point to with true propositions. We take those times to be ordered, whether we think of that as the "real" order of time or only our subjective experience. What we can examine are different assumptions about the ordering. Until now we have assumed that time is linear.

The reversibility of time
Can time be reversed? If we were to make a video of (a portion of) the world during some period of time, could it be run backwards and still "make sense"? Obviously not. We know that Dick threw the ball and Spot chased it, not that Spot ran backward from the ball which then flew into Dick's hand.

The question of whether time is reversible is meant as asking whether the laws of science are applicable if the ordering of time (that we experience) is reversed. That's a question as much about the nature of the laws we think are accurate as about our conception of time.

In our logic, the ordering of the timeline in a model establishes a direction for time. Reversing that ordering yields a model, too. Given any model $\langle \upsilon, \langle T, < \rangle, t \rangle$, if we take $a <^* b$ iff $b < a$, then $\langle \upsilon, \langle T, <^* \rangle, t \rangle$ is a model.

Time is finite, or time is infinite, or time is dense
I don't know what it means to say that time is finite or that time is infinite. That time is potentially infinite makes some sense to me: there will always be a later time. I have no good reason to believe that nor to think it is false.

We saw in Chapter 7 that finite models suffice for our logic of linear time. But that does not mean we've captured the notion of finite time in our work. The tautologies and finite semantic consequence of our logic **TPC-linear** can equally be characterized by infinite models. To see this let's first make some definitions.

Discrete models and dense models   In an ordered collection:

b is an *immediate successor* of a iff a < b and there is no c such that a < c < b

b is an *immediate predecessor* of a iff b < a and there is no c such that b < c < a.

A model M is *discrete* iff the ordering < on T is discrete: for every a,
if there is a b such that a < b, then there is an immediate successor of a in T;
and if there is a b such that b < a, there is an immediate predecessor of a in T.

A model M is *dense* iff the ordering < on T is dense:
for every a, b in T such that a < b there is a c in T such that a < c < b.

I'll let you show that every dense linear ordering is infinite.

Theorem 1 For every wff A,
 a. $\vDash_{\text{TPC-linear}} A$ iff for every discrete M, $M \vDash_{\text{TPC-linear}} A$.
 b. $\vDash_{\text{TPC-linear}} A$ iff for every countable dense M, $M \vDash_{\text{TPC-linear}} A$.

Proof Part (a) comes from Corollary 3 of Chapter 7, since every finite model is discrete.

For part (b), if $\vDash A$, then A is true in every dense model. So suppose that $\nvDash A$. Then by Corollary 2 of Chapter 7, there is a finite model $M = \langle v, <T, <>, t\rangle$ such that $M \nvDash A$. Let the linear ordering of that model be $b_T < c_1 < \ldots < c_n < e_T$, where n may be 0. Define a model M' by:

$v(p) = v'(p)$

$T' = \{ q : q$ is a rational number and $0 \leq q \leq 1\}$

If $v(p) = T$, then

$$t'(p) = \begin{cases} T' & \text{if } n = 0; \text{ otherwise} \\ \{q : i/_{n+1} \leq q \leq j/_{n+1}\} & \text{if } n > 1, \\ & \text{where } c_i \text{ is the least element and } c_j \text{ is the largest element in } t(p) \end{cases}$$

This identifies c_i with $i/_{n+1}$ when $n > 1$. Then M is dense. I'll let you show that for any wff B, $M' \vDash_{\text{TPC-linear}} B$ iff $M \vDash_{\text{TPC-linear}} B$. Hence, $M' \nvDash A$. ∎

Corollary 2 $\vDash_{\text{TPC-linear}} A$ iff for every infinite M, $M \vDash_{\text{TPC-linear}} A$.

Note that in proving that infinite models suffice, we took instants for timelines that are not identified with the beginning or ending point of any interval of time.

No wff can distinguish between finite and infinite models, though a simple infinite collection can.

Theorem 3 Every model of $\Gamma = \{(p_i)_1 <_{\text{eb}} (p_{i+1})_1$ for $i \geq 0\}$ is infinite.

No collection of wffs has only dense models because given any dense model we can always add "irrelevant" points to the timeline of a model to get a model in which the same wffs are true yet which is not dense. However, we can characterize reduced dense models.

Theorem 4 There is a computable collection of wffs Γ such that
if $M \vDash_{\text{TPC-linear}} \Gamma$, then M_R is dense.

Proof Let f be a computable function that enumerates the rational numbers between 0 and 1 without repetitions.[6] That is, for each n, $f(n) = s$, where s is a rational number such that $0 \leq s \leq 1$, and for every rational number s there is one and only one n such that $f(n) = s$. Set:

$\Gamma = \{p_n \wedge_{\text{bb}} p_m : f(n) < f(m)\} \cup \{p_n \approx_{\text{be}} p_n\}$

I'll let you show that if $M \vDash_{\text{TPC-linear}} \Gamma$ then $\langle T_r, <_r \rangle$ is order isomorphic to the rational numbers between 0 and 1, and hence is dense. ∎

---

[6] See, for example, Walter Carnielli's and my *Computability*.

Non-linear time

For **TPC-linear** we required that in every model the timeline is linear. This is not to say we assumed that time is linear. We simply imposed on ourselves the requirement that the times of any two atomic propositions can be compared.

So the time of "Tom played football" begins before the time of "Puff scratched Zoe", or the time of "Puff scratched Zoe" begins before the time of "Tom played football", or they begin at the same instant. But what if we don't know which began first? We make different models, one in which the time of "Tom played football" begins before the time of "Puff scratched Zoe", one in which it begins after, and one in which they begin at the same time, hoping in this way to decide which is correct.

Alternatively, we could allow for models in which time is not linear, either because we believe or want to investigate the idea that time "really is" not linear or because we want to factor our ignorance into our models. Then we could have an ordering and time assignments in a model that look like:

(1)   Puff scratched Zoe 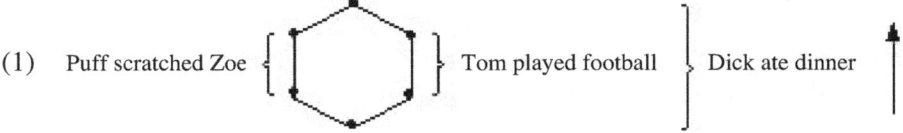 Tom played football   Dick ate dinner

In this picture, the interval assigned to "Dick ate dinner" includes all the points assigned to both "Puff scratched Zoe" and to "Tom played football".

To modify what we have done to get a logic that allows for non-linear times, recall first the definition of an interval with beginning and ending points:

> An *interval* in an ordering $\langle T, \lozenge \rangle$ is a non-empty subset I of T such that for any a, b, c in T, if a and b are in I and $a < c < b$, then c is in I. It has *endpoints* if there are c and d in T such that one of the following holds:
>
> a. $I = \{x : c < x < d\}$ c. $I = \{x : c \leq x < d\}$
>
> b. $I = \{x : c \leq x \leq d\}$ d. $I = \{x : c < x \leq d\}$

If the timeline is not linearly ordered, this could be an interval:

But it is not one with an ending point. The following could also be an interval:

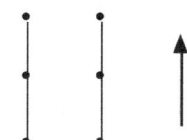

But this has no ending point and no beginning point. However, (1) could be an interval with beginning and ending point.

We can retain the definition of an interval with endpoints to serve for non-linear time. I'll show how to modify what we've done to get a logic that allows for non-linear timelines.

Syntax

The definitions in **TPC-linear** of $p \approx_{bb} q$, $p \approx_{ee} q$, and $p \approx_{be} q$ (p. 22) depend on time being linear to be evaluated correctly. So we have to take those connectives as primitive if we are to allow models in which time is not linear. There is no need to take $\approx_T$ as primitive since if $p \approx_{bb} q \wedge p \approx_{ee} q$, then the interval of p is the same as the interval of q. Hence, our language will be the following.

$L(\neg, \rightarrow, \wedge, \vee, \wedge_b, \wedge_{ee}, \wedge_{be}, \wedge_{eb}, \approx_{bb}, \approx_{ee}, \approx_{be}, p_0, p_1, \ldots)$

i. If q is a sentence-type symbol, each *indexed* version of each sentence-type variable q is a wff of length 1: $(q)_1, (q)_2, (q)_3, \ldots$.

ii. For each $i, j = 0, 1, 2, \ldots$ each of the following is a wff of length 2:

$(p_i \wedge_{bb} p_j)$ $\quad$ $(p_i \approx_{bb} p_j)$

$(p_i \wedge_{ee} p_j)$ $\quad$ $(p_i \approx_{ee} p_j)$

$(p_i \wedge_{be} p_j)$ $\quad$ $(p_i \approx_{be} p_j)$

$(p_i \wedge_{eb} p_j)$

iii. If A and B are wffs and the maximum of the lengths of A and B is n, then each of the following is a wff of length $n + 1$:

$(\neg A) \quad (A \wedge B) \quad (A \rightarrow B) \quad (A \vee B)$

iv. A concatenation of symbols is a *wff* iff it is a wff of length n for some $n \geq 1$.

We define: $p \approx_{eb} q \equiv_{Def} q \approx_{be} p$.

Semantics

We delete the restriction that the ordering of the timeline of a model is linear. We evaluate the new connectives:

$(p \approx_{bb} q) = T$ iff $\mathsf{v}(p) = T$ and $\mathsf{v}(q) = T$ and $b_p = b_q$

$\mathsf{v}(p \approx_{ee} q) = T$ iff $\mathsf{v}(p) = T$ and $\mathsf{v}(q) = T$ and $e_p = e_q$

$\mathsf{v}(p \approx_{be} q) = T$ iff $\mathsf{v}(p) = T$ and $\mathsf{v}(q) = T$ and $b_p = e_q$

The logic TPC The logic of *temporal propositional connectives*, **TPC**, is the formal language and the collection of these models with their tautologies and semantic consequence relation.

Axiomatization

In the strong completeness proof of the axiomatization of **TPC-linear**, the trichotomy of the timeline we constructed for the model in the proof of Lemma C follows from the definitions of $\approx_{bb}$, $\approx_{ee}$, and $\approx_{be}$. By taking those connectives as primitive, trichotomy is not forced. But we have to add axioms to ensure that the new primitive connectives are evaluated as equivalence relations:

i. $p \approx_{bb} p$

ii. $p \approx_{bb} q \rightarrow q \approx_{bb} p$

iii. $(p \approx_{bb} q \wedge q \approx_{bb} r) \rightarrow p \approx_{bb} r$

iv. $p \approx_{ee} p$

v. $p \approx_{ee} q \rightarrow q \approx_{ee} p$

vi. $(p \approx_{ee} q \wedge q \approx_{ee} r) \rightarrow p \approx_{ee} r$

vii. $p \approx_{be} p$

viii. $p \approx_{be} q \rightarrow q \approx_{be} p$

ix. $(p \approx_{be} q \wedge q \approx_{be} r) \rightarrow p \approx_{be} r$

I'll let you show that if we add these axioms to the axioms for **TPC-linear** we get a strongly complete axiomatization of **TPC**.

Time is branching

Time, some say, is branching.[7] At some instants time might not continue in a unique way. In this conception, time still has an arrow, there is a direction of before and after, but the ordering is not linear.

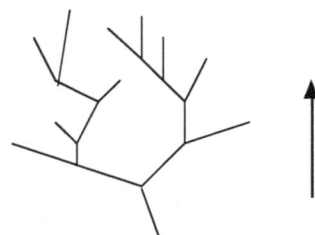

In Chapter 22 I discuss how we can characterize branching orderings, and in Appendix B I discuss what the view of time being branching might mean for our conception of the world. Here it's enough to note that we can't give a syntactic characterization of models whose timeline is branching because that would require quantifying over propositions.

---

[7] See, for example, the discussion in Nicholas Rescher and Alasdair Urquhart, *Temporal Logic*.

11 Past, Present, and Future

We can formalize

as: Spot will bark and then Dick yelled.

(Spot will bark)$_1$ $\wedge_{eb}$ (Dick yelled)$_3$

This can be true in a model since we allow any intervals to be assigned to the atomic propositions of a realization. Yet we know the example can't be true because the tenses indicate that "Dick yelled" is about a time in the past and "Spot will bark" is about a time in the future, and the past comes before the future.

Central to our conception of time as English speakers is the division of past, present, and future. Every sentence in English is tensed via its verbs. If we want to formalize reasoning from English, we must account for the past-present-future divide. Yet not every language makes the division of past, present, and future, and some have no tenses at all.[8] So we don't want to modify our logic to account for that divide. Instead, we can introduce conventions on formalizing that will allow us to apply the logic of relative times to reasoning in English.

In making the division of past, present, future, it seems we assume that time is linear. But that is not needed, as the division could work for branching time. Nonetheless, in what follows I'll assume that we are working in **TPC-linear**.

In our reasoning we talk of the present as any time that's used to split propositions into those meant to be about a time before and those meant to be about a time after. The present, no matter when we take it to be, is an interval, not a collection of disjoint times. In a narrative we can usually identify one or several atomic propositions as picking out the "now" of it. Suppose we want "(Spot will bark)$_1$" and "(Dick yelled)$_3$" together to pick out all the time of the present. Then we need that $t((\text{Spot will bark})_1) \cup t((\text{Dick yelled})_3)$ is an interval, and that is the case if "(Spot will bark)$_1$ $\wedge_O$ (Dick yelled)$_3$" is true. Then any other proposition picks out a time within the present if and only if its time is within that interval. So "(Suzy opened the gate)$_7$" picks out a time within the present iff

$t((\text{Suzy opened the gate})_7) \subseteq t((\text{Spot will bark})_1) \cup t((\text{Dick yelled})_3)$

Using the defined connective W from p. 42, this holds iff the following is true in the model:

W((Suzy opened the gate)$_7$, (Spot barked)$_1$, (Dick yelled)$_3$)

We can choose any two true atomic propositions q and r such that their times overlap and define:

---

[8] See the the essays by Dorothy Lee, Benjamin Lee Whorf, and the discussion of time in "Language and the World" in *Language and the World: Essays New and Old*.

Chapter 11

Present (p) $\equiv_{Def}$ W(p, q, r)

$\vee$(Present (p)) = T iff $\vee$(p) = $\vee$(q) = $\vee$(r) = T and
t(q) $\cup$ t(r) is an interval, and t(p) $\subseteq$ t(q) $\cup$ t(r)

For the past and the future, we first define what we mean by one interval coming before another:

I < J $\equiv_{Def}$ all a in I, all b in J, a < b

Then if we take q and r to define the present in a model:

Past (p) $\equiv_{Def}$ (p $\wedge_{eb}$ q) $\wedge$ (p $\wedge_{eb}$ r)

$\vee$(Past (p)) = T iff $\vee$(p) = T and t(p) < (t(q) $\cup$ t(r))

Future (p) $\equiv_{Def}$ (q $\wedge_{eb}$ p) $\wedge$ (r $\wedge_{eb}$ p)

$\vee$(Past (p)) = T iff $\vee$(p) = T and (t(q) $\cup$ (t(r)) < t(p))

Note that Past (p) means that the time of p is in the past, not that the time of p is all of the past. When we say "the past" or "the future" we needn't be construed as talking about all of the past or all of the future as if those were some things. Moreover, some propositions in a model may be about neither the past, nor the present, nor the future, but about a time that overlaps two or more of those intervals.[9]

We could extend this approach to allow for any number of propositions together to constitute "the present" of a model. But to give the general form of that is complicated: to say in the semi-formal language that the times of four propositions, together form an interval requires 12 separate clauses of the form p $\wedge_O$ q. So for the examples here, I'll assume just two atomic propositions are chosen to determine what we will call the present: (Puff scratched Zoe)$_{26}$ and (Zoe winced)$_{85}$. Then the interval of the present is:

N = t((Puff scratched Zoe)$_{26}$) $\cup$ t((Zoe winced)$_{85}$)

Example 1 *Tom had a dog.*

Past ((Tom had a dog)$_1$)

Analysis Because of the tensed form of "to have" we know that this is meant to be about some time in the past. For the formalization to be true, "(Tom had a dog)$_1$" has to be true.

Example 2 *Tom will have a dog.*

Future ((Tom will have a dog)$_{14}$)

Analysis From the tense of the example, we recognize that it's meant to be about the future.

---

[9] These definitions bear a superficial resemblance to operators used in the tradition of Arthur Prior, but that approach is quite different, as I explain in Appendix A.

Example 3 Tom has a dog.

Analysis The example establishes some time beginning in the past and continuing through the present, which we can formalize using the defined connectives for making temporal comparisons of conjunctions (p. 41):

(Tom has a dog)$_6$ $\wedge_{bb}$ ((Puff scratched Zoe)$_{26}$ $\wedge$ (Zoe winced)$_{85}$)

$\wedge \neg$ ((Puff scratched Zoe)$_{26}$ $\wedge$ (Zoe winced)$_{85}$) $\wedge_{ee}$ (Tom has a dog)$_6$)

Example 4 Spot will bark, and then Dick yelled.

((Spot will bark)$_1$ $\wedge_{eb}$ (Dick yelled)$_3$)

$\wedge$ Future ((Spot will bark)$_1$) $\wedge$ Past ((Dick yelled)$_3$)

Analysis Recognizing the tenses in the atomic propositions for this formalization, we have an anti-tautology. In a model we can't have $N < t$((Spot will bark)$_1$), and t((Dick yelled)$_3$) $< P$, and t((Spot will bark)$_1$) $< t$((Dick yelled)$_3$).

Example 5 Dick had eaten when Zoe arrived.

Past ((Dick had eaten)$_2$) $\wedge_{eb}$ Past ((Zoe arrived)$_3$)

Analysis The past perfect in English is used to indicate that "an action has been completed" relative to another "action" in the past. That is a bad way to say that the time of the proposition is (meant to be) completely before the time of the other proposition, both of which are in the past. After all "Zoe had been a waitress" is not an "action". The formalization then follows.

Example 6 If Spot barks, then Dick will yell. Spot barked. So Dick yelled.

Analysis The first premise is meant to talk about any time at which Spot might bark and Dick might yell, which we can't formalize (compare Example 34 of Chapter 9).

Note that in the first premise the present is meant to indicate the future, and in the last two premises, the past tense has to be construed as the future of the first premise, but the past relative to the speaker.

Summary

We're investigating how to take account of time in our reasoning. We've made no large assumptions about the nature of time. We assume only that, for whatever reason, we are able to make judgments about true atomic propositions describing the world as related by before and after.

We found, however, that "before" is ambiguous. We distinguished four readings of it, each based on comparing beginnings and endings of "what happened". Those beginnings and endings could be conceived of as real in the world or only distinctions we adopt. We then added to classical propositional logic formal equivalents of those four readings as propositional connectives. We gave semantics by assuming some instants of time as ordered, which seemed to require a substantial assumption about the nature of time. But we were able to show through a completeness theorem that all we needed were the beginnings and endings of the times established by true atomic propositions.

Though most of our analyses assumed that times are ordered linearly, we saw we could modify the formal systems to allow for non-linear conceptions of time, too.

With this temporal propositional logic we can formalize a great deal more of ordinary reasoning than we could with classical propositional logic.

Relative time, as formalized here, seems basic to any conception of how we or anyone can conceive of time in their lives and reasoning. Only judgments of before and after of true descriptions are needed.

QUANTIFYING
over
RELATIVE TIMES

12 Quantifying over Relative Times via Indices

Suppose we want to formalize

(1) There was a time when Spot barked after Suzy opened the gate.

All we can do is choose some indices and assert:

(Suzy opened the gate)$_{14}$ $\wedge_{eb}$ (Spot barked)$_{31}$

In building a model, we require that this is true.

But what if we have a model already? To see if (1) is true we would have to consider each pair of indexed sentence-types. If there are only a few of them, we could formalize (1) as:

(2) $\mathcal{W}_{i,j}$ ((Suzy opened the gate)$_i$ $\wedge_{eb}$ (Spot barked)$_j$)

Here $\mathcal{W}_{i,j}$ indicates the disjunction of the indexed wffs, associating to the left.

Suppose we want to formalize:

(3) Every time that Spot barks, Dick yells afterwards.

If there are only a few indices that are used for these sentence-types, we could formalize this as:

(4) $\mathcal{M}_i$ ((Spot barked)$_i$ $\to$ $\mathcal{W}_j$ ((Spot barked)$_i$ $\wedge_{eb}$ (Dick yelled)$_j$))

Here $\mathcal{M}_i$ indicates the conjunction of the indexed wffs, associating to the left.

There may, however, be very many or potentially infinitely many pairs of indexed sentence-types used in the model. Then we cannot use these formalizations. We need to have a way to quantify over times.

In our propositional logic of relative times, we take account of time but do not talk of times. The only times we take into consideration are those established with true atomic propositions. So to quantify over times, we should quantify over true atomic propositions. But we don't need to quantify over all true atomic propositions to formalize (1) and (3) because we're concerned only with propositions of the specific sentence-types that appear in them. We can talk of all those by quantifying over their indices. We can use for (1):

$\exists i \, \exists j$ ((Suzy opened the gate)$_i$ $\wedge_{eb}$ (Spot barked)$_j$)

For (3) we can use:

$\forall i \, \exists j$ ((Spot barked)$_i$ $\to$ (Spot barked)$_i$ $\wedge_{eb}$ (Dick yelled)$_j$)

This is a new kind of quantification, not quantifying over things but over propositions, and not quantifying over all propositions but only indexed ones via their indices. We need to see that we can do this formally.

A formal language

$L(\neg, \rightarrow, \wedge, \vee, \wedge_{bb}, \wedge_{ee}, \wedge_{be}, \wedge_{eb}, p_0, p_1, \ldots, \forall, \exists, i_0, i_1, \ldots)$

The symbols $p_0, p_1, \ldots$ are called *sentence-type symbols*.
The symbols $i_0, i_1, \ldots$ are called *index variables*.

i. If q is a sentence-type symbol, each *indexed* version of it is a wff of length 1: $((q)_1), ((q)_2), ((q)_3), \ldots$. The subscript is the *index*.

ii. If q is a sentence-type symbol, each *variable indexed* version of it is a wff of length 1: $((q)_{i_0}), ((q)_{i_1}), ((q)_{i_2}), \ldots$.
The occurrence of the index variable in these is *free*.

iii. If p and q are wffs of length 1, then each of the following is a wff of length 2:

$(p \wedge_{bb} q) \quad (p \wedge_{ee} q) \quad (p \wedge_{be} q) \quad (p \wedge_{eb} q)$

If an index variable appears in p or q, it is free in each of these wffs.

iv. If A and B are wffs and the maximum of the lengths of A and B is n, then each of the following is a wff of length $n + 1$:

$(\neg A) \quad (A \vee B) \quad (A \wedge B) \quad (A \rightarrow B)$

An occurrence of an index variable in $(\neg A)$ is free iff it is free in A.
An occurrence of an index variable in $(A \rightarrow B), (A \vee B)$, or $(A \wedge B)$ is free iff the corresponding occurrence of the variable in A or in B is free.

v. If A is an wff of length n in which some occurrence of the index variable i is free in A, then each of $(\forall i\, A)$ and $(\exists i\, A)$ is a wff of length $n + 1$.
An occurrence of an index variable in either $(\forall i\, A)$ or $(\exists i\, A)$ is free iff the variable is not i and the corresponding occurrence in A is free.

vi. A concatenation of symbols is a *wff* iff it is a wff of length n for some $n \geq 1$.

Wffs of length 1 are *atomic*; all others are *compound*.
A wff is *closed* iff there is no index variable free in it. Otherwise, it is *open*.
In $(\forall i\, A)$ the initial $\forall i$ has *scope* A and *binds* each free occurrence of i in A, and similarly for $(\exists i\, A)$. We write $A(i)$ to mean the index variable i is free in A.

In addition to our usual meta-variables, we'll use the following:

index variables: $i, j, k, j_0, j_1, \ldots, k_0, k_1, \ldots$

counting numerals: $n, m, n_0, n_1, \ldots, m_0, m_1, \ldots$

Substituting for index variables

An index variable j is *free for an occurrence* of an index variable i in A (that is, free to replace it) iff both:

58   *Chapter 12*

- The occurrence of i is free.
- The occurrence does not lie within the scope of an occurrence of $\forall j$ or $\exists j$.

An index variable j is *free for i* in A iff it is free for every free occurrence of i in A, and we write $A(j|i)$ for A with every free occurrence of i in A replaced with j.

A numeral n is *free for an occurrence* of an index variable i in A if the occurrence of i is free. We write $A(n|i)$ for A with every free occurrence of i replaced with n, and $A(n|i)$ is a *substitution instance* of B. A *complete substitution instance* of A is the result of replacing every free index variable in A by a numeral, the same numeral always replacing the same index variable.

For example, consider:

(Suzy opened the gate)$_i$ ∧ $_{eb}$ (Spot barked)$_j$

This is an open wff. Some substitution instances of it are:

(Suzy opened the gate)$_3$ ∧ $_{eb}$ (Spot barked)$_j$

(Suzy opened the gate)$_i$ ∧ $_{eb}$ (Spot barked)$_7$

(Suzy opened the gate)$_3$ ∧ $_{eb}$ (Spot barked)$_7$

The last is a complete substitution instance.

An example of a substitution instance of

is $\forall i$ ((Suzy opened the gate)$_i$ ∧ $_{eb}$ (Spot barked)$_j$)

$\forall i$ ((Suzy opened the gate)$_i$ ∧ $_{eb}$ (Spot barked)$_{482}$)

Let $i_{d_1}, \ldots, i_{d_s}$ be a list of all index variables that occur free in A such that $d_1 < \cdots < d_n$. The *universal closure* of A is:

$\forall \ldots A \equiv_{Def} \forall i_{d_1} \cdots \forall i_{d_n} A$

For example, the universal closure of

is $((\text{Spot barked})_{i_3} \rightarrow ((\text{Spot barked})_{i_{17}} \wedge_{eb} (\text{Dick yelled})_{i_{32}}))$

$\forall i_3 \forall i_{17} \forall i_{32} ((\text{Spot barked})_{i_3} \rightarrow ((\text{Spot barked})_{i_{17}} \wedge_{eb} (\text{Dick yelled})_{i_{32}})$

Conventions for abbreviating wffs

- The parentheses around an atomic sentence-type can be deleted.
  The parentheses around an atomic wff can be deleted.
  The outer parentheses around an entire wff can be deleted.

- Parentheses between successive quantifiers at the beginning of a wff along with the corresponding right-hand parentheses may be deleted.

- The following bind most strongly in the order below, where those on the same line bind equally strongly:

$\approx_{bb}$ $\approx_{ee}$ $\approx_{be}$ $\approx_{eb}$
¬
∧ ∨
→
∀i ∃i

- When an atomic sentence-type is written without an index, it's meant that the index is 1.

In what follows, I'll consider only linear time as in **TCL**. So we can adopt the definitions of $\approx_{bb}, \approx_{ee}, \approx_{be}, \approx_{eb}$, and $\approx_T$ from p. 22. Modifications for other conceptions of time can be made according to what we did previously for **TC** in Chapter 10.

Realizations and semantics
Realizations and semi-formal languages are defined as for **TCL**.
A *model* is defined as for **TCL**. To define truth in a model, we extend the definition for **TCL** by adding the following clauses:

$M \vDash \forall i\, A$ iff for every numeral n, $M \vDash A(n \mid i)$.

$M \vDash \exists i\, A$ iff for some numeral n, $M \vDash A(n \mid i)$.

I'll let you show by induction on the length of wffs that for every wff A, either $M \vDash A$ or $M \nvDash A$. Then $v(A) = T$ in M iff $M \vDash A$.

The logic **QiTCL** The logic of *quantifying over relative times via indices for linear time* is the formal language and the collection of these models with their tautologies and semantic consequence relation.

An axiomatization
Here L will stand for the formal language of **QiTCL**.

Propositional axioms

  Classical propositional axioms
  The axiom schemes of classical propositional logic (p. 26), where A, B, C are replaced by wffs of L and the universal closure is taken, e.g., $\forall \ldots (\neg A \to (A \to B))$.

  Temporal propositional axioms
  The axiom schemes of **TPC-linear** (pp. 27–28) with the universal closure taken. e.g., $\forall \ldots ((p)_{i_0} \approx_{be} (q)_{i_4} \land (q)_{i_4} \approx_{ee} (r)_{i_8} \to (p)_{i_0} \approx_{be} (r)_{i_8})$.

  Axioms for index variables
  1. a. $\forall \ldots \forall' \ldots (\forall i\, (A \to B) \to (\forall i\, A \to \forall i\, B))$ *distribution of* $\forall$
      if i is free in both A and B

b. $\forall\ldots\forall'\ldots(\forall i\,(A\to B) \to (\forall i\,A \to B))$
   if i is free in A and not free in B

c. $\forall\ldots\forall'\ldots(\forall i\,(A\to B) \to (A \to \forall i\,B))$
   i is free in B and not free in A

2. $\forall\ldots\forall'\ldots(\forall i\,\forall j\,A \to \forall j\,\forall i\,A)$ *commutativity of $\forall$*

3. $\forall\ldots\forall'\ldots(\forall i\,A(i) \to A(j|i))$ *universal instantiation*
   where j is an index or an index variable free for i in A

4. a. $\forall\ldots\forall'\ldots(\exists i\,A \to \neg\forall i\,\neg A)$
   b. $\forall\ldots\forall'\ldots(\neg\forall i\,\neg A \to \exists i\,A)$

Rules

$$\frac{A,\ A\to B}{B}$$ where A and B are closed formulas   *modus ponens*

$$\frac{\text{For every n},\ A(n/i)}{\forall i\,A(i)}$$ where i is the only variable free in A   *index-all*

I'll let you check that the axiom system is sound: If $\Gamma \vdash A$, then $\Gamma \vDash A$.

For the completeness of this system, we begin with lemmas that are the same as for the completeness proof for classical predicate logic (Appendix 5 of Volume 0), and are proved in much the same way.

Lemma   *The Syntactic Deduction Theorem*
$\Gamma, A \vdash B$ iff $\Gamma \vdash A \to B$.
$\Gamma \cup \{A_1, \ldots, A_n\} \vdash B$ iff $\Gamma \vdash A_1 \to (A_2 \to (\cdots \to (A_n \to B)\cdots))$.

Lemma   If $\vdash A \to B$ and $\vdash B \to C$, then $\vdash A \to C$.

Lemma   a. $\vdash \forall\ldots(A \to B) \to (\forall\ldots A \to \forall\ldots B)$
       b. Generalized *modus ponens* $\{\forall\ldots(A \to B),\ \forall\ldots A\} \vdash \forall\ldots B$

Theorem   Let Γ be a consistent set of closed wffs of L.
Then there is a collection of closed wffs Σ such that:

a. $\Gamma \subseteq \Sigma$.

b. Σ is a complete and consistent theory.

c. If $\exists i\,B \in \Sigma$, then for some m, $B(m|i) \in \Sigma$.

d. For every wff $B(i)$ in $L(w_0, w_1, \ldots)$ with one free index variable, if for each m, $B(m|x) \in \Sigma$, then $\forall i\,B(i) \in \Sigma$.

Proof   Let $A_0, A_1, \ldots$ be a numbering of the closed wffs of the language. Define Σ by stages:

$\Sigma_0 = \Gamma$

Σ_{n+1} is defined by cases:

Quantifying over Relative Times via Indices 61

If $\Sigma_n \vdash \neg A_n$, then $\Sigma_{n+1} = \Sigma_n \cup \{\neg A_n\}$.

If $\Sigma_n \nvdash \neg A_n$, then:

    If A_n is not $\exists i\, B$, then $\Sigma_{n+1} = \Sigma_n \cup \{A_n\}$.

    If A_n is $\exists i\, B$, then for the least such m such that $B(m\,|\,i)$ is consistent with Σ, $\Sigma_n \cup \{\exists i\, B, B(m\,|\,i)\}$.

$\Sigma = \bigcup_n \Sigma_n$

We first need to show that if A_n is $\exists i\, B$, then there is some m such that $B(m\,|\,i)$ is consistent with Σ. If there were none, then for each m, $\Sigma \vdash \neg B(m\,|\,i)$. Hence by the index-all rule, $\Sigma \vdash \forall i\, \neg B(i)$. Hence, by axiom scheme 4.a and classical propositional logic, $\Sigma \vdash \neg \exists i\, B(i)$, which contradicts that $\Sigma_n \nvdash \neg A_n$.

Part (a) then follows by construction, and the proof of part (b) follows as in the proof of completeness for classical propositional logic.

Part (c) follows by construction, and part (d) follows by the index-all rule. ∎

Theorem Every consistent collection of closed wffs in L has a model.

Proof Let Γ be a consistent collection of closed wffs of L. Let $\Sigma \supseteq \Gamma$ be as in the previous theorem. We begin with the model of Σ as in the proof of Theorem 2 for **TPC-linear** (p. 29) and add the clauses for the evaluation of the quantifiers to get a model M of **Q*i*TCL**. We extend the induction proof that $M \vDash B$ iff $B \in \Sigma$ from that previous proof to now include wffs that include quantifiers as follows:

Suppose that all wffs B of length less than that of $\forall i\, A$ are such that $M \vDash B$ iff $B \in \Sigma$.

$M \vDash \forall i\, A$ iff   for every numeral n, $M \vDash A(n\,|\,i)$
               iff   for every numeral n, $A(n\,|\,i) \in \Sigma$
               iff   by the index-all rule, $\forall i\, A \in \Sigma$

Similarly, since $\exists i\, A$ has the same length as $\forall i\, A$,

$M \vDash \exists i\, A$ iff   for some numeral n, $M \vDash A(n\,|\,i)$
               iff   it's not the case that for every numeral n, $\neg A(n\,|\,i) \in \Sigma$
               iff   it's not the case that $\forall i\, \neg A(i) \in \Sigma$ by universal instantiation
               iff $\neg \forall i\, \neg A(i) \in \Sigma$ by the completeness of Σ
               iff $\exists i\, A \in \Sigma$ by axiom scheme 4.b.

Hence for every wff B, $M \vDash B$ iff $B \in \Sigma$, and we are done. ∎

Then strong completeness then follows in the usual way.

The *index-all* rule is infinitary because it requires infinitely many premises. I do not have a proof that the index-all rule cannot be deleted or replaced by a finitary rule for a strongly complete axiomatization of **Q*i*TCL**. This matters because if infinitary, we cannot prove a compactness theorem that for every Γ, if every finite subset of Γ has a model, then Γ has a model.

Aside: *The index-all rule compared to the ω-rule*

The ω-rule for formal systems of arithmetic is:

$$\frac{\text{For every } n, \text{A}(n)}{\forall x \, \text{A}(x)} \quad \text{where } n \text{ stands for } 0, \text{S}0, \text{SS}0, \ldots$$

This is very different from the index-all rule. The symbols $0, \text{S}0, \text{SS}0, \ldots$ are names. They are meant to stand for objects of the universe, and in any model of the formal system the universe is infinite. In the index-all rule the indices do not stand for anything; they are not names.

For the same reason, formal systems in which quantification over names is allowed as presented in my *Classical Mathematical Logic* are quite different from quantification over indices.

13 Examples of Formalizing

I'll let i stand for i_{17}, i' for i_{800}, j for j_{32}, and j' for j_{4318}. In the first six examples I won't take account of tenses.

Example 1 $\forall i$ (Spot barked)$_i$

Analysis This is true iff each indexed instance of "Spot barked" is true and hence establishes a time. Those times might not be distinct.

In this system we quantify over indexed instances of sentence-types, and those may be false. So we are not quantifying over times, for times are established by true atomic propositions only.

Example 2 *Spot barked twice.*

$\exists i \, \exists j \, [\,$(Spot barked)$_i \wedge$ (Spot barked)$_j$
$\wedge \neg ($ (Spot barked)$_i \approx_T$ (Spot barked)$_j) \,]$

Analysis For this to be true, there have to be two distinct times established by indexed instances of "Spot barked" (compare Example 30 of Chapter 9).

We can formalize in this way "Spot barked three times", "Spot barked four times", ... But we can't formalize "Spot barked (potentially) infinitely many times". It might seem we could by requiring that for each i, "(Spot barked)$_i$", and "$\neg$ ((Spot barked)$_i \approx_T$ (Spot barked)$_{i+1}$". But that would take us into considering the internal structure of the indices, which would mean that using numerals for indices would be essential rather than just a convenience.

Of course we don't think that Spot barked infinitely many times, not even potentially infinitely many times. But we might want a model in which potentially infinitely many indexed instances of "An electron changed energy level in a radon atom" are true.

Example 3 *Spot barked just twice.*

$\exists i \, \exists j \, [\,$ (Spot barked)$_i \wedge$ (Spot barked)$_j$
$\quad \wedge \neg ($(Spot barked)$_i \approx_T$ (Spot barked)$_j)$
$\quad \wedge \forall i' \, ($ (Spot barked)$_{i'} \rightarrow ($ (Spot barked)$_{i'} \approx_T$ (Spot barked)$_i)$
$\quad\quad \vee ($ (Spot barked)$_{i'} \approx_T$ (Spot barked)$_j)$

Analysis We could not formalize this before (Example 31 of Chapter 9).

Example 4 $\exists i \, ($(Spot barked)$_i \wedge_{eb}$ (Dick yelled)$_i)$

Analysis Nothing significant is conveyed by the use of the same index variable.

Example 5 $\forall i \, \exists j \, ($ (Spot barked)$_i \wedge_{eb}$ (Dick yelled)$_j \,)$

Analysis For this to be true, each indexed instance of "Spot barked" must be true.

Example 6 Dick always yelled after Spot barked.

$\forall i \, \exists j \, [\, (\text{Spot barked})_i \rightarrow [\, (\text{Spot barked})_i \wedge_{eb} (\text{Dick yelled})_j] \,)$

Analysis This is how we quantify over the times that Spot barked. But it could be true if Spot never barked.

For the following examples, assume we have a proposition or conjunction of propositions to pick out the present for formalizing tenses as in Chapter 11.

Example 7 Spot barked before Dick yelled.

or
$\quad$ Past $(\,(\text{Spot barked})_2) \wedge$ Past $((\text{Dick yelled})_{18})$

$\exists i_{17} \, \exists j_{32} \, (\text{Past} \, ((\text{Spot barked})_{i_{17}}) \wedge \text{Past} \, ((\text{Dick yelled})_{j_{32}}) \,)$

Analysis The first follows what we did in the previous section. The particular indices matter. The second is how we can formalize the example in our new system. It formalizes the past tense as "some time in the past". The particular index variables do not matter.

Example 8 Just once Dick yelled before Spot barked.

Analysis It might seem we could formalize this as:

$\exists i \, \exists j \, (\text{Past} \, ((\text{Dick yelled})_i) \wedge \text{Past} \, ((\text{Spot barked})_j)$

$\wedge \, (\forall i' \, \forall j' \, \neg \, ((\text{Dick yelled})_{i'} \wedge_{eb} (\text{Spot barked})_{j'})$

$\vee \, (\, (\text{Dick yelled})_i \approx_T (\text{Dick yelled})_{i'} \wedge (\text{Spot barked})_j \approx_T (\text{Spot barked})_{j'}) \,) \,)$

But suppose Spot barked 100 times this year, and after the first time he barked, Dick yelled. Say we take "(Dick yelled)$_1$" for that. But then "(Dick yelled)$_1 \wedge_{eb}$ (Spot barked)$_j$)" would be true for all the j that establish one of the last 99 times Spot barked, even though after each of those 99 times, Dick yelled after Spot barked.

The problem here, as discussed in Example 39 of Chapter 9, is that we need a way to formalize "shortly before" and then construe the example as "Just once did Dick yell shortly before Spot barked and that wasn't when Dick yelled shortly after Spot barked."

Example 9 There was a time when both Dick yelled and Spot barked.

$\exists i \, \exists j \, [\, ((\text{Dick yelled})_i \wedge_O (\text{Spot barked})_j)$

$\wedge \, \text{Past} \, ((\text{Dick yelled})_i)) \wedge \text{Past} \, ((\text{Spot barked})_j) \,]$

Analysis This better formalizes the idea that when we use the past tense we are saying only that there is some time(s) in the past, no longer being forced to specify which time(s) (compare Example 14 of Chapter 9).

Example 10 Dick talked only when Spot wasn't barking.

$\exists i \, (\text{Dick talked})_i \wedge \, \forall i' \forall j \, (\neg \, (\text{Dick talked})_{i'} \wedge_O (\text{Spot barked})_j) \,)$

Analysis Now we can formalize this (compare Example 36 of Chapter 9).

Example 11 Caesar was assassinated (*only once*).

$\exists i\, [\, $(Caesar was assassinated)$_i \wedge \forall j\, (\, $(Caesar was assassinated)$_j \rightarrow$
((Caesar was assassinated)$_i \approx_T$ (Caesar was assassinated)$_j)\,]$

Analysis Now we can formalize this (compare Example 37 of Chapter 9).

Example 12 Spot didn't bark when Caesar was assassinated.

$\forall i\, \forall j\, (\, $(Caesar was assassinated)$_i \rightarrow$
$\neg\, ((\text{Spot barked})_j \wedge {}_O$ (Caesar was assassinated)$_i)\,]$

Analysis Now we can formalize this (compare Example 38 of Chapter 9).

Example 13 When Zoe cooks, Dick washes up.

Not formalizable.

Analysis We can now talk about any time that Zoe cooks (compare Example 34 of Chapter 9). But we still can't formalize this example because for it to be true, Dick has to wash up shortly after Zoe cooks and he washes the same dishes.

SUMMARY

In our formal systems of temporal propositional logic, talk of some time or all times is about the times established by true atomic propositions in a model. To say that some time Spot barked is to say that some proposition of the form "(Spot barked)$_1$", or "(Spot barked)$_2$", or . . . is true. To say that every time Spot barks Dick yells afterwards, is to say that there for every proposition of the form "(Spot barked)$_1$", "(Spot barked)$_2$", . . . that is true, there is some proposition of the form "(Dick yelled)$_1$", or "(Dick yelled)$_2$", or . . . that is true and whose time is after that. We can formalize such talk in our reasoning by quantifying over indexed instances of atomic sentence-types via their indices. If for each sentence-type there are only finitely many indexed instances in the semi-formal language, such quantification can be seen as just a useful abbreviation, existential quantification being shorthand for a finite disjunction and universal quantification for a finite conjunction. If, though, there are infinitely many instances of even one sentence-type in the semi-formal language, the quantifiers are essential, and to axiomatize the formal system we use an infinitistic inference rule. With this system we can formalize much more.

In a supplement to this text at <www.AdvancedReasoningForum.org> I show how to extend this system to allow for quantification over all atomic propositions rather than just indexed versions.

It is possible to incorporate temporal propositional connectives into classical predicate logic, as I do in a supplement to this text at <www.AdvancedReasoning Forum.org>. But when adopting the metaphysics of things needed for classical predicate logic it seems more apt to view times as things that we can quantify over, as we do in our ordinary speech and reasoning.

TIMES and LOCATIONS as THINGS

TIME in PREDICATE LOGIC

14 The Timelessness of Classical Predicate Logic

The following is often taken as an archetype of a valid inference:

>All men are mortal.
>Socrates is a man.
>Therefore, Socrates is mortal.

Moreover, it's said, the premises are true. So the conclusion is true. Yet Socrates does not exist. That doesn't matter. We use the simple present tense to indicate that we're not talking about any time in particular nor about all times, for this inference would be valid even if there were no men anymore.

We don't often reason like this. Mostly we talk like this when we're doing science, or mathematics, or metaphysics, trying to find out what's true of things without any consideration of time: electrons have spin; dogs bark; 2 + 2 = 4; God is omniscient; the world is made up of individual things.

Predicate logic is designed to formalize such reasoning. No account is taken of the times at which the things in a universe of a model are meant to exist. They all exist in a timeless status, their coming into existence and going out of existence is of no concern. The only existence we reason about in classical predicate logic is that which is suitable to allow for a thing to be the value given to a variable by an assignment of references.[10] We say that we can assign Socrates to x, though Socrates does not exist now. This is not the issue of whether we have the ability to pick out a thing, for we think that we can pick out Socrates among all other things to be the reference of a variable.

So identity is timeless in predicate logic. Though Norma Jeane Mortensen adopted the name "Marilyn Monroe" only when she was an adult, "Marilyn Monroe ≡ Norma Jeane Mortensen" is true: the names pick out the same thing.

Quantification is timeless in predicate logic. For an existential quantification to be true there must be something in the universe that can be assigned to the variable that makes the resulting proposition true, and that assignment is timeless. For a universal quantification to be true, each thing in the universe, regardless of any considerations of time, must satisfy the predicate.

So consider:

(1) All dogs bark.
     Spot is a dog.
     Therefore Spot barks.

---

[10] This is the way to understand W. V. O. Quine's memorable phrase "To be is to be the value of a variable" in "Designation and Existence", p. 708. Quine didn't mean that mud doesn't exist since it can't be the value of a variable, and he amended that dictum to read on p. 13 of "On What There Is": "To be assumed as an entity is, purely and simply, to be reckoned as the value of a variable". In *Predicate Logic* I suggest that predicate logic as a whole characterizes our notion of thing.

In classical predicate logic we formalize this as:

(2) $\forall x\,((\text{— is a dog})\,(x) \to (\text{— barks})\,(x))$
     $(\text{— is a dog})\,(\text{Spot})$
     Therefore, $(\text{— barks})\,(\text{Spot})$

The formal inference is valid. We justify that by saying that if the collection of all things that are dogs is within the collection of all things that bark, then if Spot is a dog, he barks. It is remarkably unclear what we mean by this. The reading of the predicates isn't atemporal nor omnitemporal. It's more like ascribing essential attributes or permanent capabilities or dispositions to things. If Spot is a dog, that is, if he has that attribute without any consideration of time, then Spot barks, without any consideration of time. But that Spot barks is not an essential attribute of Spot. Or perhaps we need to think that it is in order to use classical predicate logic to formalize (1). And perhaps that is indeed what we mean by (1). This is how we must understand the wffs at (2): they are true because being a dog and barking are attributes we ascribe to an object independent of time.

But suppose we wish to use both "— is a puppy" and "— is a dog" in a semi-formal language. These are related in meaning: a puppy is an immature dog. So we should adopt a meaning axiom to codify that:

(3) $\forall x\,((\text{— is a puppy})\,(x) \leftrightarrow ((\text{— is a dog})\,(x) \wedge \neg(\text{— is mature})\,(x)))$

What does this mean? It's said to be atemporal, but nothing is a puppy atemporally. The truth-conditions for (3) are: something is in the collection of things that are puppies if and only if it is in the collection of things that are not mature dogs. Suppose, then, that my dog Birta is in the universe of a model. Is she in the collection of mature dogs or is she in the collection of puppies? To use (3), we must choose, but that means we cannot formalize a true proposition about Birta when she is in the other.

Whatever properties we ascribe to an object in a model must be unchanging. That does not preclude, however, having all of the following true in a model:

$(\text{— was a puppy})\,(\text{Birta})$
$(\text{— is a dog})\,(\text{Birta})$
$(\text{— is mature})\,(\text{Birta})$

Such a model amounts to setting out what is true of certain objects at one particular time. In 2009 both "$(\text{— is a dog})\,(\text{Birta})$" and "$(\text{— was a puppy})\,(\text{Birta})$" are true. Yet if that's how we interpret our models, then (3) does not ensure that the following is true in the model.

$(\text{— was a dog})\,(\text{Birta}) \wedge \neg(\text{— was mature})\,(\text{Birta})$

We would have to add as well:

$\forall x\,((\text{— was a puppy})\,(x) \leftrightarrow ((\text{— was a dog}) \wedge \neg(\text{— was mature})\,(x)))$

But that is false, for Birta was both a puppy and was a mature dog, just not at the same time. We'd need to find a better way to relate "— was a dog" to "— is a dog", and "— was a woman" to "— is a woman", and . . . . To reason about things in time in classical predicate logic, about how objects have different properties at different times, we would have to adopt *ad hoc* meaning axioms governing every atomic predicate.

Classical predicate logic is useful for formalizing reasoning about things outside of time or about essential attributes or permanent attributes of things that are in time. What will occupy us now is how or whether we can modify classical predicate logic to reason about things in time, for it is not clear that the semantic assumptions of classical predicate logic are compatible with taking account of when things exist.

15 Time and Reference in Predicate Logic

If we are to take account of time in classical predicate logic, then the objects we talk about, the things that can be values of variables, will be things in time. Those objects have to have sufficient stability that we can identify and re-identify them, at least in theory. For example, "Birta" refers to a dog—*that*. Yes "she" changed: she was a puppy, now she's a mature dog; she shed hair; she's grown hair. Yet we say that there is one object we pick out with the name "Birta" that persisted through all those "changes". The name "Birta" picks out an object that is not atemporal but supratemporal: not of one time but across many times. It is a difficult question how and when we are justified in making such identifications, which we'll consider more in Chapter 19. To start, it is enough that in practice we can make judgments that are more or less generally accepted.

So when I say that Birta is the value of the variable x, that is a supratemporal reference; not timeless, but ignoring any particular times at which she existed. That is, reference to an object is independent of any time or times. And it is also independent of any other object. Birta was adopted by me, and perhaps the only way we can pick her out is with a description that talks of other objects and times, such as "the brown dog that Arf adopted at the animal pound on August 23, 2001". But once the reference is established, we need not consider those objects and times. It's enough to say that Birta is the value of x.[11]

But when we as speakers of English divide time into past, present, and future, it seems that when a thing exists is important for how we refer to it.

There are some future-tense propositions we have good reason to believe. I have good reason to believe "No dog lives more than 30 years" and "Chocolate is a dog", so I can conclude "Chocolate will be dead 31 years from now". Every scientific law is meant as true of all times, including the future. For instance, "Electrons have spin" is meant not as a proposition that is true of the present or timelessly, but true for all times, including the future where it serves as a prediction: any electron in the future will have spin.

On the other hand, now as I'm writing we have no reason to believe that "Eduardo Ribeiro will have a cold two years from now" is true nor to believe it is false. When we do not have good reason to believe that a future-tense sentence is true or that it is false, we can treat it as we treat any other proposition whose truth-value we do not know: we consider a model in which it is true and a model in which it is false in order to investigate its consequences and perhaps learn whether it is true or whether it is false. But that doesn't help us with the problem of how to use variables in such reasoning. We need a notion of pointing and naming.

---

[11] As I write this now in 2017, sadly Birta no longer is alive. But she continues to exist in the sense needed to reason about her in predicate logic.

Suppose we have a realization and wish to take for its universe all creatures that are living, have lived, or will live. What can it mean to have in the universe some thing that has never lived but will live three years from now? If we wish to use predicate logic, we are under the obligation to explain what we mean when we say that we can pick out and re-identify any object in the universe to serve as a reference for a variable.

If everything in the universe is picked out with a name from our ordinary speech, there is no problem; we can assume that we know how to use ordinary names well enough. But for a universe of all mountain lions in New Mexico it would be too difficult to name each one of them. Yet we understand well enough what it means to say "that mountain lion" when pointing at one, and we take that as our explanation of how we'll use variables.

For a universe of all creatures that have ever lived, it is more difficult to say what we mean by pointing to a donkey that lived in Julius Caesar's time. But unless we have some notion of what we mean by that, the semantics and inductive definition of truth in a model for predicate logic are an empty formalism. We can pick out such an animal by linguistic means by saying "the donkey that is referred to by — in —". But of all the creatures that have lived, the existence of only a few were ever noted in writing. Yet we want to say that there must have been a donkey alive when the first humans lived in Africa. How do we know that such a donkey lived? We infer its existence from evidence we now have.

To pick out things by evidence is to use an informal logic to establish the semantics of our formal logic. It is an informal logic that must remain informal on pain of an infinite regress on what we mean by pointing and naming. Yet we can say that we know well enough what it means to point and name for donkeys, or for any creature that ever lived. The problem is that we can't be there to do the pointing and naming, just as we can't point and name every mountain lion now. Our notion is clear enough; it's just that we're not able to do it. And, we could argue, it is only the method not the execution that is needed for the use of the semantics of predicate logic. We need not be complete empiricists to use predicate logic.

For a universe that includes all creatures that will live any time up to 30 years from now the problem of pointing and naming seems quite different. To say that the baby that will be born to Tom and Suzy three years from now is the reference of x seems to invoke a determinism, an assumption that such a thing will exist, though it doesn't exist now and we have no evidence that it will exist. We can't invoke evidence to use a variable in that way.

The problem would be simpler if we could confine our reasoning to only things in the future that are named. We could treat "Humbert", "Lucilinda", "Meribel" as names for things that are not alive now but will be alive within 30 years from now in the same way we treat fictional names, as we did in Volume 1. A proposition such as "Meribel will be a girl" is assigned a truth-value, and we reason with that without assuming there is an object that "Meribel" refers to. But we want to analyze inferences about the future that are not confined to talk about things we provide names for now. It seems clear that the following is valid:

> If Tom and Suzy have a baby, it will be born in a hospital.
> If a baby is born in a hospital, the parents will have to pay a doctor.
> Therefore, if Tom and Suzy have a baby, they will have to pay a doctor.

If we wish to use predicate logic to reason about future-tense sentences, it is not enough to say that such sentences are propositions because we can treat them as if they are true or false. We need to explain what we mean by pointing and naming. To do that I see no way except to invoke the general method we use for the kind of objects we are talking about that do exist. We talk about all creatures that will live 30 years from now by invoking the method of pointing and naming we use for creatures that are alive now. We survey different universes to determine what is valid: one in which there is a baby born to Tom and Suzy three years from now, one in which there is a baby born to Tom and Suzy eight years from now, one in which there is no baby born to Tom and Suzy, ... For each of those we claim that the notion of pointing and naming is clear (enough).[12]

If you agree that these comments are adequate justification for our use of predicate logic to analyze reasoning with future-tense sentences as propositions, perhaps our work here will be useful to you. If you disagree, then you can excise all such applications to allow for formalizations of how to reason only about the present and the past. If you disagree further that we can't use the semantics of predicate logic for past-tense sentences as propositions, then the applications that follow will be at best curious, an attempt to convert a logic of timeless propositions into a logic of temporal propositions.

Aside: *Waismann on future tense propositions and determinism*

We have made no judgment here about whether the future is determined.[13] Given a future-tense proposition, such as "Richard L. Epstein will die in 2029", we consider one model in which it is true and one in which it is false. We see the consequences of each choice, but we do not assume or know which is the "correct" model. This is consonant with the analysis that Friedrich Waismann gives in "How I See Philosophy", pp. 8–10:

---

[12] Nicholas Rescher and Alasdair Urquhart in *Temporal Logic*, p. 236, take a similar approach to resolving issues of temporal reference:

> Let us, for reasons such as this, eschew a rigorous nominalism, in the context of the present discussion, and be prepared to accept collections that are given, not by an extensionalistic display, but in terms of an abstract criterion of membership, and so be prepared to accept future individuals of a given type (e.g. liars) which, *ex hypothesi* lies beyond the reach of any ostensive procedure such as an enumeration or labeling.

[13] See Richard Taylor, "The Problem of Future Contingencies" for a discussion that touches on many issues in what follows.

This doubt has taken many different forms, one of which I shall single out for discussion—the question, namely, whether the law of excluded middle, when it refers to statements in the future tense, forces us into a sort of logical Predestination. A typical argument is this. If it is true now that I shall do a certain thing tomorrow, say, jump into the Thames, then no matter how fiercely I resist, strike out with hands and feet like a madman, when the day comes I cannot help jumping into the water; whereas, if this prediction is false now, then whatever efforts I may make, however many times I may nerve and brace myself, look down at the water and say to myself, "One, two, three—", it is impossible for me to spring. Yet that the prediction is either true or false is itself a necessary truth, asserted by the law of excluded middle. From this the startling consequence seems to follow that it is already now decided what I shall do tomorrow, that indeed the entire future is somehow fixed, logically preordained. Whatever I do and whichever way I decide, I am merely moving along lines clearly marked in advance which lead me towards my appointed lot. We are all, in fact, marionettes. If we are not prepared to swallow *that*, then—and there is a glimmer of hope in the "then"—there is an alternative open to us. We need only renounce the law of excluded middle for statements of this kind, and with it the validity of ordinary logic, and all will be well. Descriptions of what will happen are, at present, neither true nor false. (This sort of argument was actually propounded by Łukasiewicz in favour of a three-valued logic with "possible" as a third truth-value alongside "true" and "false".)

The way out is clear enough. The asker of the question has fallen into the error of so many philosophers: of giving an answer before stopping to ask the question. For is he clear what he is asking? He seems to suppose that a statement referring to an event in the future is at present undecided, neither true nor false, but that when the event happens the proposition enters into a sort of new state, that of being true. But how are we to figure the change from "undecided" to "true"? Is it sudden or gradual? At what moment does "it will rain tomorrow" begin to be true? When the first drop falls to the ground? And supposing that it will not rain, when will the statement begin to be false? Just at the end of the day, at 12 p.m. sharp? Supposing that the event *has* happened, then the statement *is* true, will it remain so for ever? If so, in what way? Does it remain uninterruptedly true, at every moment of day and night? Even if there were no one about to give it any thought? Or is it true only at the moments when it is being thought of? In that case, how long does it remain true? For the duration of the thought? We wouldn't know how to answer these questions; this is due not to any particular ignorance or stupidity on our part but to the fact that something has gone wrong with the words "true" and "false" applied here.

If I say, "It is true that I was in America", I am saying that I was in America and no more. That in uttering the words "It is true that—" I take responsibility upon myself is a different matter that does not concern the present argument. The point is that in making a statement prefaced by the words "It is true that" I do not *add* anything to the factual information I give you. *Saying* that something is true is not *making* it true: cf. the criminal lying in court, yet every time he is telling a lie protesting, his hand on his heart, that he is telling the truth.

What is characteristic of the use of the words "true" and "false" and what the pleader of logical determinism has failed to notice is this. "It is true" and "It is false", while they certainly have the force of asserting or denying, are not descriptive. Suppose that someone says, "It is true that the sun will rise tomorrow" all it means is that the sun will rise tomorrow: he is not regaling us with an extra-description of the trueness of what he says. But supposing that he were to say instead, "It is true *now* that the sun will rise tomorrow", this would boil down to something like "The sun will rise tomorrow now"; which is nonsense. To ask, as the puzzle-poser does, "Is it true or false *now* that such-and-such will happen in the future?" is not the sort of question to which an answer can be given: which *is* the answer.

16 Times as Things

One way we can take account of time in predicate logic is suggested by how we talk.

> Some time before Spot barked, Suzy opened the gate.
> Dick never pets Puff.
> Flo spilled a drink three times yesterday.
> Harry always calls before he visits.

We talk about how many times, counting them: some, none, three, all.

If we're going to use the methods of classical predicate logic to quantify over times, we'll have to treat times as things, for it is things we quantify over in classical predicate logic. That seems to be a big assumption.

Time, in our experience, is not a collection of times. Time is a mass. Every part of time is time, just as every part of mud is mud. There are, in our experience, no smallest parts of time, just as there are no smallest parts of mud. The time when Spot barked contains the time when Suzy was startled. The time when Suzy was startled contains the time when Suzy flinched. The time when Suzy flinched contains the time when Suzy blinked. We have no experience of smallest times, so how can times be things?

Water is a mass, too, pervading the world, not coming in bits that are things.[14] Yet we can talk of this cup of water, of that lake of water, of that bathtub of water, treating them, conceiving of them, experiencing them as things. So, too, with time. We can use containing descriptions: the time when Birta barked at Buddy, the time when I slipped getting out of the shower. I can focus my attention on a part of the mass of time, treating that part as a thing, and hope to direct your attention to that part of time by using a description. Time does not come packaged as things as dogs come to us as individual things. The parts of time, the things that are times are what we pay attention to by containing descriptions.

Some disagree. Time, they say, is composed of instants, indivisible, like points on a line. Each instant has no duration, as each point on a line has no breadth. The time that Birta barked is a collection of those instants. Instants, like points on a line, are real, independent of us and our descriptions. Times are things: instants and collections of instants.

This is to make into things the result of our abstracting. A point is a part of space we mark or describe that is sufficiently small for our purposes at hand to treat as having no dimension: the point we make with a pencil when we wish to draw a line to saw a piece of wood; a distant star as a "point mass" in calculations. An instant is a part of time we describe that is sufficiently small for our purposes at hand to treat as having no duration, such as the time that Suzy blinked.

---

[14] To assert that H_2O molecules are the smallest bits of water is to mistake an abstraction for our experience, as I explain in "Models and Theories" in *Reasoning in Science and Mathematics*.

If time were the totality of durationless instants, we would have to explain how it is that many of those together can make an interval of time that does have duration. Some say it's the same as with points of space: uncountably many together can give an interval that has length. But that would be to postulate a metaphysics that is ungrounded in experience, mistaking our abstracting—paying attention to some times as if they were indivisible for our purpose—for a reality. It would leave us in a mathematical wonderland with no justification for how to analyze simple temporal inferences.

Let us base our work here on experience. We can treat times as things, each of which can contain other times until we no longer wish to consider smaller parts of time. We pick them out, as we must in order to use them in the semantics of predicate logic, by descriptions.

But what of times we name, like "September 20, 2009", "1985", "5:02 p.m. March 4th, 1831"? Surely those pick out times without any description: the times are there, independent of us and any way we may choose to talk of them. When I say, "I fed the sheep at 7:09 p.m. Saturday, April 7, 2018", the time name is shorthand for a description: when Arf looked at his watch and it said "7:09 p.m. Saturday, April 7, 2018". A scientist who uses the time name "April 8, 2018 at 6:04.0002113 p.m." is using a description: when the standard clock indicated "April 8, 2018 at 6:04.0002113 p.m.". Our clocks are used to describe parts of time.[15]

To treat times as things in the semantics of predicate logic, we will need to assume that we have some way of picking out times, of specifying a reference, a value, for a variable. The specific method of picking out times, whether with verbal descriptions of "what happened" or pointing to a clock, are part of how we specify a model, as the ways we pick out dogs or numbers are part of how we specify a model.

As with our talk of individual things, we can treat a reference to a time as independent of other times and things. That's not to say that a particular description of an object or a time does not and need not involve talk of any other object or other time—that would be to invoke some kind of abstract pure reference, as discussed in Chapter 5.H of Volume 0. It's that we can focus on the individual thing-in-time or time-as-a-thing in our reasoning independently of what things or times are used in a particular description.

Things in Time, the World, and Propositions The world is made up at least in part of individual things that exist in time. Times are also (can be conceived of as) individual things. Reference to each of those things-in-time is independent of any time or any other individual thing. Reference to each of those times-as-things is independent of any individual thing or other time.

The only propositions in which we are interested in are those that are about things-in-time and/or times-as-things.

---

[15] Time names have to specify a particular place on earth for which the times are marked. Here let's assume that's the center of the universe, Dogshine, the home of the Advanced Reasoning Forum.

Chapter 16

A time can be a value of a variable. But we don't want "$\forall x_1$ (— is a dog) (x_1)" to be false because yesterday isn't a dog. So we'll adopt new variables $t_0, t_1, t_2, \ldots$ to stand for times. These new variables are meant to take values from a universe of times T in a model that is distinct from the universe U of things from which the usual variables $x_0, x_1, x_2, \ldots$ take values.

As noted above, we also use names for times, such as "5:02 p.m. March 4th, 1831", which we'll call *fixed time markers*. We needn't limit ourselves to just those names, however. I might want to use "BB" for the time that my dog Birta was born, since I don't know the date. So let's adopt the symbols $b_0, b_1, b_2, \ldots$ that can be realized as names of times from T, different from the name symbols $c_0, c_1, c_2, \ldots$ that can be realized as names of things from U.

I'll use $t, t', w, w', w_1, w_2, \ldots$ to stand for time variables or time name symbols.

Aside: Time as things and time as later than

In English it seems natural to speak of times as things. But there are languages in which there is no or only a quite secondary notion of individual thing and which do not talk of times. Speakers of those "reckon time" using a relation of later than, as we do with "Dick yelled after Spot barked". Vera da Silva Sinha in *Linguistic and Cultural Conceptualisations of Time in Huni Kui, Awety´, and Kamaiurá Communities in Brazil* shows this for some Amazonian tribes, and Benjamin Lee Whorf in "The Relation of Habitual Thought and Behavior to Language" makes this point comparing how Hopi and Western European speakers talk.

17 Basic Assumptions about Times

We are not giving a theory of time. Still, we must begin with some assumptions about times if we are to have any theory that incorporates talk of times.

Orderings of times
Central to our conception of time is that we understand times to be ordered by before and after. The time when Spot barked is before the time when Dick yelled. The time when I slipped getting out of the shower is after the time when I moved to Dogshine.

There is an ordering of times In a model, there is an ordering on the universe of times. That is, there is a relation < on T that satisfies:

> Not t < t. < is *anti-reflexive*
>
> If t < w and w < t´, then t < t´. < is *transitive*

It follows that every ordering satisfies:

> If t < w, then not w < t. < is *anti-symmetric*

That's because if we were to have both t < w and w < t, then by transitivity t < t, which would contradict that the ordering is anti-reflexive.

Let's adopt "— $<_{time}$ —" as a predicate which we'll interpret as the ordering of times in a model. We want it to be available in every semi-formal language, so we'll put it in the formal language. Hence, we'll be able to use it in formalizations even though we'll allow different interpretations of it.[16]

Two abbreviations will be useful:

t ≤ w ≡$_{Def}$ t < w or t = w

w_1 < w_2 < w_3 ≡$_{Def}$ w_1 < w_2 and w_2 < w_3
 and similarly for ≤

We do not assume that the times are all lined up in order. Each of the following could be an ordering of times in a model (the arrow shows the direction of the ordering):

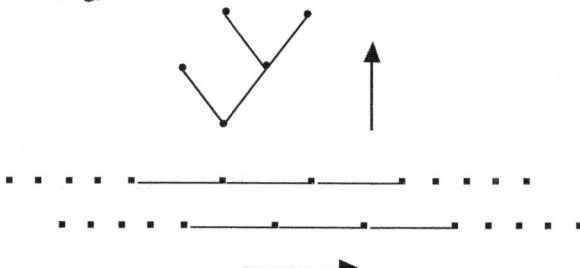

---

[16] Compare the use of "∈" in formal theories of collections.

Times within times

Central to our conception of time as described in the last chapter is that there can be a time within a time. So in every model we will have a binary relation W on the universe of times that we'll understand as one time being within another. The relation will not be total: the time I am typing this is neither contained in nor contains the time I fed the sheep this morning.

We're not consistent in our usual speech about whether a time is a part of itself. Let's stipulate that it is: for every t, $W(t,t)$.

Two distinct times cannot each be within the other, for otherwise the relation would be circular. So for every t and w, if $W(t,w)$ and $W(w,t)$, then $t = w$.

The time t_1 I've been typing this chapter today is within the time t_2 I've been at my computer this afternoon: $W(t_1, t_2)$. The time t_2 I've been at my computer this afternoon is within the time t_3 I've been in my office this afternoon: $W(t_2, t_3)$. So the time I've been typing this chapter this afternoon is within the time I've been in my office this afternoon: $W(t_1, t_3)$. The within-relation is transitive.[17]

The part-whole relation on times In every model, there is a binary relation W on the universe of times T that satisfies:

$W(t,t)$	W is *reflexive*
If $W(t,w)$ and $W(w,t)$, then $t = w$.	W is *anti-symmetric*
If $W(t_1, t_2)$ and $W(t_2, t_3)$ then $W(t_1, t_3)$.	W is *transitive*

Suppose now that the time from 12:16 p.m. until 12:18 p.m. on January 6, 2001 has exactly the same times in it as the time during which my dog Birta was born. What, then, could distinguish those two times? To assume that we could distinguish them is to assume some further property of times that we haven't taken account of. So let's proceed on the assumption that there isn't anything that distinguishes them.

Parts determine times $t = t'$ iff (for all w, $W(w,t)$ iff $W(w,t')$)

Let's adopt "$W_{time}(-,-)$" as a predicate that we'll interpret as the relation of one time being within another in a model. This, too, will be in the formal language, though we'll allow different interpretations of it.

Overlapping times

Maria went to Tom's football game Saturday, but she had to leave early to go to work. At work she listened to the end of the game on radio. The time of the game overlapped, from before, the time Maria was working. The week starting March 20, 2009 overlaps, from before, the week starting March 23, 2009. We can define a relation of one time overlapping another time from before.

---

[17] These three assumptions are enough to guarantee that whatever relation W we choose is a part-whole relation as characterized by Roberto Casati and Achille Varzi in *Parts and Places*.

$O_<(w_1, w_2) \equiv_{Def}$ there is some t such that $W(t, w_1)$ and $W(t, w_2)$,
and there is some t' such that $W(t', w_1)$ and
for all w such that $W(w, w_2), t' < w$.

The last clause ensures that neither $W(w_1, w_2)$ nor $W(w_2, w_1)$, so $w_1 \neq w_2$.

To define that w_1 overlaps w_2 afterwards, we can use $O_<(w_2, w_1)$. More generally, we can say that two times overlap if there is some time in both.

Overlapping times $O(w_1, w_2) \equiv_{Def}$ for some t, $W(t, w_1)$ and $W(t, w_2)$.

By this definition, every time overlaps itself since $W(w, w)$.

Is the time of Tom's football game before or after the time when Maria worked on Saturday? It's sort of before—not really before but it started before. Sort of before is not before; nor is it after. Is March 21, 2009 before or after March, 2009? Is the time Dick cooked dinner before or after the time he was heating up the frying pan? Neither. Overlapping times are neither before nor after one another.

Overlapping times are not related in the ordering
   If $O(w_1, w_2)$, then neither $w_1 < w_2$ nor $w_2 < w_1$.

The time t_1 that I've been typing this chapter this afternoon is within the time t_2 I've been in my office: $W(t_1, t_2)$. The time I've been in my office is before the time t_3 I'll feed grain to my sheep: $t_2 < t_3$. So it follows that the time I've been typing this afternoon is before the time I'll feed grain to my sheep: $t_1 < t_3$. A part of a time is related to other times in the ordering as the whole is related.

Parts of times are related to other times as the whole is related
   If $W(t_1, t_2)$ and $t_2 < t_3$, then $t_1 < t_3$.
   If $W(t_1, t_2)$ and $t_3 < t_2$, then $t_3 < t_1$.

Intervals of time
Yesterday at 3 p.m. and today at 9 a.m. are not one time. The time I was in my office last Monday and the time I was feeding my sheep a couple hours later are not one time. Nor are all the times I fed my sheep last week one time—indeed, we say "all the times". We don't conceive of just any collection of times as a whole. To be a whole, to be a time that has parts, we expect that the times in it are sequentially connected. Thus, March 2009 is a whole, though it has parts like March 20, 2009. It might seem that we could define connectedness by assuming that for every time w, if $W(t_1, w)$ and $W(t_2, w)$ and $t_1 < t_2$, then for every t_3 such that $t_1 < t_3 < t_2$, $W(t_3, w)$. That's certainly needed to ensure that there are no gaps, but it's not enough, for it would allow the following to qualify as sequentially connected:

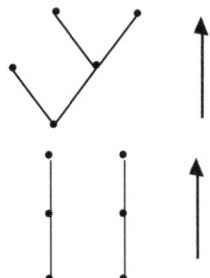

We need in addition that that the times within a time are all connected: if there are two times in the time, then either they overlap or one is less than the other.

Intervals A collection of times **I** is an *interval* iff
if t_1 and t_2 are in **I**, then
$O(t_1, t_2)$ or $t_1 < t_2$ or $t_2 < t_1$, and
if $t_1 < t_2$, then for every t_3 such that $t_1 < t_3 < t_2$, t_3 is in **I**.

And from what we said above, we assume the following.

Times are intervals Every time is an interval.

Should we further assume that every interval is a time? We could have a timeline with just four times:

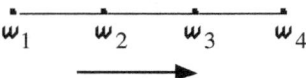

Each time is an interval: itself. There are no other times that are intervals. There are no overlapping times. Though we can talk of the interval that is w_1 and w_2, it isn't a time. Though the interval of w_1, w_2, and w_3 overlaps the interval of w_2, w_3, and w_4, neither of those intervals are times. This seems odd. We can specify the interval, pick it out as if it were a time. But that's only in the metalogic. There might be no way to pick out these intervals in the semi-formal language. To say that every interval of time is a time we would need to quantify over all intervals, and that means quantifying over collections of times, which would require second-order logic.

Suppose we have a model in which **T** is just those times that are picked out by fixed time markers. In **T** there is an interval from 11:21 a.m. November 26, 2020 to 11:23 a.m. November 26, 2020. But that is not a time in **T**. And all times after 11:23 a.m. November 26, 2020 constitute an interval, but that is not a time in **T**. To include each interval in **T** as a time would expand **T** greatly. So I'll not assume that in every model every interval is a time, though you can add that condition for particular models.

18 Times and Predications

Syntax

(1) Zoe sang at 4:15 p.m. November 4, 2003.

This proposition is about Zoe and about 4:15 p.m. November 4, 2003. So to formalize it, we need a blank to be filled with a term for a time along with a blank to be filled with a term for an individual thing.

$\quad$ (— sang —$_{time}$) (Zoe, 4:15 p.m. November 4, 2003)

But this won't do because the predicate has a time indication built into it with the use of the past tense. We need a tenseless form of the predicate. In English we have that with the infinitive "to sing":

$\quad$ (— to sing —$_{time}$) (Zoe, 4:15 p.m. November 4, 2003)

$\quad$ But consider:

(2) Zoe is a woman.

The predicate in (2) is "— is a woman". The infinitive form of that is "— to be a woman". That seems to mean that being a woman is a process rather than a classification, as if we were to say that Zoe is being a woman at July 29, 2017. There seems to be a big difference between a process predicate, such as "— sings", and a classification predicate, such as "— is a woman". When we assert "Zoe sings" we are asserting that Zoe is doing something, whereas when we assert "Zoe is a woman" we are asserting that Zoe is something. The latter is static.

$\quad$ To say that "— is a woman" is static and "— sings" is process is to divide our predicates according to duration. When we assert "Zoe is a woman" we assume a longer duration than for "Zoe sings". If we mean, on the other hand, that right now Zoe is a woman, then that can't be distinguished in terms of duration from right now Zoe sings. Certainly there was a time when Zoe wasn't a woman, and then she became a woman over a period of time. It's just that the time she remains being a woman is much longer than the time she sings—or at least we expect it to be. But she could have sung for a whole week, hit puberty and then died the next day. Moreover, there are plenty of process predicates whose duration is long, such as "— breathes". Dick breathes now and his whole life, so "— breathes" is even more static than "— is a man". I can see no way to justify a distinction between what is static and what is process. The bad versions of Aristotelian logic tried to make everything into static conditions: "Plato is running" was converted to "Plato is a running being".[18] That seems far from our experience. Converting all to process,

---

[18] A similar view of all as static is the basis of attempts to use events as a semantic foundation of predicate logic as discussed in Appendix 2 of Volume 1. See also the discussion of the dynamic versus the static in Appendix 3 of Volume 1.

however, seems more a stressing of one view we hold but ignore a great deal of the time. So let's formalize (2) as:

(— to be a woman —$_{time}$) (Zoe, July 29, 2017)

Ordinary simple temporal atomic predicates Given an *n*-ary ordinary language predicate P that we could take as a simple predicate in classical predicate logic, we can convert it to an infinitive phrase I$_P$. Then the *ordinary simple temporal predicate* we derive from it is the *n* + 1-ary predicate with *n* unmarked blanks: I$_P$ (—, ..., —, —$_{time}$).

Only terms for individuals can fill the unmarked blank(s) in an ordinary simple atomic predicate, and only a time term can fill the time-marked blank. We allow no predication about things in time that does not take account of time. Infinitive predicates are atemporal in the sense that we understand whether they apply to an object or objects at any given time.

In our semi-formal languages, let's place the blanks for the individual terms as we derive them from our ordinary speech and add the blank for a time term at the end. We'll leave out the commas in time names that are dates so that commas will have just the role of separating terms and blanks in wffs. So we'll have atomic wffs:

(— to bark —$_{time}$) (Spot, 4:15 p.m. November 4 2003)

(— to be a dog —$_{time}$) (Dick, t_4)

(— to be a dog —$_{time}$) (x_7, BB)

(— to be taller than —, —$_{time}$) (Socrates, Plato, 398 B.C.E.)

In our ordinary speech we incorporate times into predicates in other ways:

Tom dated Suzy during 1999.

Suzy will be a cheerleader in 2017.

Flo cried throughout the night of September 18, 2001.

Should we adopt different time-marked blanks to formalize "during", "in", and "throughout"? We'll see in examples of formalizing that with just the one time-marked blank meant to formalize "at" we can formalize uses of these.

Semantics

Suppose Zoe says:

Spot barked at 12:19 a.m. August 18, 2017.

Does this mean he barked at just this time? He could have started at 12:18 a.m. and continued barking until 12:24 a.m. until Dick shouted at him to stop.

If we say that "Spot barked" is true of the minute 12:19 a.m., August 18, 2017, does that mean he never stopped barking for even a second to take a breath or look around? Did sound have to keep coming out of his mouth the entire minute? It takes less than a whole minute for Spot to bark, but how much less? Could Spot have barked at a millisecond within 12:19 a.m.? Doesn't there have to be some minimal interval of time for barking to occur?

If Spot barked at 12:19 a.m., then throughout that entire time (minute) he barked, for barking continues during each millisecond. If Spot was a dog at July, 2017, then he was a dog throughout all that time. There are no gaps. What we mean by a predicate being true of an object at a time has to be taken as primitive, open to many interpretations. But we can agree that if a predication applies to an object or objects at a time, then it applies to that object or objects at every part of that time. That is, *the truth of a predication is closed downwards in time*. This does not involve us picking out beginnings and endings or postulating that for every true predication there is a minimal time for it.

Does the converse hold? Suppose Spot barked at every minute in the hour of 10 p.m. April 3, 2017. Does this mean that he barked at the hour of 10 p.m.? If not, what would distinguish barking at that interval from barking at every time within that interval? After all, we've assumed that a time is completely characterized by the times within it. If a predication applies to an object or objects at all times within a time, then it applies to that object or objects at the larger time. That is, *the truth of a predication is closed upwards in time*.

Downward and upward closure of truth in time

$I(-,\ldots,-,-_{time})(x_1,\ldots,x_n,t)$ is true of $\mathsf{o}_1,\ldots,\mathsf{o}_n,\mathsf{t}$ iff
for all t' such that $W(\mathsf{t}',\mathsf{t})$, $I(-,\ldots,-,-_{time})(x_1,\ldots,x_n,t)$
is true of $\mathsf{o}_1,\ldots,\mathsf{o}_n,\mathsf{t}'$.

Aside: Time indications as adverbs?

Someone might say this approach misconstrues the role of the time phrases in (1) and (2). They're adverbial.

Some think that time adverbial phrases can be treated as propositional operators. But we don't do that with other adverbs: it makes no sense to read "Juney barked loudly" as "loudly (Juney barked)". Attempts that have been made to treat times as propositional operators are not coherent, as I explain in Appendix A.

In Volume 1, though, we saw how to formalize adverbs as part of the internal structure of predicates. Why not do that here? Compare:

Spot barked loudly.
Spot barked near Dick.
Spot barked at 4:05 p.m. April 3, 2017.

We formalize the first two by taking "loudly" and "near Dick" as restrictors of the predicate "— barked". Perhaps we could treat "at 4:05 p.m. April 3, 2017" by viewing "at —" as a variable restrictor:

(*)   ((— barked) / at (4:05 p.m. April 3 2017)) (Spot)

If "at (—)" is a variable restrictor, then from (*) we could conclude:

(— barked) (Spot)

But this uses an unanalyzed time marking for the predicate. Perhaps we could use:

((— to bark) / at (4:05 p.m. April 3 2017)) (Spot)

But if "at (—)" is a variable restrictor, we could conclude :

(— to bark) (Spot)

This would be a timeless application of "to bark", and no predicate is meant to apply timelessly due to our assumption *Things in Time, the World, and Propositions* (p. 79).

19 Identity and the Equality Predicate

In classical predicate logic we can assert:

   Marilyn Monroe ≡ Norma Jeane Mortensen

This is true in a model just in case the name "Marilyn Monroe" picks out the same thing as the name "Norma Jeane Mortensen" regardless of any considerations of time. Reference is atemporal. We've assumed the same here: reference does not depend on time, though it picks out objects in time.

There is a tension in our system: we consider time in every predication, but we take objects to be in some way supratemporal. We take a universe of a model to be a collection of things in time. Yet everything that exists in time changes.

Suppose I had pointed out my dog Birta to you twelve years ago. Then you visit and I say, pointing, "That's Birta". You're surprised. "That's the same dog I saw all those years ago?" She looks a lot different—she was just a puppy then. There was a thing twelve years ago; now there's this thing. I say they're the same. You doubt me. And in one sense you're right. I know they're not identical; I am drawing an equivalence. The equivalence is justified, implicitly, by our sharing a notion of what it means for things at different times to be judged to be the same dog.

We use equivalences built on implicit understandings of what "the same" means relative to any particular kind of thing. We look at an apartment house we both knew as a single-family home four years ago. We don't say "That's the same home that was here four years ago"; we do not say "That's the same apartment house that was here four years ago". We say "That's the same building that was here four years ago". We change the kind relative to which we judge whether things are the same, and with that we change the implicit conditions for whether two things are the same. There can never be "true identity" of things that are in any way different.

An object can never change: it is of that time, and what is true of it at that time is all that is true of it. Yet we believe that things persist in time and do change. We resolve this tension between our belief that all is flux and our need to find continuity in our experience by using criteria of identity over time.

We establish criteria for what counts as the same thing by how we describe the universe of a model. If we say that the universe is all dogs, then the criteria for two things-in-time to be the same is that they are the same dog. If we say that the universe is all living creatures and all buildings, then the criteria for two things-in-time to be the same is that they are the same living creature or the same building. Crucial to this is that we don't use a description of a universe that gives competing criteria for whether two things-in-time are the same. Crucial also is that our description is clear enough that it provides a way to distinguish whether two descriptions of things-in-time pick out the same object.

Chapter 19

An often discussed conundrum about identity is the ship of Theseus. Theseus had a ship, and it started to weather badly. So to preserve it, he replaced first one plank, then another plank, then later a gunwale, then a mast, until finally there was no single bit that was the same as when it was built. Looking at the object then and thinking of the object when the replacements began, someone might ask, "Is it the same thing?". That's a loaded question: it assumes there is a notion of identity of things independent of what kinds of things and independent of human purposes. That assumption is very much at odds with human practice and with any use to which we can put predicate logic. Considering the variety of ways that different languages classify as things, it is more plausible to suppose that our notion of thing is a way we divide up experience codified in and shaped by our language: only relative to the idea of a particular kind of thing do we have a notion of identity. What counts as a name, whether "Birta", or "Arf", or "the ship of Theseus", depends on what kind of things are being considered. If we're considering ships that persist in time as objects to be sailed, then the object now is the same as the object before. If we're considering ships as built from the same planks, gunwales, masts, and so on, then there are two different things. The conundrum comes from trying to use competing criteria to evaluate an identity claim.

It might seem that we could avoid the tension between things-in-time and things as supratemporal by requiring that the things in a universe are things-at-times. So pointing out my window, there is only this thing here now that is a dog. To say that it was sleeping two hours ago, we would have to say that there was a thing two hours ago that was sleeping and that thing-at-a-time is equivalent to this thing-at-a-time. We would need no supratemporal objects, only our notions of equivalence of kinds of things.

But really we should say that the thing that was is equivalent to the thing that was when I pointed out my window, for the pointing is now in the past. I can't continue to point at the same thing. Each pointing is to a different thing-at-a-time because each pointing occurs at a different time. We couldn't use a name such as "Birta" to unify all the pointings or possible pointings. There would be no stability. To say that "$\exists x\, (-\text{ is a dog})\,(x)$" is true would be to say that for some time t there is a thing-at-that-time such that "$(-\text{ to be a dog})\,(-, -_{\text{time}})$" is true of that thing and time. We could even delete the time-marked blank because the time is encoded into the thing: "$(-\text{ is a dog})$" applies to that thing-at-a-time already says what time. There is the thing-at-a-time, but no need for times. Predications are of things that are already at a time. A time is picked out, perhaps, by what things-at-a-time there are at that time. There is not one universe but myriad universes: one for each time. They are unified by our method of classifications of different things-at-a-time as the same under a predicate, such as "$-$ is a dog" or "$-$ is sleeping".

One of the motivations for using classical predicate logic to reason about things in time is that we expect it to give simple formalizations of propositions such as

"Sometimes Spot barks when Suzy opens the gate" using variables and quantifiers. But the use of quantifiers in classical predicate logic depends on reference being timeless. If we have only temporal objects and temporal reference, there is no simplicity but only enormous complications, trying to replace the stability of reference we find in our language with the multiplicity of reference we see that we need when talking of things-at-times.

We use names, "Birta", "Dick", "Spot", that reflect our belief that things are stable despite the "changes" they go through. That is the basic metaphysics of English. It is the metaphysics that lies behind the syntax and semantics of classical predicate logic. To reason about only things-at-times is to abandon that metaphysics for the metaphysics of a different language, a mass-process language that we will see in Volume 3. Perhaps we could modify the syntax and semantics of classical predicate logic to talk of only things-at-times. But that would not reconcile our conception of supratemporal things with things in time; it would be to abandon the conception of supratemporal things and with it the metaphysics of English and classical predicate logic.

So in our work here, the things in a universe are things in time but not things at times. With one pointing we can use a name to refer to a thing across all times that "it" exists. So we will take the equality predicate "$- \equiv -$" to be unmarked for time. It will be part of our logical vocabulary. Not atemporal, not omnitemporal, but supratemporal. In this way we can investigate to what extent the metaphysics of English from which we build classical predicate logic is compatible with a conception of time and change. We evaluate the equality predicate as before:

$(- \equiv -)(u, v)$ is true iff u refers to the same object in U as v

We also want to be able to assert that times are the same, so we will add "$- \equiv_{time} -$" to our vocabulary where only time terms may fill the blanks. We evaluate that as:

$(- \equiv_{time} -)(t, w)$ is true iff t refers to the same time in T as w

We will continue to restrict our reasoning to extensional predicates. That now requires that it doesn't matter how we refer to a time. So the following will be equivalent:

$(-$ to bark $-_{time})$ (Birta, June 3 2001)

$(-$ to bark $-_{time})$ (Birta, t_7) t_7 is assigned reference June 3, 2011

$(-$ to bark $-_{time})$ (x_{32}, Axmer) x_{32} is assigned reference Birta, "Axmer" refers to June 3, 2011

Aside: Things-at-times

In some languages people do talk about things-at-times rather than supratemporal things. Here is what Richard M. Gale says in *The Philosophy of Time*:

> It is true that Indo-European languages make [past-present-future] determinations through the grammatical device of tensed verbs. But why should this particular grammatical device for conveying such temporal information be sacrosanct? Eskimo and many American Indian languages seem to be able to express the pastness, presentness, or futurity of a reported event or state of affairs by altering the end of a substantive. In certain African languages, like Mende and Hausa, pronouns are inflected, rather than verbs or nouns . . . . "The chief went" would be translated by words meaning "Chief he-past go." The latter sentence translated into Mende or Hausa obeys the same rules of use as the English sentence "The chief went," in that both can be made to make a true statement only if uttered or inscribed later than the chief's leaving. These sentences are subject to the same temporal restrictions in their use for making true statements. pp. 302–303

Otto Jespersen in *The Philosophy of Grammar* says:

> In some far-off languages tense-distinctions of substantives are better represented. Thus, in the Alaska Eskimo we find that *ningla* "cold, frost," has a preterit *ninglithluk* and a future *ninglikak*, and from *puyok* "smoke" is formed a preterit *puyuthluk* "what has been smoke," and a future *puyoqkak* "what will become smoke," an ingenious name for gunpowder . . . Similarly in other American languages. Thus the prefix *-neen* in Athapascan (Hupa) denotes past time both in substantives and verbs . . . . pp. 282–283

We'll see in Volume 3, *Reasoning about the World as the Flow of All*, that such talk can be modeled in a logic of the world as the flow of all rather than a logic of things.[19]

Arthur Prior in "Time, Existence, Identity" remonstrates against the idea of a thing-at-a-time by asserting that there are only supratemporal things:

> . . . the successive phases of the history of a thing are in no sense parts of the thing itself; it is one and the same thing (the whole thing, so far as talk of parts and wholes is appropriate here) which at one time does or undergoes this and at another does or undergoes that, and at one time stands here and at another time—by which time it (the same thing) has *stood* here—stands there. p. 101

---

[19] George Myro in "Identity and Time" sketches a way to discuss these issues in an extension of classical predicate logic with times used as propositional operators. But see Appendix A here.

20 When Things Exist

The existence predicate
When classical predicate logic was first developed it was considered a major advance that assertions about existence were to be made only with the existential quantifier. Existence as a predicate was expunged from formal logic and with it anomalous sentences such as "There is something that does not exist". The idea of different existences, of subsisting versus existing, was considered at best too confused to require formalization.

In classical predicate logic, and here, reference is timeless. So the existential quantifier is for timeless existence: there is something that can be taken to be the value of a variable. But we are talking about things in time, and they exist at some times and not at others. For example, the following is true:

(1) Juney was a dog that no longer exists.

We're not talking about a different kind of existence; we're concerned about when a thing exists. To talk about when things exist, let's adopt the *existence predicate*:

$(-$ to exist $-_{\text{time}})$

Then to formalize (1), taking the past to be relative to when I am writing, we can use:

(2) $\exists x \, \exists t \, (\, (-$ to be a dog $-_{\text{time}}) \, (\text{Juney}, t) \wedge \, (t <_{\text{time}} \text{March 31, 2018}) \,)$
$\wedge \, \neg \, (\, (-$ to exist $-_{\text{time}}) \, (\text{Juney}, \text{March 31, 2018}) \,)$

We want the existence predicate to have the same meaning in every model. And we want to be able to use it in formalizations of wffs in which no phrase equivalent to "exists" appears. So it will be part of the syncategorematic vocabulary. What semantic conditions should we adopt for it?

Things exist in time
Our first constraint is that the things in the universe of a model are in time, so each exists at some time.

Things exist in time Given any object o in U, there is a time t in T such that $(-$ to exist $-_{\text{time}}) \, (x, t)$ is true of o at time t.

Hence we have the tautology:

$\exists x \, (x \equiv \text{Socrates}) \, \leftrightarrow \, \exists t \, (-$ to exist $-_{\text{time}}) \, (\text{Socrates}, t)$

This is only a truth-value equivalence. The two sides cannot be semantically equivalent as the left-hand side is atemporal and the right-hand side is temporal.

Everything exists at some time. At every time does something exist? Should we assume the following as a tautology?

$\forall t\, \exists x\, (\text{— to exist —}_{\text{time}})\,(x, t)$

That would be a substantial assumption we should be able to investigate rather than assume in the logic we're developing.

Consider now:

$(\text{— to exist —}_{\text{time}})\,(\text{Birta}, 2009)$

This is true iff $(\text{— to exist —}_{\text{time}})$ applies to Birta at 2009. We know what that means: Birta exists at 2009. At what times does Birta exist? Did she exist an hour after she died before I buried her on January 18, 2016? Or do we say that she existed only at times when she was alive? When did the rock that I tripped over on January 6, 2017 exist? The answers depend on what we count as the criteria for identity of a dog and the criteria for identity of a rock. We can only assume that we have agreed-on ways of identifying what is the same dog and what is the same rock. The times when Birta exists and the times when the rock exists depend on those.

Downward and upward closure of existence in time
Birta existed at 2009. So she existed at December 18, 2009 and at every time in 2009. Conversely, if she existed at every time in 2009, she existed in 2009. The principles of the downward and upward closure of truth of a predication apply to the existence predicate.

Downward and upward closure of existence in time For any object o in the universe and time t, $(\text{— to exist —}_{\text{time}})\,(x, t)$ is true of o at t iff for all w such that $W(w, t)$, $(\text{— to exist —}_{\text{time}})\,(x, t)$ is true of o at w.

Continuity of existence in time
Birta existed in 2001. She also existed in 2010. So she existed in 2004. It's not possible for something to exist, go out of existence, then exist again. Whatever ceased to exist is not the same thing as that which exists later. But . . . Dick throws a ball for Spot to catch. It goes behind a tree, then comes out the other side and Spot gets it. Did "it" exist when no one—not Dick, not Spot, no one—saw it? Was "it" behind the tree? We know it's the same ball: it's the same size; it's the same color; it has the same scuff marks; it came out from behind the tree on the same trajectory; it smelled the same to Spot.

I meet a friend I haven't seen for years. He's older, grey, has a paunch now, but I recognize him. We recall some old times, he tells me he has grandchildren now and that he recently retired. We smile, then go on our way. I know he's the same person I knew all those years ago. But did he exist during all those years we had no contact?

Of course the ball existed when it went behind the tree. Of course my friend existed all those years. Our criteria of identity do not prove that but depend on the assumption that there are no gaps between times when an individual thing exists. That is a large metaphysical assumption, yet one that does not seem to matter, for we identify the ball, the person regardless. But it does matter, for it is how we link together our experiences: the same ball, the same person, continuous.

As part of our program of trying to understand how the metaphysics embedded in our language and our way of living can or should affect how to reason well, I am going to assume the continuity in time of when things exist. This assumption constrains our notion of an individual thing in time. It constrains our notion of naming: we cannot use the same name for what once existed and what now exists if there was a gap in time when there was no thing for which we could use the name.

No gaps doesn't mean that we can't have a model like:

(3)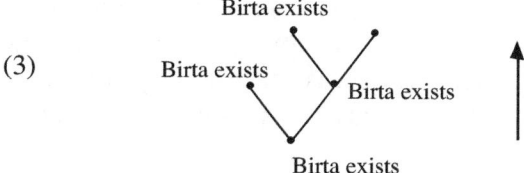

In Appendix B I suggest that this is just what we want if we wish to trace how choices affect our lives.

But no gaps does mean we won't allow a model in which we have either:

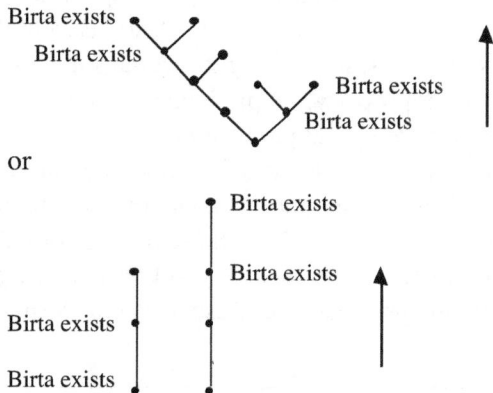

or

The following formal assumption rules these out but allows for (3).

Continuity of existence in time
If $(-$ to exist $-_{time})$ (x, t) is true of $\mathbf{o}$ at time t_1, and is true of $\mathbf{o}$ at time t_2, then:
  i. there is some w such that $w \leq t_1$ and $w \leq t_2$ and $(-$ to exist $-_{time})$ (x, t) is true of $\mathbf{o}$ at w, and
  ii. if $t_1 < t_2$, then $(-$ to exist $-_{time})$ (x, t) is true of $\mathbf{o}$ at all times t_3 such that $t_1 < t_3 < t_2$.

Predicates are positive for existence

Dick tells Suzy that Socrates was a philosopher in 400 B.C. So she knows that Socrates was alive then: you can't be a philosopher when you don't exist. On the phone Zoe tells Dick that Spot barked really loud a few minutes ago. Dick knows—he doesn't have to think about it—that Spot existed then. How could Spot have barked if he didn't exist?

A predicate is *positive for existence* means that if the predicate is true of an individual thing at a particular time, then that thing existed at that time.[20] For example, "— to be a stone —$_{time}$" and "— to be three feet long —$_{time}$" are positive for existence. The binary predicate "(— and — to be twins, —$_{time}$)" is positive for existence: if true of some individuals at a time, each of the individuals exists at that time. Is every atomic predicate positive for existence?

Certainly "— to exist —$_{time}$" is positive for existence. On the other hand, the question doesn't even arise for "— $<_{time}$ —" and "$W_{time}(-, -)$", since neither can be predicated of an individual thing. The issue is whether every ordinary atomic predicate we'll want to use in a semi-formal language is positive for existence. The examples suggest so.

But consider:

(4) Socrates is dead.

This is true, but Socrates does not exist now. Yet how can "Socrates" currently have some property when he no longer exists? Socrates was a philosopher, he was a man, he was an Athenian, but no longer is he any of those. He's dead. Except that in our logic we're taking reference to be timeless, so pointing to Socrates in a timeless way, we can say that he does have a current property. In particular, he's dead—now. This is to mix badly timeless reference with temporal predication.

It seems to me that if we construe "— to be dead" or any other atomic predicate as not positive for existence, we will have conundrums about the mixing of timeless reference and temporal predication. In Chapter 27 we'll see how to formalize (4) with the predicate "—to die", which is positive for existence. So I propose we adopt the assumption that every ordinary atomic predicate we'll use in a semi-formal language is positive for existence.

Ordinary atomic predicates are positive for existence
If A is an ordinary atomic predicate, and $A(-, \ldots, -, -_{time})(x_1, \ldots, x_n, t)$ is true of $a_1, a_2, \ldots, a_n$ at time t, then for each i, $(-$ to exist $-_{time})(x_i, t)$ is true of a_i at time t.

This is a simplification, similar to what we do when we agree to use only extensional predicates in classical predicate logic. If further study shows that there

---

[20] In *Some Uses of Type Theory in the Analysis of Language*, p. 128, M. K. Rennie says, "We may define an *acutalizer* as an "existence-entailing" property."

are more than a few atomic predicates that are not positive for existence yet are otherwise not problematic for applications of our logic, we could use the condition for being positive for existence to formulate a meaning axiom for some but not all ordinary atomic predicates.

Now we have enough to set out a formal logic for quantifying over times that extends classical predicate logic.

Aside: *On death and existence*
When rangers from the New Mexico Department of Game and Fish killed a mountain lion on my ranch they brought the body to my house and said, "This is the mountain lion that was eating your sheep". For things that live, we cannot identify existing with being alive. Yet if the rangers had pointed to the body eight weeks after they had killed the mountain lion, I'm pretty sure they would have said, "This *was* the mountain lion that was eating your sheep". After some length of time, a dead mountain lion is no longer a mountain lion, a dead person is no longer a person, a dead cockroach is no longer a cockroach. That length of time depends not only on the kind of thing but also on the state of the remains. If someone were to kill a mountain lion with a trap loaded with high explosives, immediately after the explosion there would be nothing we could point to and say, "That is the mountain lion", yet we could say, "The mountain lion that was killing your sheep is dead". For things that can live, the relation between being alive and existing involves complex metaphysical and moral issues that are too much to investigate here.[21]

---

[21] See https://youtu.be/watch?v=vnciwwsvNcc.

21 The Logic QT for Quantifying over Times in Classical Predicate Logic

The formal language

The letters i, j, k, n stand for counting numerals.

Vocabulary

    infinitive symbols I_j^n for $j \geq 0$ and $n \geq 1$, where n indicates the arity

    time order predicate $-<_{time}-$

    time part predicate $W_{time}(-,-)$

    existence predicate $(-$ to exist $-_{time})$

    equality predicate $(-\equiv-)$

    time equality predicate $(-\equiv_{time}-)$

    individual name symbols $c_0, c_1, \ldots$ ⎱ *individual terms*
    individual variables $x_0, x_1, \ldots$ ⎰

    time name symbols $b_0, b_1, \ldots$ ⎱ *time terms*
    time variables $t_0, t_1, \ldots$ ⎰

    propositional connectives $\neg, \rightarrow, \wedge, \vee$

    quantifiers $\forall, \exists$

Punctuation

    parentheses $(\ )$

    comma   ,

    blank   —

    time-marked blank $-_{time}$

Wffs

  i. If I is an n-ary infinitive symbol, then $I(-,\ldots,-,-_{time})$ with n unmarked blanks is a *formal ordinary atomic predicate*.

    If $u_1, \ldots, u_n$ are individual terms and t is a time term, then the following is an *ordinary* wff of *length* 1:

$$(I(-,\ldots,-,-_{time})(u_1,\ldots,u_n,t))$$

    The terms $u_1, \ldots, u_n$ *fill* the unmarked blanks in that order; the time term t *fills* the time-marked blank.

ii. If t and w are time terms, each of the following is a wff of length 1:
$$((-<_{\text{time}}-)(t,w))$$
$$(W_{\text{time}}(-,-)(t,w))$$
$$((-\equiv_{\text{time}}-)(t,w))$$

iii. If u and v are individual terms, the following is a wff of length 1:
$$((-\equiv-)(u,v))$$

iv. If u is an individual term and t is a time term, the following is a wff of length 1:
$$((-\text{ to exist }-_{\text{time}})(u,t))$$

Each occurrence of each variable in a wff of length 1 is *free*.

v. If A is a wff of length n, then $(\neg A)$ is a wff of length $n+1$.
An occurrence of a variable in $(\neg A)$ is free iff it is free in A.

vi. If A and B are wffs, and the maximum of the lengths of A and B is n, then each of $(A \rightarrow B)$ and $(A \wedge B)$ and $(A \vee B)$ is a wff of length $n+1$.
An occurrence of a variable in $(A \rightarrow B)$ is free iff the corresponding occurrence of the variable in A or in B is free, and similarly for $(A \wedge B)$ and $(A \vee B)$.

vii. If A is a wff of length n and some occurrence of an individual variable x is free in A, then each of $(\forall x A)$ and $(\exists x A)$ is a wff of length $n+1$.
An occurrence of a variable in either $(\forall x A)$ or $(\exists x A)$ is free iff the variable is not x and the corresponding occurrence in A is free.

viii. If A is a wff of length n and some occurrence of a time variable t is free in A, then each of $(\forall t A)$ and $(\exists t A)$ is a wff of length $n+1$.
An occurrence of a variable in either $(\forall t A)$ or $(\exists t A)$ is free iff the variable is not t and the corresponding occurrence in A is free.

ix. A concatenation of symbols is a *wff* iff it is a wff of length n for some $n \geq 1$.

A wff of length 1 is *atomic*; all other wffs are *compound*.
A wff is *closed* if there is no occurrence of a variable free in it; otherwise it is *open*.

In $(\forall x A)$ the initial $\forall x$ has *scope* A and *binds* each free occurrence of x in A, and that occurrence is *bound* by that quantifier, and similarly for $(\exists x A)$

In $(\forall t A)$ the initial $\forall t$ has *scope* A and *binds* each free occurrence of t in A, and that occurrence is *bound* by that quantifier, and similarly for $(\exists t A)$.

I'll leave to you to prove the unique readability of wffs (compare the proof for classical predicate logic in Volume 0).

Substituting for variables
An individual term u is *free for an occurrence of an individual variable x in A* (that is, free to replace it) iff both:

- The occurrence of x is free.
- If u is a variable, the occurrence does not lie within the scope of some occurrence of $\forall u$ or $\exists u$.

An individual term u is *free for x in A* iff it is free for every free occurrence of x in A, and then we write $A(u/x)$ to mean the wff that results from replacing every free occurrence of x in A with u.

A time term w is *free for an occurrence of a time variable t in A* (that is, free to replace it) iff both:

- The occurrence of t is free.
- If w is a variable, the occurrence does not lie within the scope of some occurrence of $\forall w$ or $\exists w$.

A term w is *free for t in A* iff it is free for every free occurrence of t in A, and then we write $A(w/t)$ to mean the wff that results from replacing every free occurrence of t in A with w.

Abbreviations

$A, B, C, A_0, A_1, B_0, B_1, C_0, C_1, D_0, D_1, \ldots$ stand for wffs.

$(u \equiv v)$	stands for	$((- \equiv -)(u, v))$
$(u \not\equiv v)$	stands for	$\neg((- \equiv -)(u, v))$
$(t \equiv_{time} w)$	stands for	$((- \equiv_{time} -)(t, w))$
$(t \not\equiv_{time} w)$	stands for	$\neg((- \equiv_{time} -)(t, w))$
$(t <_{time} w)$	stands for	$((- <_{time} -)(t, w))$
$(W_{time}(t, w))$	stands for	$(W_{time}(-, -)(t, w))$
$(t \leq_{time} w)$	stands for	$(t <_{time} w) \vee (t = w)$

$(w_1 <_{time} w_2 <_{time} w_3)$ stands for $(w_1 <_{time} w_2) \wedge (w_2 <_{time} w_3)$, and similarly for $\leq$

$O_{time}(w_1, w_2) \equiv_{Def} \exists t ((W_{time}(t, w_1) \wedge W_{time}(t, w_2))$

$O_<(w_1, w_2) \equiv_{Def} \exists t ((W_{time}(t, w_1) \wedge W_{time}(t, w_2)) \wedge$
$\quad\quad\quad\quad\quad\quad \exists t' (W_{time}(t', w_1) \wedge \forall w (W_{time}(w, w_2) \rightarrow t' <_{time} w))$

We use the same conventions on deleting parentheses as in classical predicate logic, adding that $\forall x, \exists x, \forall t, \exists t$ bind equally strongly.

The universal closure of a wff
Let $x_{e_1}, \ldots, x_{e_r}$ be a list of all the individual variables that occur free in A such that $e_1 < \cdots < e_r$. Let $t_{d_1}, \ldots, t_{d_s}$ be a list of all the time variables that occur free in A such that $d_1 < \cdots < d_s$. The *universal closure* of A is:
$$\forall \ldots A \equiv_{\text{Def}} \forall x_{e_1} \cdots \forall x_{e_n} \forall t_{d_1} \cdots \forall t_{d_s} A$$

Realizations and semi-formal languages
An ordinary language name or predicate is *simple* iff it contains no part we could formalize as a name, predicate, propositional connective, variable, quantifier, or a combination of those. A *simple infinitive* is any simple predicate with the verb in it converted to an infinitive; it has the same arity as the predicate from which it is derived.

A *realization* of the formal language is an assignment of simple names to none, some, or all of the name symbols (individual or time) and an assignment of simple infinitives to none, some, or all of the infinitive symbols. The *realization of a formal wff* is what we get when we replace the formal symbols in it with the parts of ordinary language that are assigned to them; it is a *semi-formal wff*. The *semi-formal language* for a realization is the realizations of formal wffs.

The simple infinitives, names, W_{time}, and $<_{\text{time}}$ are *categorematic*. The rest of the vocabulary is *syncategorematic*. Categorematic parts of the semi-formal language joined by punctuation or logical vocabulary are categorematic.

Models
Given a semi-formal language, we adopt two universes for it:

A universe U which is a non-empty collection of *things that are in time*.

A universe T which is a non-empty collection of things that are *times*.

There is no thing that is in both U and T.

There is a binary relation < on T.

There is a binary relation W on T.

These relations satisfy the following conditions.

Reflexivity of W
$W(t, t)$

Anti-symmetry of W
If $W(t, w)$ and $W(w, t)$, then $t = w$.

Transitivity of W
If $W(t_1, t_2)$ and $W(t_2, t_3)$ then $W(t_1, t_3)$

Parts determine times
$t = t'$ iff (for all w, $W(w, t)$ iff $W(w, t')$)

Anti-reflexivity of <
  not t < t

Transitivity of <
  If t < w and w < t', then t < t'.

Parts of times are related to other times as the whole is related
  If W(t$_1$, t$_2$) and t$_2$ < t$_3$ then t$_1$ < t$_3$.

  If W(t$_1$, t$_2$) and t$_3$ < t$_2$ then t$_3$ < t$_1$.

We write O for the interpretation of O.

We write O$_<$ for the interpretation of O$_<$.

Times are intervals
  If W(t$_1$, w) and W(t$_2$, w) then either O(t$_1$, t$_2$) or t$_1$ < t$_2$ or t$_2$ < t$_1$, and if t$_1$ < t$_2$, then for every t$_3$, such that t$_1$ < t$_3$ < t$_2$, W(t$_3$, w).

Overlapping times are not related in the ordering
  If O(t, w), then neither t < w nor w < t.

Assignments of references An assignment of references σ assigns:

To each individual variable x an object σ(x) from U.

To each individual name c an object σ(c) from U such that for every assignment of references τ, σ(c) = τ(c).

To each time variable t a time σ(t) from T.

To each time name b a time σ(b) from T such that for every assignment of references τ, σ(b) = τ(b).

Completeness of the collection of assignments of references

There is at least one assignment of references.

For every assignment of references σ, and every individual variable x, and every object in U, either σ assigns that object to x or there is an assignment τ that differs from σ only in that it assigns that object to x.

  We write τ ~$_x$ σ to mean that τ differs from σ at most in what it assigns x.

For every assignment of references σ, and every time variable t, and every time in T, either σ assigns that time to t or there is an assignment τ that differs from σ only in that it assigns that time to t.

  We write τ ~$_t$ σ to mean that τ differs from σ at most in what it assigns t.

Satisfaction of atomic wffs

For each assignment of references σ, there is a valuation $\vee_\sigma$ that assigns to each atomic wff a truth-value T or F. The valuation of A with respect to σ is a *predication*. If A is an ordinary atomic wff and the individual terms appearing in A are $u_1, \ldots, u_n$ in order reading from the left (including repetitions), and the time term appearing in A is t, we write $\vee_\sigma(A) = T$ or $\vee_\sigma \vDash A$ to mean that A is (taken to be) *true of* or *applies to* the objects $\sigma(u_1), \ldots, \sigma(u_n)$ in that order *at the time* $\sigma(t)$.

The valuations of the atomic wffs satisfy the following conditions.

$\vee_\sigma \vDash (w <_{time} t)$ iff $\sigma(w) < \sigma(t)$.

$\vee_\sigma \vDash W_{time}(w, t)$ iff $W(\sigma(w), \sigma(t))$.

$\vee_\sigma \vDash u \equiv v$ iff $\sigma(u)$ is the same object as $\sigma(v)$.

$\vee_\sigma \vDash w \equiv_{time} t$ iff $\sigma(w)$ is the same time as $\sigma(t)$.

The extensionality condition

Let $u_1, \ldots, u_n$ be the individual terms that appear in A (if any) and w_1, w_2 the time terms in A (if any). For any σ and τ and individual terms $v_1, \ldots, v_n$ and time terms w'_1, w'_2, if for all i, $\sigma(u_i) = \tau(v_i)$, and $\sigma(w_i) = \tau(w'_i)$,

$\vee_\sigma \vDash A$ iff $\vee_\tau \vDash A(v_1/u_1, \ldots, v_n/u_n, w'_1/w_1, w'_2/w_2)$

Downward and upward closure of truth in time

If A is an ordinary atomic wff and t the time term appearing in it,

$\vee_\sigma \vDash A(t)$ iff for all τ such that $\tau \sim_w \sigma$,
if $\vee_\tau \vDash W_{time}(w, t)$, then $\vee_\tau \vDash A(w/t)$.

Things exist in time

For every σ there is a $\tau \sim_t \sigma$ such that $\vee_\tau \vDash (-$ to exist $-_{time}) (x, t)$.

Downward and upward closure of existence in time

$\vee_\sigma \vDash (-$ to exist $-) (x, t)$ iff for all τ such that for all $\tau \sim_w \sigma$,
if $\vee_\tau \vDash W_{time}(w, t)$, then $\vee_\tau \vDash (-$ to exist $-) (x, w)$.

Continuity of existence in time

If $\vee_\sigma \vDash (-$ to exist $-) (x, t_1)$ and $\vee_\sigma \vDash (-$ to exist $-) (x, t_2)$, then
i. there is some τ such that $\tau \sim_{t_3} \sigma$ and $\vee_\tau \vDash (t_3 \leq_{time} t_1)$
and $\vee_\tau \vDash (t_3 \leq_{time} t_2)$, and $\vee_\tau \vDash (-$ to exist $-) (x, t_3)$, and
ii. if τ is such that $\tau \sim_{t_3} \sigma$ and $\vee_\tau \vDash (t_1 <_{time} t_3 <_{time} t_2)$,
then $\vee_\tau \vDash (-$ to exist $-) (x, t_3)$.

Ordinary atomic predicates are positive for existence

If A is an ordinary atomic wff, and $\vee_\sigma \vDash A(-, \ldots, -, -_{time}) (x_1, \ldots, x_n, t)$,
then for each i, $\vee_\sigma \vDash (-$ to exist $-_{time}) (x_i, t)$.

Satisfaction of compound wffs
The valuations for all assignments of wffs are extended to all wffs simultaneously:

$v_\sigma(\neg A) = T$ iff $v_\sigma(A) = F$

$v_\sigma(A \wedge B) = T$ iff $v_\sigma(A) = T$ and $v_\sigma(B) = T$

$v_\sigma(A \vee B) = T$ iff $v_\sigma(A) = T$ or $v_\sigma(B) = T$

$v_\sigma(A \rightarrow B) = T$ iff $v_\sigma(A) = F$ or $v_\sigma(B) = T$

$v_\sigma(\exists x\, A) = T$ iff   for some τ such that $\tau \sim_x \sigma$, $v_\tau(A) = T$

$v_\sigma(\forall x\, A) = T$ iff   for every τ such that $\tau \sim_x \sigma$, $v_\tau(A) = T$

$v_\sigma(\exists t\, A) = T$ iff   for some τ such that $\tau \sim_t \sigma$, $v_\tau(A) = T$

$v_\sigma(\forall t\, A) = T$ iff   for every τ such that $\tau \sim_t \sigma$, $v_\tau(A) = T$

We write $v_\sigma \vDash A$ to mean that $v_\sigma(A) = T$ and $v_\sigma \nvDash A$ to mean $v_\sigma(A) = F$.

The *model* M is the realization, universe of individuals, universe of times, the complete collection of assignments of references, valuations for atomic wffs satisfying the conditions, the extension of the valuations to all wffs by the inductive definition, and the valuation on all closed wffs.

A closed wff B of the semi-formal language is *true in* M iff for every σ, $v_\sigma(A) = T$; in that case we write $M \vDash B$. It is *false in* M otherwise, in that case we write $M \nvDash B$.

A formal wff A is *valid* or a *tautology* iff in every model its realization is true; in that case we write $\vDash A$. The formal inference Γ therefore A is *valid*, written $\Gamma \vDash A$, means there is no model in which the realizations of all the wffs in Γ are true and the realization of A is false; this is the *semantic consequence relation*. These definitions are extended to semi-formal wffs via formal wffs of which they are realizations.

> *Sufficiency of the collection of models*
> Given any realization, collection of things in time, collection of times, and complete collection of assignments of references, any assignment of truth-values to the atomic predications satisfying these semantic conditions defines a model.

Classical predicate logic with quantifying over times The formal language, definitions of models, tautology, and semantic consequence together constitute *classical predicate logic with quantifying over times*, **QT**.

The pure language of time and pure logic of time
The *pure language of time* is the formal language defined above without infinitive symbols, existence predicate, equality predicate, individual variables, and individual name symbols. A *model* of the pure language of time is defined as above except

Aside: *What is true at a particular time*

Given a model of **QT**, each time in the universe of times determines a model of classical predicate logic with universe those things that exist at that time and predications true iff true of those things at that time.

Let M be a model for a semi-formal language L with universe U and collection of times T. An *object* o *in the universe of* M *exists at time* t iff there is some x and some t and an assignment of references σ for M such that $σ(t) = t$ and $σ(x) = o$ and $v_σ \vDash (-\text{ exists }-_{\text{time}})(x, t)$. Let L_t be the language L such that all names that do not refer to an object in U at time t are deleted, the time vocabulary is deleted, and predicates are not converted to infinitives. The *model* M_t *of what is true at time* t *in* M is defined:

The universe of M_t is the collection of objects of M that exist at time t.

The assignments of references $σ_t$ of M_t are those of M restricted to this universe with assignments of times deleted.

Given any $σ_t$, let σ be any assignment of references of M that agrees with $σ_t$ on the variables and the names of L_t and $σ(t) = t$. Then for any atomic wff A,
$v_{σ_t} \vDash A$ iff $v_σ \vDash_M A$.

The true wffs of M_t constitute *what is true at time* t *in* M.

From a collection of models of classical predicate logic can we derive a model of **QT**? We would need to derive a timeline from the collection. That would require being able to discern from what is true and what is false in one model whether that model is or should be of a later time, or within the time, of another model.

An axiom system for QT

Propositional axioms

The axiom schemes of classical propositional logic, **PC** (Chapter 2 of Volume 0), where A, B, and C are any wffs of L and the universal closure is taken.

$\forall \ldots (\neg A \to (A \to B))$

$\forall \ldots (B \to (A \to B))$

$\forall \ldots ((A \to B) \to ((\neg A \to B) \to B))$

$\forall \ldots ((A \to (B \to C)) \to ((A \to B) \to (A \to C)))$

$\forall \ldots (A \to (B \to (A \wedge B)))$

$\forall \ldots ((A \wedge B) \to A)$

$\forall \ldots ((A \wedge B) \to B)$

$\forall \ldots (A \to (A \vee B))$

$\forall \ldots (B \to (A \vee B))$

$\forall \ldots ((A \to C) \to ((B \to C) \to ((A \vee B) \to C)))$

Axioms governing $\forall$

1. a. $\forall \ldots (\forall x\, (A \to B) \to (\forall x\, A \to \forall x\, B))$
      if x is free in both A and B
   b. $\forall \ldots (\forall x\, (A \to B) \to (\forall x\, A \to B))$
      if x is free in A and not free in B
   c. $\forall \ldots (\forall x\, (A \to B) \to (A \to \forall x\, B))$
      if x is free in B and not free in A
2. $\forall \ldots (\forall x\, \forall y\, A \to \forall y\, \forall x\, A)$
3. $\forall \ldots (\forall x\, A(x) \to A(u/x))$
   where u is free for x in A
4. a. $\forall \ldots (\forall t\, (A \to B) \to (\forall t\, A \to \forall t\, B))$
      if t is free in both A and B
   b. $\forall \ldots (\forall t\, (A \to B) \to (\forall t\, A \to B))$
      if t is free in A and not free in B
   c. $\forall \ldots (\forall t\, (A \to B) \to (A \to \forall t\, B))$
      if t is free in B and not free in A
5. $\forall \ldots (\forall t\, \forall w\, A \to \forall w\, \forall t\, A)$
6. $\forall \ldots (\forall t\, A(t) \to A(w/t))$
   where w is free for t in A
7. $\forall \ldots (\forall x\, \forall t\, A \to \forall t\, \forall x\, A)$
8. $\forall \ldots (\forall t\, \forall x\, A \to \forall x\, \forall t\, A)$

Axioms governing the relation between $\forall$ *and* $\exists$

9. a. $\forall \ldots (\exists x\, A \to \neg \forall x\, \neg A)$
   b. $\forall \ldots (\neg \forall x\, \neg A \to \exists x\, A)$
10. a. $\forall \ldots (\exists t\, A \to \neg \forall t\, \neg A)$
    b. $\forall \ldots (\neg \forall t\, \neg A \to \exists t\, A)$

Axioms for equality and extensionality

11. $\forall x\, (x \equiv x)$
12. $\forall \ldots \forall x\, \forall y\, (x \equiv y \to (A(x) \to A(y/x)))$
    where A is atomic and y replaces some
    but not necessarily all occurrences of x in A
13. $\forall t\, (t \equiv_{\text{time}} t)$
14. $\forall \ldots \forall t\, \forall w\, (t \equiv_{\text{time}} w \to (A(t) \to A(w/t)))$
    where A is atomic and w replaces some
    but not necessarily all occurrences of t in A

Axioms for time

W_{time} *is a part-whole relation*
$\forall t_1 \, W_{time}(t_1, t_1)$
$\forall t_1 \, \forall t_2 \, (W_{time}(t_1, t_2) \land W_{time}(t_2, t_1) \rightarrow (t_1 \equiv_{time} t_2))$
$\forall t_1 \, \forall t_2 \, \forall t_3 \, (W_{time}(t_1, t_2) \land W_{time}(t_2, t_3) \rightarrow W_{time}(t_1, t_3))$

Parts determine times
$\forall t_1 \, \forall t_2 \, ((t_1 \equiv_{time} t_2) \leftrightarrow (\forall t_3 \, (W_{time}(t_3, t_1) \leftrightarrow W_{time}(t_3, t_2))))$

$<_{time}$ *determines an ordering*
$\forall t_1 \, \neg (t_1 <_{time} t_1)$
$\forall t_1 \, \forall t_2 \, \forall t_3 \, ((t_1 <_{time} t_2) \land (t_2 <_{time} t_3) \rightarrow (t_1 <_{time} t_3))$

Parts of times are related to other times as the whole is related
$\forall t_1 \, \forall t_2 \, \forall t_3 \, (W_{time}(t_1, t_2) \land (t_2 <_{time} t_3)) \rightarrow (t_1 <_{time} t_3))$
$\forall t_1 \, \forall t_2 \, \forall t_3 \, (W_{time}(t_1, t_2) \land (t_3 <_{time} t_2)) \rightarrow (t_3 <_{time} t_1))$

Times are intervals
$\forall t_1 \, \forall t_2 \, \forall t_3 \, (W_{time}(t_1, t_3) \land W_{time}(t_2, t_3) \rightarrow$
$\quad (O_{time}(t_2, t_1) \lor (t_1 <_{time} t_2) \lor (t_2 <_{time} t_1)$
$\quad \land ((t_1 <_{time} t_2) \rightarrow \forall t_4 \, ((t_2 <_{time} t_4 <_{time} t_3) \rightarrow W_{time}(t_4, t_1)))))$

Overlapping times are not related in the ordering
$\forall t_1 \, \forall t_2 \, (O_{time}(t_1, t_2) \rightarrow (\neg (t_1 <_{time} t_2) \land \neg (t_2 <_{time} t_1)))$

Downward and upward closure of truth in time
$\forall \ldots (A(x_1, \ldots, x_n, t_1) \leftrightarrow \forall t_2 \, (W_{time}(t_2, t_1) \rightarrow A(x_1, \ldots, x_n, t_2))$
for every ordinary atomic predicate A

Things exist in time $\forall x \, \exists t \, (- \text{ to exist } -_{time}) \, (x, t)$

Downward and upward closure of existence in time
$\forall x_1 \, \forall t_1 \, ((- \text{ to exist } -_{time}) \, (x_1, t_1) \leftrightarrow$
$\quad \forall t_2 \, (W_{time}(t_2, t_1) \rightarrow (- \text{ to exist } -_{time}) \, (x_1, t_2)))$

Continuity of existence in time
$\forall x \, \forall t_1 \, \forall t_2 \, (((- \text{ to exist } -_{time}) \, (x, t_1) \land (- \text{ to exist } -_{time}) \, (x, t_2)) \rightarrow$
$\quad (\exists t_3 \, ((t_3 \leq_{time} t_1) \land (t_3 \leq_{time} t_2) \land (- \text{ to exist } -) \, (x, t_3)) \land (t_1 <_{time} t_2)$
$\quad \rightarrow \forall t_4 \, ((t_1 <_{time} t_4 <_{time} t_2) \rightarrow (- \text{ to exist } -_{time}) \, (x, t_4))))$

Ordinary atomic predicates are positive for existence
$$\forall x_1 \ldots \forall x_n, \forall t \, (\, A(-, \ldots, -, -_{time}) (x_1, \ldots, x_n, t)$$
$$\rightarrow (- \text{ to exist } -_{time}) (x_i, t) \,)$$

Rule $\dfrac{A, A \rightarrow B}{B}$ where A and B are closed formulas

    The axioms not including those for time and those for existence constitute a strongly complete axiomatization of classical logic with two sorts (kinds) of variables. That can be proved via a translation of classical predicate logic with two sorts of variables to classical predicate logic with just one kind of variables, as given in Chapter XIV.H of *Classical Mathematical Logic*. Or that can be shown by a modification of the completeness proofs for classical predicate logic with one sort of variable (Appendix 5 of Volume 0). Adding the axioms governing time and existence then yields a strongly complete axiom system for **QT**.

    For any collection of wffs Γ and wff A, $\Gamma \vdash_{\mathbf{QT}} A$ iff $\Gamma \vDash_{\mathbf{QT}} A$.

22 Examples of Formalizing: The Nature of Time

In this chapter we'll see how we can incorporate or compare different conceptions of time using the pure language and logic of time.

I'll use x for x_1, y for x_{32}, t for t_7, t' for t_{83}, and w for t_{13}.

Example 1   There is more than one time.

$\exists t \exists w \, (t \neq_{\text{time}} w)$

Analysis   Nothing we've done so far requires that there is more than one time in a model. A model with only one time would be like confining classical predicate logic to what is true at a particular time (see Chapter 14 and the *Aside* on p. 105).

Example 2   There is a smallest unit of time.

$\exists t \neg \exists w \, ((w \neq_{\text{time}} t) \wedge W_{\text{time}}(w, t))$

Analysis   There is a smallest unit of time iff there is some time such that no other time is contained in it—in the model. We're not talking of "all time" here, if that phrase even makes sense.

This example is compatible with the view that time is composed of instants and that any interval is just a collection of those instants. It does not seem to be compatible with the view of time as a mass: every part of a mass is a mass of the same kind, and masses have no smallest parts. But the example could be true in a model on the understanding of time as a mass since we could have a time in the universe of times without any of its parts. Let's define:

instant $(t) \equiv_{\text{Def}} \neg \exists w \, ((w \neq_{\text{time}} t) \wedge W_{\text{time}}(w, t))$

Example 3   2:49.000000001 p.m. May 3, 2004 is an instant of time.

instant (2:49.000000001 p.m. May 3 2004)

Analysis   If this is true, the preceding example is true, too. But nothing in this or the preceding example requires that all smallest units of time be the same length or even that there is more than one smallest unit of time.

Example 4   There are only instants of time.

$\forall t \, \text{instant}\,(t)$

Analysis   In a model of this, the relation W would be the identity.

Example 5   Every time contains an instant of time.

$\forall t \exists w \, (W_{\text{time}}(w, t) \wedge \text{instant}\,(w))$

Analysis   This is true if there are instants of time and some times are intervals of those instants. However, it is also true in a model that has only one time.

110   Chapter 22

Example 6   There are smallest units of time, and every interval of time is a collection of those and nothing more.

$\exists t \, (\text{instant}(t)) \land$
$\forall t_1 \, \forall t_2 \, ((t_1 \equiv t_2) \leftrightarrow (\forall t \, (\text{instant}(t) \rightarrow (W_{\text{time}}(t, t_1) \leftrightarrow W_{\text{time}}(t, t_2)))))$

Analysis   If this is true in a model, "anything else" in a time besides instants is irrelevant. This conception of time is the standard in treatments of time devised by mathematicians.

Example 7   There is a time which contains no smallest unit of time.

$\exists t \, \forall w \, (W_{\text{time}}(w, t) \rightarrow \neg \, \text{instant}(w))$

Analysis   This is consistent with the assumption that there are smallest units of time, for it could be that instants are contained only within particular times. For this to be true in a model, there must be (potentially) infinitely many times in the universe of times, along with some method of picking out those.

Example 8   There are no smallest units of time.

$\neg \, \exists t \, \text{instant}(t)$

Analysis   This is the negation of Example 2. It is part of the conception of time as mass but is incompatible with the conception of time as made up of instants.

Example 9   There are parallel timelines.

Analysis   We could have a model with two timelines:

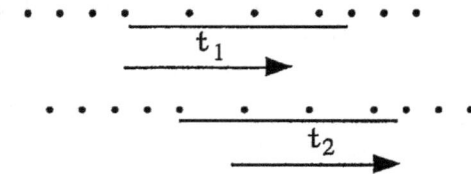

Parallel timelines can be characterized as having two times that are completely incomparable: nothing above or below the one is above or below the other.

Parallel timelines

$\exists t_1 \, \exists t_2 \, ((\neg O_{\text{time}}(t_1, t_2) \land \neg (t_1 <_{\text{time}} t_2) \land \neg (t_2 <_{\text{time}} t_1))$
$\land \, \forall w \, (((w \leq_{\text{time}} t_1) \lor (t_1 \leq_{\text{time}} w) \lor O_{\text{time}}(t_1, w)) \rightarrow$
$(\neg (t_2 \leq_{\text{time}} w) \land \neg (w \leq_{\text{time}} t_2) \land \neg O_{\text{time}}(t_2, w))))$

Example 10   Branching time.

Analysis   Suppose we know that Suzy visited Dick before Dick ate dinner. And we know that both Puff scratched Zoe and Tom sprained his ankle before Dick ate dinner and after Suzy visited Dick. But we might not know and might never know whether Puff scratched Zoe before or after Tom sprained his ankle.

(1)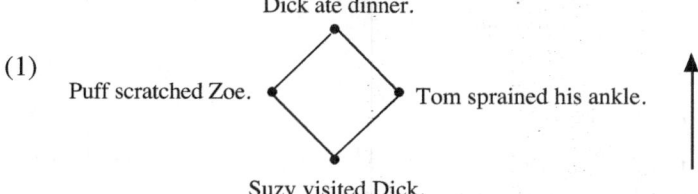

To say that nonetheless Puff scratched Zoe either before Tom sprained his ankle or after Tom sprained his ankle or at the same time as Tom sprained his ankle is to assume an ordering of those times that is not only independent of us but without any basis in our experience. Our logic allows for models in which the timeline branches, either in reality or due to our ignorance, as in (1) and:

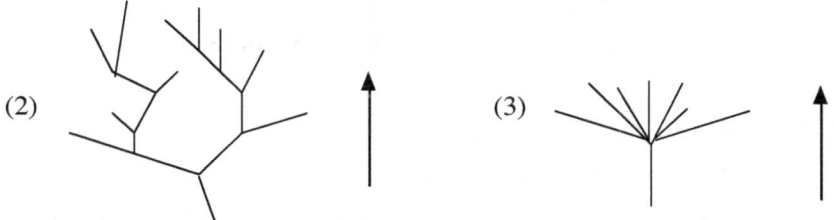

In these, given any two non-overlapping times, there is a time less than or equal to both. But that won't do as a characterization of branching timelines since it would exclude:

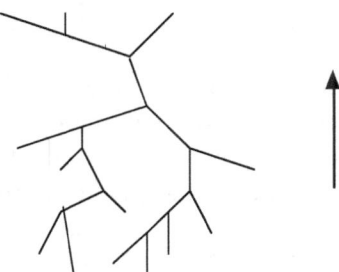

So let's say that given any two non-overlapping times there is a time less than or equal to both or greater than or equal to both. That rejects the following though each part is itself branching:

But it does allow a timeline with a single branch:

112   Chapter 22

To exclude a single timeline, we add to the characterization that there are two non-overlapping times that are not comparable.

Branching times

$\forall t_1 \forall t_2 ((\neg O_{time}(t_1, t_2) \rightarrow$
$\exists w ((w \leq_{time} t_1) \wedge (w \leq_{time} t_2)) \vee \exists w ((w \geq_{time} t_1) \wedge (w \geq_{time} t_2)))$
$\wedge \exists t_1 \exists t_2 (\neg O_{time}(t_1, t_2) \wedge \neg (t_1 <_{time} t_2) \wedge \neg (t_2 <_{time} t_1)))$

This allows for (1) to be characterized as branching.

In Appendix B I discuss what the view of time being branching might mean for our conception of the world.

Example 11   Every time is contained in some other time.

$\forall t \exists w (t \not\equiv_{time} w \wedge W_{time}(t, w))$

Analysis   For this to be true, there must be (potentially) infinitely many times in a model.

Example 12   There are largest units of time.

$\exists t \forall w (w \not\equiv_{time} t \rightarrow \neg W_{time}(t, w))$

Analysis   This could be true while there is no single largest unit of time that contains all times. It could be true in a linear timeline so long as intervals of times are not themselves times.

Example 13   There is no time that contains all other times.

$\neg \exists t \forall w W_{time}(w, t))$

Analysis   To adopt this as a condition on our models would not be to deny that there is some totality of times. Indeed, the universe of times plays just that role in our models. But it is to deny that the totality of times is itself a time. This would be consistent with the adoption of "all" fixed-time markers in a semi-formal language.

Example 14   Time has no beginning and no ending.

$\neg \exists w \forall t (w \leq_{time} t) \wedge \neg \exists w \forall t (t \leq_{time} w)$

Analysis   For this to be true, there must be (potentially) infinitely many times in the model.

There being no latest time does not mean there is not a limit to all times. There could be later and later times, but the intervals between those grow smaller so that there is a limit to all times, but not one that is ever attained.

Example 15   Characterizing equality for times

$\forall t \forall w ((t \equiv_{time} w) \leftrightarrow \forall t' (W_{time}(t', t) \leftrightarrow W_{time}(t', w)))$

Analysis   This is one of our axioms. We could use it as a definition of "$\equiv_{time}$", adjusting the statement of anti-symmetry for W_{time} accordingly.

Example 16 Every two times are within some time.

$$\forall t\,\forall t'\,\exists w\,(\,W_{\text{time}}(t,w) \land W_{\text{time}}(t',w)\,)$$

Analysis This would be true if there is only a single branch for the timelines where every interval is a time.

Example 17 The ordering of times is dense.

$$\forall t\,\forall w\,(\,t <_{\text{time}} w \rightarrow \exists t'\,(t <_{\text{time}} t' <_{\text{time}} w)\,)$$

Analysis This does not assert that between any two times there is a third time, for between May 3, 1996 and May 1996 there can be no time since neither is less than the other. Rather, if two times are related in the ordering then there is a third time between them. Note that this could be true in a model with dense timelines in parallel or in which each branching is dense.

Example 18 There are times other than those named by fixed time markers.

    Not formalizable.

Analysis We could formalize this example if we allow for quantification over time names, though that theory would not be axiomatizable.[22]

Example 19 Everything happens at once.

Analysis Nothing in what we've done precludes having a model in which there is only one time at which atomic wffs are true. For example, we could allow for only wffs about December 26, 2009 to be true in our model. There could be lots of other times in the model, but they would all fall into two classes: those within December 26, 2009, and those not within December 26, 2009. All of the latter would be equivalent as far as which ordinary atomic predications are true at them: none. It's not that in such a model everything happens at once. It's that we are concerned about only one time at which anything happens. We can pay attention to as much or as little as we want in our reasoning.

Example 20 Time flows.

    Not formalizable.

Analysis Many people say this and believe it. Yet what they mean by it is at best obscure. Unless they can clarify it, we cannot formalize it. Perhaps, in the end, the example is not a proposition but a guide to thinking about time, a way of conveying a viewpoint rather than a truth.

---

[22] Compare Chapter XV.F of *Classical Mathematical Logic*.

23 The Internal Structure of Atomic Predicates

In *The Internal Structure of Predicates and Names* (Volume 1) we saw how to extend the scope of what we can formalize in classical predicate logic by taking account of the internal structure of atomic wffs. In the same way we can include in our language of temporal predicate logic:

- predicate restrictors
- predicate negators
- conjunctions of terms
- conjunctions of restrictors
- conjunctions of predicates
- disjunctions of predicates

The semantic conditions for these will be the same as in Volume 1 with two modifications.

- The conditions are for ordinary atomic predicates only.
- Each ordinary atomic predicate has a time term.

For example, consider:

Spot barked loudly at 1 p.m. May 3, 1998.
Therefore, Spot barked at 1 p.m. May 3, 1998.

This is valid, and we want its formalization to be valid, too:

$$\frac{((-\text{ to bark } -_{\text{time}})/\text{loudly}) (\text{Spot}, 1 \text{ p.m. May 3 1998})}{(-\text{ to bark } -_{\text{time}}) (\text{Spot}, 1 \text{ p.m. May 3 1998})}$$

The condition for restrictors in classical predicate logic is:

RES   If $\upsilon_\sigma \vDash (Q/R)(\vec{d})$, then $\upsilon_\sigma \vDash Q(\vec{d})$.

Here Q is an atomic predicate, R is a restrictor, and $\vec{d}$ is a sequence of terms or conjunctions of terms of the correct arity for the predicate. Now we adopt for any such R and $\vec{d}$, and Q an ordinary atomic predicate:

RES-time   If $\upsilon_\sigma \vDash (Q/R)(\vec{d}, t)$, then $\upsilon_\sigma \vDash Q(\vec{d}, t)$.

Recall that there is one *distinguished variable restrictor* "obj(—)" we use *to formalize direct objects of verbs*. For example, in classical predicate logic we formalize "Spot chased Puff" as "((— chased)/obj(Puff)) (Spot)".

Similarly, the condition for negators N in classical predicate logic is:

NEG   If $\upsilon_\sigma \vDash (Q/N)(\vec{d})$, then $\upsilon_\sigma \nvDash Q(\vec{d})$.

Now we adopt:

NEG-time If $\upsilon_\sigma \vDash (Q/N)(\vec{d},t)$, then $\upsilon_\sigma \nvDash Q(\vec{d},t)$.

There is one case, though, where a stronger principle applies. For conjunctions of restrictors there is no notion of two restrictors acting together beyond applying simultaneously. Hence, we can now draw an equivalence between:

(1) $((-$ to be a dog $-_{time})/$ small & cute) (Bruno, March 1 2001) $\leftrightarrow$
$((-$ to be a dog $-_{time})/$ small) (Bruno, March 1 2001)
$\wedge$ $((-$ to be a dog $-_{time})/$ cute) (Bruno, March 1 2001)

& *is equivalent to time-indexed* $\wedge$
If Q is an ordinary atomic predicate and R, R´ are restrictors, then:
$\upsilon_\sigma \vDash (Q/(R \& R´))(\vec{d},t)$ iff $\upsilon_\sigma \vDash (Q/R)(\vec{d},t)$ and $\upsilon_\sigma \vDash (Q/R´)(\vec{d},t)$

We still need to be able to form the left-hand side of (1) so we can modify the conjunction of the restrictors, as in:

$(((-$ to be a dog $-_{time})/$ small & cute)/large) (Bruno, March 1, 2001)

The principles of the downward and upward closure of truth in time were formulated for ordinary atomic predicates that have no internal structure. The same arguments apply for atomic predicates that have internal structure. Hence, those principles need not be modified.

The existence predicate and modifiers
Should we allow the existence predicate to be modified? Consider:

(2) $((-$ to exist $-_{time})/$ tall) (Zoe, January 18 2016)

To evaluate this we first consider all things that exist, and then restrict to those that exist in a tall way, and if Zoe is among those at January 18, 2016, the wff is true. But that's just to say that "Zoe is tall at January 18, 2016" is true. In Volume 1 we rejected "Zoe is tall" as too vague: "tall" is an adjective that can be used only relative to a predicate. To allow for (2) or any other modification of an existence predicate is to return to the view of propositions as subject-copula-predicate. We won't allow the existence predicate to be modified.

The pure negator
Compare:

(3) Spot barked at 7:12 am May 5th, 2005.

(4) Spot didn't bark at 7:12 am May 5th, 2005.

Both of these are about Spot. In (3) we're told he barked. In (4) we're not told what he did except that he didn't bark. If (3) is true, Spot existed at 11:15 a.m., May 5th, 2005. Equally, if (4) is true, Spot existed at that time. So we can't formalize (4) as:

¬ (((— to bark —$_{time}$) (Spot, 7:12 am May 5 2005))

This could be true if Spot didn't exist at that time. Using the syncategorematic *pure negator* ~ , which can modify only atomic predicates and restrictors (as in Chapter 21 of Volume 1), we can formalize "not" in (4) as negating the predicate:

(5)   (~(— to bark —$_{time}$)) (Spot, 7:12 am May 5 2005)

Here "~(— to bark —$_{time}$)" is an ordinary atomic predicate and hence positive for existence. So "(— to exist —$_{time}$) (Spot, 7:12 am May 5 2005)" follows from (5). We have the equivalence:

(~ (— to bark —$_{time}$)) (Spot, 11:15 a.m. 7:12 am May 5 2005) ↔
(¬ ((— to bark —$_{time}$) (Spot, 7:12 am May 5 2005))
∧ (— to exist —$_{time}$) (Spot, 7:12 am May 5 2005))

More generally, the following is a tautology:

$\forall x \forall t$ (~ Q(x, t) ↔ (¬ Q(x, t) ∧ (— to exist —$_{time}$) (x, t)))

So the pure negator modifying a predicate directly could be taken as an abbreviation in context. Can we eliminate it in other contexts?

Consider:

Spot is barking not loudly at 7:12 am May 5 2005.

With the pure negator modifying a restrictor we can formalize this as:

((— to bark —$_{time}$)/(~ loudly)) (Spot, 7:12 am May 5 2005)

But this is equivalent to:

(— to bark —$_{time}$) (Spot, 7:12 am May 5 2005)
∧ ¬((— to bark —$_{time}$)/loudly) (Spot, 7:12 am May 5 2005)

More generally, the following is a tautology:

$\forall x \forall t$ ((Q/~ R) (x, t) ↔ (Q(x, t) ∧ ¬(Q/R) (x, t)))

So, it seems, the pure negator modifying a restrictor could be taken as just an abbreviation in context. But now consider:

(6)   Tom spoke carefully not clearly October 9. 2021.

Tom was speaking not clearly. And he was doing so carefully. That is, "carefully" modifies "not clearly". With the pure negator, we can formalize (6) as:

(7)   ((— to speak —$_{time}$)((~ clearly)/carefully)) (Tom, October 9. 2021)

Here there doesn't seem to be a way to eliminate the pure negator with a contextual definition.

In order to allow for formalizations such as (6) and to avoid using two different abbreviations for the use of ~ depending on context, I propose we allow the syncategorematic pure negator to be part of our formal logic, modifying a predicate directly or modifying a restrictor to create a new restrictor.

Now I'll set out an extension **QT+internal** of **QT** that allows for parsing the internal structure of atomic predicates as described above.

The formal language

Vocabulary

We add to the vocabulary of **QT**.

predicate modifier symbols

> *restrictor symbols* $R_0, R_1, \ldots$
>
> *variable restrictor symbols* R_n^i for $n \geq 0$ and $i \geq 1$
>
> *logical variable restrictor* $\mathrm{obj}(-)$
>
> *negator symbols* $N_0, N_1, \ldots$
>
> *logical negator* $\sim$

internal logical symbols

> *term conjoiner* $\wedge$
>
> *predicate conjoiner* $+$
>
> *modifier conjoiner* $\&$
>
> *predicate disjoiner* $\cup$

The propositional connectives and quantifiers are now called *external logical symbols*.

Individual terms and formal conjunctions of individual terms

i. Each individual variable and each individual name symbol is an individual term of degree 1.

ii. If d is an individual term or a formal conjunction of individual terms of degree n, and d' is an individual term or formal conjunction of individual terms of degree m, then $(d \wedge d')$ is a formal conjunction of individual terms of degree $n + m$.

iii. A concatenation of symbols is a formal conjunction of individual terms iff it is a conjunction of individual terms of degree n for some $n > 1$.

Formal modifiers

i. Every predicate restrictor symbol is a formal restrictor of degree 1.

   If R is a k-ary variable predicate restrictor symbol and $d_1, \ldots, d_k$ are individual terms or conjunctions of individual terms, then $R(d_1, \ldots, d_k)$ is a formal restrictor of degree 1.

ii. If R is a formal restrictor of degree n, then (~R) is a formal restrictor of degree $n + 1$.

iii. If R is a formal restrictor of degree n, and N is a formal negator, then (R/N) is a formal restrictor of degree $n + 1$.

iv. If N and N´ are formal negators, then (N/N´) is a formal restrictor of degree 2.

v. If R and R´ are formal restrictors of degree n and degree m respectively, then (R & R´) is a formal restrictor of degree $n + m$. It is a *conjunction of restrictors*.

vi. A concatenation of symbols is a formal modifier iff it is a formal restrictor or formal negator of degree n for some $n \geq 1$.

Formal modifiers of degree > 1 are *complex*.

Formal ordinary atomic predicates

i. If I is a k-ary infinitive symbol, then $I(-, \ldots, -, -_{time})$ with k unmarked blanks is a *formal ordinary atomic predicate* of degree 0. It is a *simple* formal atomic predicate.

ii. If $Q(-, \ldots, -, -_{time})$ is a k-ary formal ordinary atomic predicate of degree n, then $(Q/\sim)(-, \ldots, -, -_{time})$ is a k-ary formal ordinary atomic predicate of degree $n + 1$. It is a *pure negated predicate*.

iii. If $Q(-, \ldots, -, -_{time})$ is a k-ary formal ordinary atomic predicate of degree n, and M is a formal modifier other than ~, and Q is not a pure negated predicate, then $(Q/M)(-, \ldots, -, -_{time})$ is a k-ary formal ordinary atomic predicate of degree $n + 1$.

iv. If $Q_1(-, -_{time})$ and $Q_2(-, -_{time})$ are unary formal ordinary atomic predicates of degrees n and m, respectively, and at least one of them is not a pure negated predicate, then $(Q_1 + Q_2)(-, -_{time})$ is a unary formal ordinary atomic predicate of degree $n + m + 1$. It is an *internally conjoined* formal atomic predicate.

v. If $Q_1(-, -_{time})$ and $Q_2(-, -_{time})$ are unary formal ordinary atomic predicates of degrees n and m, respectively, and neither is a pure negated predicate, then $(Q_1 \cup Q_2)(-, -_{time})$ is a unary formal ordinary atomic predicate of degree $n + m + 1$. It is an *internally disjoined* formal atomic predicate.

vi. A concatenation of symbols is a formal atomic predicate iff it is a formal atomic predicate of degree n for some n.

Formal atomic predicates of degree > 0 are *modified formal atomic predicates*.

Wffs

We modify the definition in **QT** for the first clause only.

i. If $A(-, \ldots, -, -_{time})$ is an n-ary formal ordinary atomic predicate, and $u_1, \ldots, u_n$ are individual terms, and t is a time term, then the following is an *ordinary* wff of *length* 1:

$$(A(-, \ldots, -, -_{time})\,(u_1, \ldots, u_n, t))$$

The terms $u_1, \ldots, u_n$ *fill* the unmarked blanks in that order, and the term t *fills* the time-marked blank.

Realizations

An ordinary language name, or predicate, or modifier is *simple* iff it contains no part we could formalize as a name, predicate, propositional connective, variable, quantifier, internal conjunction, internal disjunction, predicate, predicate modifier, or a combination of those. A *simple infinitive* is a simple predicate with the verb in it converted to an infinitive; it has the same arity as the predicate from which it is derived.

A *realization* of the formal language is an assignment of:

- Simple names to none, some, or all of the individual or time name symbols.
- Simple infinitives to none, some, or all of the infinitive symbols.
- Simple non-variable restrictors to none, some, or all of the non-variable restrictor symbols.
- Simple variable restrictors to none, some, or all of the variable restrictor symbols.
- Simple negators to none, some, or all of the negator symbols.

The definition of the semi-formal language then proceeds as for **QT**.

Models

Conditions on valuations of atomic wffs

The valuations of atomic wffs now satisfy the conditions for **QT** plus the conditions for classical predicate logic with predicate modifiers and internal conjunctions and disjunctions (Chapter 31 of Volume 1) adapted to the formal language here:

- The conditions are for only formal ordinary atomic predicates.
- Each formal ordinary atomic predicate has a time term.

(These conditions are given explicitly in Chapter 45 below for the logic of quantifying over both times and locations.) We have in addition the following four new conditions, where are R, R´ are any restrictors, N is any negator, Q is any atomic predicate, and $\vec{d}$ is a sequence of terms or conjunctions of terms of the correct arity for the predicate.

Chapter 23

& *is equivalent to time-indexed* ∧

$v_\sigma \vDash (Q/(R \& R'))(\vec{d}, t)$ iff $v_\sigma \vDash (Q/R)(\vec{d}, t)$ and $v_\sigma \vDash (Q/R')(\vec{d}, t)$.

The pure negator and negation

$v_\sigma \vDash (Q/\sim R)(\vec{d}, t)$ iff $v_\sigma \vDash Q(\vec{d}, t)$ and $v_\sigma \nvDash (Q/R)(\vec{d}, t)$

$v_\sigma \vDash \sim Q(\vec{d}, t)$ iff $v_\sigma \nvDash Q(\vec{d}, t))$ and $v_\sigma \vDash (-\text{ to exist} -_{\text{time}})(\vec{d}, t)$

If $v_\sigma \vDash (Q/N)(\vec{d}, t)$, then $v_\sigma \vDash \sim Q(\vec{d}, t)$.

Satisfaction of compound wffs
The valuations for atomic wffs are extended to compound wffs in the same way as for **QT**, and the semantic consequence relation is then defined in the usual way.

Sufficiency of the collection of models
Given any realization, collection of things in time, collection of times, complete collection of assignments of references, and assignment of truth-values to the atomic predications satisfying the conditions above they together constitute a model.

Classical predicate logic with quantifying over times taking account of the internal structure of atomic predicates

The formal language, definitions of models, tautology, and semantic consequence constitute *classical predicate logic with quantifying over times and internal structure of atomic predicates*, **QT+internal**.

Axiom system
We start with the axiom system for **QT** of Chapter 21, except that now A, B, C stand for wffs of the extended language of **QT+internal**. We then add all the axioms for classical predicate logic with predicate modifiers and internal conjunctions and disjunctions (Chapter 31 of Volume 1), adapted in the same way as the semantic conditions were adapted. Then we add the four new axioms:

& *is equivalent to time-indexed* ∧

$\forall \ldots (Q/(R \& R'))(\vec{d}, t) \leftrightarrow ((Q/R)(\vec{d}, t) \wedge (Q/R')(\vec{d}, t))$

The pure negator and negation

$\forall x \, \forall t \, ((Q/\sim R)(x, t) \leftrightarrow (Q(x, t) \wedge \neg (Q/R)(x, t)))$

$\forall x \, \forall t \, (\sim Q(x, t) \leftrightarrow (\neg Q(x, t) \wedge (-\text{ to exist} -_{\text{time}})(x, t)))$

$\forall x \, \forall t \, ((Q/N)(x, t) \rightarrow \sim Q(x, t))$

The axiomatization is complete for the same reasons as the axiomatization of **QT** is strongly complete (p. 108 above).

FORMALIZING with QUANTIFYING over TIMES

24 Quasi-Linear Time

The past-present-future division
We want to use our logic of time to formalize claims and inferences from our daily language. In English every sentence is tensed via the verbs in it. The tenses are based on the idea that time can be divided into past, present, and future. There is a now, the present, that divides all times into those before now and those after now. So to formalize reasoning from English, we'll need to choose a particular time to serve as the present, whether that be an instant or an interval.

Convention for formalizing the present
When formalizing reasoning that is based on a past-present-future tense system, in a model we choose a particular time we name *now* to serve as the present.

Quasi-linear time
The past-present-future divide is compatible with incomparable past times, as we saw at p. 111. But most of us conceive of time as spread along a single line running from the past to the future, with "now" marked on it. That's usually understood as meaning that the timeline is *linear*: any two times can be compared.

$$\forall t \, \forall w \, (\, (t <_{\text{time}} w) \vee (w <_{\text{time}} t) \vee (t \equiv_{\text{time}} w) \,) \qquad \text{trichotomy}$$

But this can hold in a model only if every time is an instant. A more general notion of linearity assumes only that any two times that are not overlapping are comparable.

Quasi-linear orderings An ordering of times in a model is *quasi-linear* if for any times t and w, either t < w or w < t or O(t, w).

$$\forall t_1 \, \forall t_2 \, (\, (t_1 <_{\text{time}} t_2) \vee (t_2 <_{\text{time}} t_1) \vee O_{\text{time}}(t_1, t_2) \,) \qquad \text{quasi-trichotomy}$$

When Spot barked
With quasi-linear time, if Spot barked at 1:03 a.m. then there is a maximal interval around 1:03 a.m. in which he barked: an interval (perhaps just that minute) during which he barked and immediately outside of which he didn't bark.

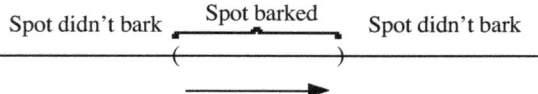

If Spot barked only once, we'd say that this interval is *the* time when Spot barked. More likely is that there are lots of times when Spot barked, and so lots of maximal intervals in which Spot barked:

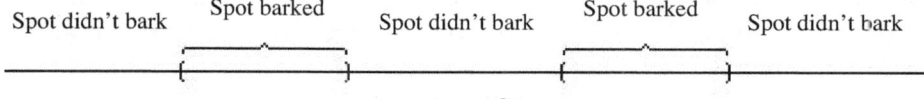

Each of these maximal intervals is *a* time at which Spot barked. Or at least it is if such intervals are themselves times.

When we say "Every time Spot barked Dick yelled" we're not talking about each instant at which Spot barked, for that would include 1:03.21 a.m. When we say "The last time Spot barked Dick didn't yell" we're not talking about the last instant at which Spot barked. We're talking about maximal intervals in which he barked. To quantify over those, we need that each such maximal interval is a time. Or at least we do if, as in the pictures above, those times are bounded. What if it's unbounded? Consider:

(1) $(\sim(-\text{ to bark }-_{\text{time}}))\,(\text{Puff}, t)$

If Puff the cat never barked, and we have a model with times named by every fixed time marker, then the interval in which (1) is true is unbounded: it's the entire (potentially) infinite collection of times. It seems at best odd to call that "the time when Puff didn't bark". So let's require that only bounded maximal intervals of true predications of ordinary wffs are times.

Maximal intervals of true predications   If $C(t)$ is an ordinary atomic wff, and w, t' are the least time variables not appearing in $C(t)$, then:

$$\text{Max } C(t) \equiv_{\text{Def}} C(t) \wedge$$
$$\neg \exists w\,((t <_{\text{time}} w) \wedge \forall t'\,((t \leq_{\text{time}} t' \leq_{\text{time}} w) \to C(t'))) \wedge$$
$$\neg \exists w\,((w <_{\text{time}} t) \wedge \forall t'\,((w \leq_{\text{time}} t' \leq_{\text{time}} t) \to C(t')))$$

Bounded maximal intervals of true predications are times
$$\forall \ldots (\,(C(t) \wedge \exists w\,((t <_{\text{time}} w) \wedge \neg C(w)))$$
$$\wedge \exists w\,((w <_{\text{time}} t) \wedge \neg C(w)))$$
$$\to \exists w\,(W_{\text{time}}(t, w) \wedge \text{Max } C(w)))$$

Though maximal intervals are bounded, we don't assume each has endpoints. When a train goes by and blasts its whistles, the sound of the whistle just trails off with no distinct end. Yet we still talk of the time when the train whistled.

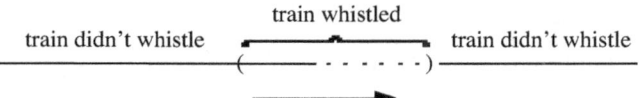

When Spot exists
By the continuity of existence in time, in any model in which "Spot" is a name there is a single maximal interval in which Spot exists. We can call that "the time when

124   Chapter 24

Spot exists". Or at least we can if the interval in which Spot exists is bounded. But Spot isn't immortal. If the interval in which an object exists is bounded, we can call it *the time when the object exists*.

If for each object in the universe of a model the interval in which it exists is bounded, then each object determines a time. It does not follow from this that every time is determined by a single thing. Birta and her tail existed always at the same time.

When Spot barked and Puff ran
Suppose Spot barked, and while he was barking, Puff began to run. Looking at maximal intervals in which Spot barked and Puff ran, we have:

(2)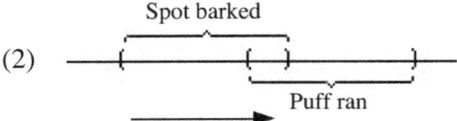

The overlap of these is what we'd say is the or a time at which Spot barked and Puff ran. So let's assume that the overlap is indeed a time.

Overlaps of maximal times of bounded true predications are times
For ordinary atomic wffs A and B, if a bounded maximal interval of time in which A is true overlaps a bounded maximal interval of time in which B is true, then that overlap is a time.

$\forall \ldots ((\text{Max } A(t) \wedge \text{Max } B(w) \wedge O_{time}(t, w)$
$\wedge \text{ bounded } A(t) \wedge \text{ bounded } B(w)) \rightarrow$
$\exists w_1 \forall w_2 (W_{time}(w_2, w_1) \leftrightarrow (W_{time}(w_2, w) \wedge W_{time}(w_2, t))))$

We could also describe in the semi-formal language the union of the maximal intervals at (2) using ∨ rather than ∧. But we don't usually talk of the time when Spot barked or Puff ran, nor have I found a need in formalizing to assume that the union of maximal intervals is a time.

Bringing together the assumptions in this chapter, we can begin formalizing propositions and inferences that involve a past-present-future division of time in the logic of quasi-linear time.

The theory of quantifying over times in quasi-linear time
The logic **QTL** is **QT+internal** plus

The ordering of times is quasi-linear.

Bounded maximal intervals of true predications are times.

Overlaps of bounded maximal times of true predications are times.

Aside: *Past-present-future*

The conception of the world as unfolding through time from the past to the future, with the present dividing the past from the future, is not universal across language cultures. Here is what Benjamin Lee Whorf says in "The Relation of Habitual Thought and Behavior to Language":

> The three-tense system of SAE [Standard Average European] verbs colors all our thinking about time. This system is amalgamated with the larger scheme of objectification of the subjective experience of duration already noted in other patterns —in the binomial formula applicable to nouns in general, in temporal nouns, in plurality and numeration. This objectification enables us in imagination to 'stand time units in a row.' Imagination of time as like a row harmonizes with a system of *three* tenses; whereas a system of *two*, an earlier and a later, would seem to correspond better to the feeling of duration as it is experienced. For if we inspect consciousness we find no past, present, future, but a unity embracing complexity. *Everything* is in consciousness, and everything in consciousness *is*, and is together. There is in it a sensuous and a nonsensuous. We may call the sensuous—what we are seeing, hearing, touching—the 'present' while in the nonsensuous the vast image-world of memory is being labeled 'the past' and another realm of belief, intuition, and uncertainty 'the future;' yet sensation, memory, foresight, all are in consciousness together—one is not 'yet to be' nor another 'once but no more.' Where real time comes in is that all this in consciousness is 'getting later,' changing certain relations in an irreversible manner. In this 'latering' or 'durating' there seems to me to be a paramount contrast between the newest, latest, instant at the focus of attention and the rest—the earlier. Languages by the score get along well with two tense-like forms answering to this paramount relation of later to earlier. We can of course *construct and contemplate in thought* a system of past, present, future, in the objectified configuration of points on a line. This is what our general objectification tendency leads us to do and our tense system confirms it. p. 209

Many peoples native to North and South America speak or spoke languages in which such objectification is not basic if available at all.[23] They use what some call "event-based time": times are established by true propositions and ordered by saying that one is before the other, with no tense marking.[24] In the first part of this book we saw how to formalize reasoning that takes account of time solely in terms of before and after.

---

[23] See the essay "Language and the World" in my *Language and the World*.
[24] See for example Vera da Silva Sinha, *Linguistic and Cultural Conceptualisations of Time in Huni Kui, Awety´, and Kamaiurá Communities in Brazil*.

25 Formalizing English Tenses

I cannot hope to uncover "the logic" of English tenses here, for the use of tenses in English is very complicated. I can only set out a method for trying out various assumptions for taking account of tenses in our reasoning. Then you and others can use the methods, if not those assumptions, to extend or remedy the work here or to apply the methods to formalizing reasoning in other languages.

Let's start with an example in the simple past:

(1)   Spot barked.

This appears to be about some time in the past, wholly in the past, and hence about some maximal interval, as suggested in the last chapter. So we can formalize it as:

(2) $\exists t \, (\, \text{Max} \, (-\text{ to bark } -_{\text{time}}) \, (\text{Spot}, t) \, \wedge \, (t <_{\text{time}} \text{now}) \,)$

All the vocabulary in (2) except "$(-\text{ to bark } -_{\text{time}})$" and "Spot" is part of the formal language and hence allowed in a formalization, with the exception of "now" which we allow as part of a convention on formalizing tenses, not an analysis before formalization.

Here is another example in the simple past:

(3)   Spot was a dog.

If we formalize this as we did (1), we get:

(4) $\exists t \, (\, \text{Max} \, (-\text{ to be a dog } -_{\text{time}}) \, (\text{Spot}, t) \, \wedge \, (t <_{\text{time}} \text{now}) \,)$

But Spot is still alive and continues to be a dog. So (4) would be false: there is no such maximal interval wholly in the past. Instead, we formalize (3) as:

(5) $\exists t \, (\, (-\text{ to be a dog } -_{\text{time}}) \, (\text{Spot}, t) \, \wedge \, (t <_{\text{time}} \text{now}) \,)$

Yet if someone were asserting that Spot was once a dog but now has become a bird, (4) would be a good formalization.

The difference between (1) and (3) is not that (1) uses a process predicate and (3) a classification predicate, for consider:

Spot was a puppy.

This we understand to be about a time wholly in the past, and so we formalize it as:

$\exists t \, (\, \text{Max} \, (-\text{ to be a puppy } -_{\text{time}}) \, (\text{Spot}, t) \, \wedge \, (t <_{\text{time}} \text{now}) \,)$

And consider:

(6)   Spot breathed. Spot is breathing. Spot will breathe.

The predicate "$(-\text{ to breathe } -_{\text{time}})$" is a process predicate. To formalize "Spot breathed" in (6), we have to ask whether in context the proposition implies that it is about a time wholly in the past; I think it doesn't. That's not because of the tense, nor whether it is a classification or process predicate, but because we know that breathing is something that a creature continues to do while it is alive, which can start in the past and continue into the future. So to formalize a proposition in the simple past, we have to ask the speaker or infer from what is said whether he or she intends it to mean about a time wholly in the past. If so, we formalize it as about a maximal interval of time before now. If not, we formalize it as about just some time before now.

Similar remarks apply to formalizing propositions in the simple future. For example, we can formalize:

Spot will bark.

as either:

$\exists t\,(\,\mathrm{Max}\,(-\text{ to bark }-_{\text{time}})\,(\text{Spot}, t) \land (\text{now} <_{\text{time}} t)\,)$

or

$\exists t\,(\,(-\text{ to bark }-_{\text{time}})\,(\text{Spot}, t) \land (\text{now} <_{\text{time}} t)\,)$

We do not usually use the simple present to make assertions about the present. For that we typically use the present progressive (present continuous), as in:

(7)   Spot is barking.

There is no suggestion here of a complete, maximal interval. We could formalize this as:

(8) $(-\text{ to bark }-_{\text{time}})\,(\text{Spot}, \text{now})$

But the present (that we're talking about) may be a very long time. Better is:

(9) $\exists t\,(\,(-\text{ to bark }-_{\text{time}})\,(\text{Spot}, t) \land \mathrm{W}_{\text{time}}(t, \text{now})\,)$

The choice between (8) and (9) depends partly on context and partly on what we take to be the present. Note that if (8) is true, so is (9).

These remarks suggest a rough guide for formalizing the simple past, present progressive, and simple future.

Conventions for formalizing simple past, present continuous, and simple future tense propositions

Given an *n*-ary ordinary language predicate P which we could take as a simple predicate in classical predicate logic, we convert it to an infinitive phrase I_P.

- If P is in the *simple past tense*, and:
    - in our judgment $P(u_1, \ldots, u_n)$ is meant as a description of what is completed wholly in the past, we formalize $P(u_1, \ldots, u_n)$ as:

        $\exists t\,(\,\mathrm{Max}\,I_P(u_1, \ldots, u_n, t) \land (t <_{\text{time}} \text{now})\,)$

128   Chapter 25

- in our judgment, $\{(u_1, \ldots, u_n)\}$ is meant as a description of what might or might not be completed wholly in the past, we formalize $P(u_1, \ldots, u_n)$ as:

  $\exists t \, (I_P(u_1, \ldots, u_n, t) \wedge (t <_{\text{time}} \text{now}))$

- If P is in the *simple future tense*, and:
  - in our judgment $P(u_1, \ldots, u_n)$ is meant as a description of what is completed wholly in the future, we formalize $P(u_1, \ldots, u_n)$ as:

    $\exists t \, (\text{Max} \, I_P(u_1, \ldots, u_n, t) \wedge (\text{now} <_{\text{time}} t))$

  - in our judgment, $P(u_1, \ldots, u_n)$ is meant as a description of what might or might not be completed wholly in the future, we formalize $P(u_1, \ldots, u_n)$ as:

    $\exists t \, (I_P(u_1, \ldots, u_n, t) \wedge (\text{now} <_{\text{time}} t))$

- If P is in the *present continuous tense*, we formalize $P(u_1, \ldots, u_n)$ as:

  $\exists t \, (I_P(u_1, \ldots, u_n, t) \wedge W_{\text{time}}(t, \text{now}))$

We extend this guide to atomic predicates with internal structure by replacing I_P with A, where A is any atomic predicate using I_P.

Aside: Choosing and talking about a time as the present

We've left the choice of what to take as the present in a model to be outside the formal system. We might choose a time we already have a name for in the semi-formal language, like "December 12, 2020". In that case, we'd add to the semi-formal language "now $\equiv_{\text{time}}$ December 12, 2020". But if we want the present to be the interval between December 9, 2020 and December 21, 2020, that might not be a time in the model even if we have "all" fixed time markers.

Commonly, we pick out times by "what happened": the time that Suzy broke her arm; the time when the dog Moya was killed by a tiger; the time when Spot caught a skunk. We talk of times relative to them by noting that "what happened" can be described with a true proposition.[25] Suppose we want the present to be when Julius Caesar lived. Then past and future tenses mark times before and after that interval. So we could formalize:

Socrates was a philosopher.

$\exists ! t \, (\text{Max} \, (- \text{ to live } -_{\text{time}}) \, (\text{Caesar}, t) \wedge$
$\quad \exists w \, (\text{Max} \, (- \text{ to be a philosopher } -_{\text{time}}) \, (\text{Socrates}, w) \wedge (w <_{\text{time}} t)))$

and

Augustine will convert.

$\exists ! t \, (\text{Max} \, (- \text{ to live } -_{\text{time}}) \, (\text{Caesar}, t) \wedge$
$\quad \exists w \, (\text{Max} \, (- \text{ to convert } -_{\text{time}}) \, (\text{Augustine}, t) \wedge (t <_{\text{time}} w)))$

---

[25] Talk of "what happened" as an event only muddles what is clear in talking of true propositions, as I explain in Appendix C here.

(Here $\exists!$ is the defined operator that formalizes "there exists exactly one"; see p. 146 of Volume 0, *An Introduction to Formal Logic*.)

Or we could take the present to be the one time that Spot barked and Puff didn't run. We can do that because an overlap of maximal predications is a time. Then we could formalize:

Suzy will jump.

$\exists!w_1 \exists!w_2 \, (\, \text{Max}\,(-\text{ to bark }-_{\text{time}})\,(\text{Spot}, w_1)$
$\quad \wedge \; \text{Max}\,(\sim (-\text{ to run }-_{\text{time}})\,(\text{Puff}, w_2)) \wedge \text{O}_{\text{time}}(w_1, w_2))$
$\quad \wedge \; \exists t \, (\text{Max}\,(-\text{ to jump }-_{\text{time}})\,(\text{Suzy}, t)$
$\quad \wedge \; \forall t' \, (\text{W}_{\text{time}}(t', w_1) \wedge \text{W}_{\text{time}}(t', w_2) \rightarrow (t' <_{\text{time}} t))))$

Formalizing this way we do not need a name "now". But even introducing abbreviations, this will be much harder to read and will obscure the main issues of the examples we'll look at. So I'll use "now" in the formalizations.

Aside: Atomic formalizations of tensed atomic wffs?
"Spot barked", "Spot is barking", and "Spot will bark" are atomic propositions in classical predicate logic. To formalize these as atomic we could introduce tenses directly into the formal language with "(— to bark-past)", "(— to bark-present)", "(— to bark-future)". Then we could formalize "Spot barked" as "(— to bark-past) (Spot)". This would elevate the past-present-future divide into a logical distinction rather than a convention added to the logic for the purposes of formalizing English propositions. We would need to relate predications using these predicates. If we also marked the present with "now", then "(— to bark-past) (Spot)" would be equivalent to (1). I prefer the metalogical approach, clearly differentiating conventions for formalizing tenses from the basic logic.

26 Examples of Formalizing: Context of Utterance

Here and in succeeding chapters, I'll use t, t', w, w', w_1, w_2 for $t_1, t_2, t_3, t_4, t_5, t_6$, and x, y, z for x_1, x_2, x_{47}.

Example 1 Spot barked. (said by Dick on May 3, 2002)

$\exists t \, (\, (t <_{\text{time}} \text{now}) \land W_{\text{time}}(\text{now}, \text{May 3 2002})$
$\quad \land \text{Max} \, (— \text{ to bark} —_{\text{time}}) \, (\text{Spot}, t) \,)$

Analysis The context of the utterance determines not the time of which the assertion is meant to be true, but a time relative to which the past tense is meant. The now of this example is when Dick spoke. We can't pick that out with the wff "$(— \text{ to speak}) (\text{Dick}, t)$" because it's not just that Dick spoke but what he said, and we don't have names for wffs in our language. Should we take "now" to be May 3, 2002? That would rule out Spot having barked earlier that day before Dick spoke. We have to take "now" as a name meant to pick out a narrow time when Dick spoke, which is within May 3, 2002.

Example 2 Spot barked. (said by Zoe on April 7, 2004)

$\exists t \, (\, (t <_{\text{time}} \text{now}) \land W_{\text{time}}(\text{now}, \text{April 7 2004})$
$\quad \land \text{Max} \, (— \text{ to bark} —_{\text{time}}) \, (\text{Spot}, t) \,)$

Analysis The words Zoe utters are the same as Dick uttered in the previous example, but it's a different proposition. It could be that what Zoe says is true and what Dick says is false if Spot never barked until 2003. Tenses are indexicals.

The formalizations are different, too, for we constrain "now" to be within the day in which Dick spoke or the day in which Zoe spoke. But then we can't formalize Examples 1 and 2 in the same semi-formal language because "now" picks out different times in them. To avoid that restriction, we could use indexed versions of "now":

^{now}Dick said on May 3, 2002, "Spot barked"

^{now}Zoe said on April 7, 2004, "Spot barked"

Such indices would be indivisible labels, reminding us how to use these time names in formalizations.

Example 3 $(— \text{ to bark} —_{\text{time}}) \, (\text{Spot}, 3{:}00 \text{ p.m. May 3 2002})$

Analysis This is an untensed example. We know the time of which it is meant to be true, but there is no indication of whether it is meant as past, present, or future.

Example 4 Suzy was a student throughout 1998.

$(— \text{ to be a student} —_{\text{time}}) \, (\text{Suzy}, 1998) \land (1998 <_{\text{time}} \text{now})$

Analysis By the downward closure of truth in time, the formalization is equivalent to:

$\forall t \, (\, (W_{\text{time}}(t, 1998) \rightarrow (— \text{ to be a student} —_{\text{time}}) \, (\text{Suzy}, t)) \land (1998 <_{\text{time}} \text{now}) \,)$

We don't need a separate time blank "$—_{\text{throughout}}$".

Example 5   When did Suzy jump? When Spot barked and Puff screeched.

Analysis   Here "and" does not mean "and then". Rather, we understand the example as "Suzy jumped during the time that both Spot barked and Puff screeched", as if there were only one such time. We can formalize that as:

$\exists! t_1 \exists! t_2 \exists t_3$ (Max (— to bark —$_{time}$) (Spot, t_1) $\wedge$ ($t_1 <_{time}$ now)

$\wedge$ Max (— to screech —$_{time}$) (Puff, t_2) $\wedge$ ($t_2 <_{time}$ now) $\wedge$ O$_{time}$(t_1, t_2)

$\wedge$ Max (— to jump —$_{time}$) (Suzy, t_3) $\wedge$ ($t_3 <_{time}$ now)

$\wedge$ $\forall w$ (W$_{time}(w, t_3)$ $\rightarrow$ W$_{time}(w, t_1)$ $\wedge$ W$_{time}(w, t_2)$))

We don't need a separate time blank "—$_{during}$".

Example 6   Tom ate lunch yesterday.

Analysis   To formalize this we need a name in the semi-formal language for the time that "yesterday" is meant to pick out. Let's use "b". Then we can formalize the example as:

$\exists t$ (Max ((— to eat lunch —$_{time}$) (Tom, t)) $\wedge$ (b $<_{time}$ now) $\wedge$ W$_{time}(t, b)$)

This formalizes the example as "Tom ate lunch sometime during yesterday". We don't use "(— to eat lunch —$_{time}$) (Tom, b)" because then by the downward closure of truth in time Tom would have to have eaten lunch at every time within yesterday.

Example 7   If Dick yelled, then Spot barked. Dick yelled. So Spot barked.

$\exists t_1$ (Max (— to yell —$_{time}$) (Dick, t_1) $\wedge$ ($t_1 <_{time}$ now)) $\rightarrow$
  $\exists t_2$ (Max (— to bark —$_{time}$) (Spot, t_2) $\wedge$ ($t_2 <_{time}$ now))

$\underline{\exists t_1 \text{ (Max (— to yell —}_{time}\text{) (Dick, } t_1\text{)} \wedge (t_1 <_{time} \text{ now))}}$

$\exists t_2$ (Max (— to bark —$_{time}$) (Spot, t_2) $\wedge$ ($t_2 <_{time}$ now))

Analysis   This is to read "then" not as temporal but as part of a conditional. The example is valid, an application of *modus ponens*.

Example 8   If Spot barks, then Dick will yell. Spot barked. So Dick yelled.

$\exists t_1$ (Max (— to bark —$_{time}$) (Spot, t_1) $\wedge$ (now $<_{time} t_1$) $\rightarrow$
  $\exists t_2$ (Max (— to yell —$_{time}$) (Dick, t_2) $\wedge$ (now $<_{time} t_2$) $\wedge$ ($t_1 <_{time} t_2$)))

$\underline{\exists t_1 \text{ (Max (— to bark —}_{time}\text{) (Spot, } t_1\text{)} \wedge (\text{now} <_{time} t_1))}$

$\exists t_2$ (Max (— to yell —$_{time}$) (Dick, t_2) $\wedge$ (now $<_{time} t_2$))

Analysis   This is to take "then" as "and then". The inference is valid. Note that the time indicated by the present tense shifts here: the past tense of the last two propositions is in the future of the "now" of the first. And the present tense in the first proposition should be understood as indicating a time in the future.

The context of an assertion is often needed to establish what we should take as the present relative to which the tenses in it are to be evaluated. But in most cases to give such a context for an example will obscure the points the example is meant to illustrate.

Aside: *Contexts of utterance and propositions as tokens*
Suppose we allow the stipulation of the context of utterance via a blank for a person and a blank for a time, separating those with square brackets, as suggested by Eduardo Ribeiro. For example:

$$((((- \text{ is a dog } -_{time}) [-, -_{time}]) (\text{Spot, May 6 2007}) [\text{Eddie, July 6 2021}])$$

How is this different from treating

$$(- \text{ is a dog } -_{time}) (\text{Spot, May 6 2007})$$

as spoken by Eddie on July 6, 2121 as a *token*? If we take seriously that any sentence, taken as a linguistic entity, is not a proposition in some abstract way but is one only when either asserted or set out as hypothetical by someone at some time and place, then propositions are tokens. Propositions as tokens, though not developed this way, is what I used in resolving the liar paradox in my paper "A Theory of Truth Based on a Medieval Solution to the Liar Paradox" (as revised in *Classical Mathematical Logic*), using a solution of Buridan. This might be a way to make clearer or at least be part of a general solution to incorporating more of context into the syntax of propositions.

27 Examples of Formalizing: Existence

Example 1 There are unicorns.
Analysis In classical predicate logic we formalize this as:
$\exists x \, (- \text{ is a unicorn}) \, (x)$

This is true if and only if there is something in the universe that is a unicorn: "there are" is read as timeless. But taking account of times, nothing exists timelessly, and we formalize the example as:
$\exists x \, \exists t \, (- \text{ to be a unicorn} -_{\text{time}}) \, (x, t)$

If we think the example is meant about the present, we should formalize it as:
$\exists x \, \exists t \, ((- \text{ to be a unicorn} -_{\text{time}}) \, (x, t) \land W_{\text{time}}(t, \text{now}))$

Example 2 There were unicorns.
$\exists x \, \exists t \, ((- \text{ to be a unicorn} -_{\text{time}}) \, (x, t) \land (t <_{\text{time}} \text{now}))$

Analysis "There are" can be understood as either the present or as timeless. But "there were" is definitely temporal.

Example 3 There is something that was a dog.
$\exists x \, \exists t \, ((- \text{ to be a dog} -_{\text{time}}) \, (x, t) \land (t <_{\text{time}} \text{now}))$

Analysis This is to read "there is" as the timeless existential quantifier. Since the thing might still be a dog, we can't use:
$\exists x \, \exists t \, (\text{Max} \, (- \text{ to be a dog} -_{\text{time}}) \, (x, t) \land (t <_{\text{time}} \text{now}))$

Example 4 There is something now that was a dog.
$\exists x \, ((- \text{ to exist} -_{\text{time}}) \, (x, \text{now}) \land$
$\quad \exists t \, ((- \text{ to be a dog} -_{\text{time}}) \, (x, t) \land (t <_{\text{time}} \text{now})))$

Analysis Note that the example does not imply that the thing is a dog now.

Example 5 There was something that was a dog.
  Formalization as for Example 3.
Analysis That the thing was follows from the predicate being positive for existence.

Example 6 Ralph does not exist.
Analysis If this is meant as saying that Ralph doesn't exist now, we formalize it as:
$\neg \exists t \, ((- \text{ to exist} -_{\text{time}}) \, (\text{Ralph}, t) \land W_{\text{time}}(t, \text{now}))$

This could be true if Ralph existed only in the past. If the example is meant as saying that Ralph doesn't exist at all, not ever, we can formalize it as:
$\forall t \, \neg \, ((- \text{ to exist} -_{\text{time}}) \, (\text{Ralph}, t))$

This is an anti-tautology because names refer and things exist in time.

Example 7 Nothing existed.
$$\forall x \, \forall t \, (\neg \, (-\text{ to exist }-_{\text{time}}) \, (x, t) \land (t <_{\text{time}} \text{now}))$$
Analysis This can be true in a model if every thing in the universe exists only within the time denoted by "now".

Example 8 Nothing exists.
$$\forall x \, \forall t \, \neg \, ((-\text{ to exist }-_{\text{time}}) \, (x, t))$$
Analysis This is an anti-tautology. It would have to be, since the semantics for predicate logic make no sense if there is nothing in the universe (p. 67 of Volume 0). Understood as a claim about the present, however, we can formalize the example as:
$$\neg \, \exists x \, (-\text{ to exist }-_{\text{time}}) \, (x, \text{now})$$
This would seem to be a self-refuting proposition, since it itself exists in the present. But that need not be if "now" picks out some time other than our current present.

Example 9 There is something that will exist.
$$\exists x \, \exists t \, ((-\text{ to exist }-_{\text{time}}) \, (x, t) \land (\text{now} <_{\text{time}} t))$$
Analysis This is to read "there is" as the existential quantifier. If what is meant is that something that exists now will exist, we formalize the example as:
$$\exists x \, \exists t \, \exists w \, ((-\text{ to exist }-_{\text{time}}) \, (x, w) \land W_{\text{time}}(w, \text{now})$$
$$\land \, (-\text{ to exist }-_{\text{time}}) \, (x, t) \land (\text{now} <_{\text{time}} t))$$

Example 10 If something existed, then something exists, and something will exist.
$$\exists x \exists t \, ((-\text{ to exist }-_{\text{time}}) \, (x, t) \land (t <_{\text{time}} \text{now})) \rightarrow$$
$$(\exists y \, \exists t_1 \, ((-\text{ to exist }-_{\text{time}}) \, (y, t_1) \land W_{\text{time}}(t_1, \text{now}))$$
$$\land \, \exists z \, \exists t_2 \, ((-\text{ to exist }-_{\text{time}}) \, (z, t_2) \land (\text{now} <_{\text{time}} t_2))))$$
Analysis This is not an assertion that one thing existed, exists, and will exist.

Example 11 There is something that always existed, exists, and always will exist.
$$\exists x \, \forall t \, (-\text{ to exist }-_{\text{time}}) \, (x, t)$$
Analysis This could be true in a model. Whether it could be true *simpliciter* is another question. Any thing that satisfies this wff is not time-bounded if the universe of times is not time-bounded.

Example 12 There is a time when nothing exists.
$$\exists t \, \forall x \, \neg \, (-\text{ to exist }-_{\text{time}}) \, (x, t)$$
Analysis Some think that the example cannot be true, since, they say, we can pick out times, we can conceive of times, only by what "happens" at them.

Example 13 Socrates did not exist when Julius Caesar did.
Analysis We can formalize this as:

$\neg \exists t \exists w (((- \text{ to exist } -_{\text{time}}) (\text{Socrates}, t) \wedge (t <_{\text{time}} \text{now})$
$\wedge (- \text{ to exist } -_{\text{time}}) (\text{Julius Caesar}, w) \wedge (w <_{\text{time}} \text{now}) \wedge O_{\text{time}}(t, w))$

Example 14 Wanda remembers Bruno.
Analysis This is true. So Wanda exists now—how could she not if she is remembering? But her dog Bruno died three years ago, so he does not exist now. So the predicate "$(- \text{ remembers } -, -_{\text{time}})$" is not positive for existence. But that's not what we use to formalize this example:

$((- \text{ to remember } -_{\text{time}}) / \text{obj}(\text{Bruno})) (\text{Wanda}, \text{now})$

The predicate "$(- \text{ to remember } -_{\text{time}})$" is positive for existence. We've made no assumption about whether a term that appears in a variable modifier refers to something at the time the predication is true.

Example 15 Spot is biting Puff. So Puff exists.

$$\frac{\exists t (((- \text{ to bite } -_{\text{time}}) / \text{obj}(\text{Puff})) (\text{Spot}, t) \wedge W_{\text{time}}(t, \text{now}))}{\exists t ((- \text{ to exist } -_{\text{time}}) (\text{Puff}, t) \wedge W_{\text{time}}(t, \text{now}))}$$

Analysis The example is valid, at least on our usual understanding of "is biting". But the semi-formal inference is not valid. We can add a meaning axiom relative to which the formalization will be valid:

$\forall t \forall x \forall y (((- \text{ to bite } -_{\text{time}}) / \text{obj}(y)) (x, t) \rightarrow (- \text{ to exist } -_{\text{time}}) (y, t))$

Example 16 Rex was the father of Bruno.
   Therefore, Bruno existed.

$$\frac{\exists t (\text{Max} (- \text{ to be a father } -_{\text{time}}) / \text{of}(\text{Bruno})) (\text{Rex}, t) \wedge (t <_{\text{time}} \text{now}))}{\exists t ((- \text{ to exist } -_{\text{time}}) (\text{Bruno}, t) \wedge (t <_{\text{time}} \text{now}))}$$

Analysis The informal inference is valid. That's not due to its form but to the meaning of "to be a father of". We can ensure that the formalization is valid relative by adopting a meaning axiom:

$\forall x \forall y \exists t (((- \text{ to be a father } -_{\text{time}}) / \text{of}(x)) (y, t) \rightarrow (- \text{ to exist } -_{\text{time}}) (x, t))$

However, the formalization does not reflect that "the" in the premise indicates there was one and only one father of Bruno, which I'll leave to you to formalize.

Example 17 Socrates is dead.
Analysis In Chapter 20 we worried that the predicate "$(- \text{ to be dead } -_{\text{time}})$" is not positive for existence. So we excluded it from our a semi-formal languages. But if Socrates is dead, then he died completely in the past, and the predicate "$(- \text{ to die } -_{\text{time}})$" is positive for existence (if something is dying, it exists). So we can formalize the example:

$\exists t (\text{Max} (- \text{ to die } -_{\text{time}}) (\text{Socrates}, t) \wedge (t <_{\text{time}} \text{now}))$

136   Chapter 27

We can add a meaning axiom that something can die only once:

$$\exists x \,(\exists t\,(-\text{ to die }-_{\text{time}})\,(x,t)) \rightarrow \exists! t\,\text{Max}\,(-\text{ to die }-_{\text{time}})\,(x,t))$$

Example 18   Socrates is famous.
Analysis   This is true. But Socrates does not exist now. So "— is famous" is not positive for existence. But that's not why we can't formalize the example here. We understand it as something like "Most people know of Socrates", and "to know of" is not clear and "most" is a vague quantifier outside the scope of our methods.

Example 19   Puff is nearly coughing. So Puff exists.

$$\frac{((-\text{ to cough }-_{\text{time}})/\text{nearly})\,(\text{Puff, now})}{(-\text{ to exist }-_{\text{time}})\,(\text{Puff, now})}$$

Analysis   The formalization is valid because the predicate in the premise is atomic and hence positive for existence.

Example 20   There was something that was not a dog.

$$\exists x \,\exists t\,(\sim (-\text{ to be a dog}-_{\text{time}})\,(x,t)\,\wedge\,(t<_{\text{time}}\text{now}))$$

Analysis   That the thing was follows from the predicate being positive for existence, which is why predicate negation is used rather than propositional negation.

Example 21   Maria used to be a student.

$$\exists t\,(\,\text{Max}\,(-\text{ to be a student }-_{\text{time}})\,(\text{Maria},t)\,\wedge\,(t<_{\text{time}}\text{now}))$$
$$\wedge\,(\sim(-\text{ to be a student }-_{\text{time}}))\,(\text{Maria, now})$$

Analysis   Here "used to" means that in the past it was true but is not true now of something that exists now. So we use predicate negation in the last conjunct.

Example 22   Bruno doesn't exist anymore.

$$\exists t\,((-\text{ to exist }-_{\text{time}})\,(\text{Bruno},t)\,\wedge\,(t<_{\text{time}}\text{now}))$$
$$\wedge\,\neg\exists w\,((-\text{ to exist }-_{\text{time}})\,(\text{Bruno},w)\,\wedge\,W_{\text{time}}(w,\text{now}))$$

Analysis   We use "not anymore" as "used to" in the previous example to indicate that a condition held in the past and no longer holds.

These examples suggest a convention for when to use predicate negation and when to use propositional negation.

Convention on formalizing "not"
Given a part of an informal proposition, we use ~ to formalize "not" in it if an existence claim follows from the negated part; otherwise we use ¬.

Example 23   Rudolfo nearly existed on February 18, 1931 at 2:46 p.m.
Analysis   Though odd, someone might say this if Rudolfo was born on February 18,

1931 at 2:48 p.m. Or perhaps someone would say this if Rudolfo was conceived on February 18, 1931 at 2:47 p.m.

The predicate "— to exist —$_{time}$" is not meant to formalize another kind of existence but only to place the existence of the existential quantifier in time. It doesn't seem to be a good idea to introduce more kinds of existence by allowing the existence predicate to be modified.

Example 24 Socrates was shorter than Julius Caesar.

Analysis This is true, yet there was no time at which both Socrates and Julius Caesar existed. So if it's true, of what time is it true? It can't be a time when Socrates existed but Julius Caesar didn't or vice-versa: there was nothing to compare then. Perhaps we could say that it's true of any time after both of them existed. But what is true of a time need not be true of all succeeding times: "Socrates was a baby" is true of some time but not true of all succeeding times. Moreover, Socrates was not shorter than Julius Caesar at the time that Julius Caesar was born. What then do we mean when we assert this example? It must be something like:

Socrates, when he was mature, was shorter than Caesar, when he was mature.

But this isn't better because there is no time when both were mature. The predicate "(— to be shorter than —, —$_{time}$)" applies at a time and is positive for existence at that time.

Nor can we take the example to be a comparison not of Socrates and Julius Caesar but of their heights, using a degree approach to comparatives, because we'd need times for those heights.

Example 25 7 is a prime number.

Not formalizable.

Analysis If numbers are abstract objects, then they are not of time: they are atemporal, not omnitemporal. In that case the example is not formalizable in our logic.

Some view numbers not as abstract things but as abstractions from experience. The experience of counting, adding, subtracting, multiplying, and dividing are in time: "2" and "7" are adjectives, meaningful only when attached to a noun, as in "2 apples" or "7 cats". But on that human-based conception of numbers, the example also is not formalizable.[26]

---

[26] Mathematicians ignore the physical and temporal aspects of what is being counted. Actually, they ignore the counting entirely. See my "Mathematics as the Art of Abstraction".

28 Examples of Formalizing: Attributes

Here and in succeeding chapters, I'll use "now" for the time of a formalization of a present tense proposition rather than "$W_{time}(t, now)$" in order to make clearer the other issues that are being discussed.

Example 1 Birta is a dog.
Analysis If this is meant as an assertion about the present, we formalize it as:

$\quad$ (— to be a dog —$_{time}$) (Birta, now)

If we understand the example as meaning that being a dog is the nature of Birta, we can formalize it as:

$\quad \forall t\,((-\text{ to exist }-_{time})\,(Birta, t) \to (-\text{ to be a dog }-_{time})\,(Birta, t)\,)$

We can use the existence predicate here because it is in the formal language.

Example 2 Birta is a dog, but she will be a princess.

$\quad (-\text{ to be a dog }-_{time})\,(Birta, now)$
$\quad \land\ \exists w\,((-\text{ to be a princess }-_{time})\,(Birta, w) \land (now <_{time} w)\,)$

Analysis Whoever asserts this does not mean by "Birta is a dog" that she always was, is, and will be a dog.

Example 3 If Birta is a dog, then Birta is a mammal.

Analysis We could understand the example as omnitemporal:

$\quad \forall t\,((-\text{ to be a dog }-_{time})\,(Birta, t) \to (-\text{ to be a mammal }-_{time})\,(Birta, t)\,)$

Or we could understand the example as saying that if Birta is a dog now, then she is a mammal now:

$\quad (-\text{ to be a dog }-_{time})\,(Birta, now) \to (-\text{ to be a mammal }-_{time})\,(Birta, now)$

Example 4 Tom had a dog that will bark.
$\quad$ Therefore, *there will be a dog that will bark.*

$\quad \exists x\,(\,\exists t\,((-\text{ to have}_p\,-, -_{time})\,(Tom, x, t)$
$\quad\quad \land\ (-\text{ to be a dog}-_{time})\,(x, t) \land (t <_{time} now)\,)$
$\quad\quad \land\ \exists w\,((-\text{ to bark }-_{time})\,(x, w) \land (now <_{time} w)\,)\,)$

$\overline{\exists x\,\exists t\,((-\text{ to be a dog}-_{time})\,(x, t) \land (-\text{ to bark }-_{time})\,(x, t) \land (now <_{time} t)\,)}$

Analysis The subscript "p" indicates that "to have" is used in the sense of possesses (see p. 21 of Volume 0). The formalization is invalid, and so is the example: the dog that Tom had could have been transformed into a prince and then barked. From the premise we can conclude only that there will be something that was a dog and that will bark. Yet we all believe that if something once was a dog, it continues to be a dog as long as it exists; it is only the stuff of fairy tales that a dog becomes a prince.[27] But some things do change: what was once a house is now an apartment building.

Example 5 Once a dog, always a dog.

Analysis We could understand the example as we do "Once a king, always a king" to mean that if an object is once a dog, it will be a dog at all times after that at which it exists, though it might not have been a dog previously. In that case we formalize the example as:

$\forall x \, \forall t \, ((- \text{ to be a dog } -_{\text{time}}) \, (x, t) \rightarrow$
$((\forall w \, (- \text{ to exist } -_{\text{time}}) \, (x, w) \wedge (t <_{\text{time}} w)) \rightarrow (- \text{ to be a dog } -_{\text{time}}) \, (x, w)))$

Alternatively, we can understand the example as meaning that if at any time an object is a dog, then at all times that it existed, exists, or will exist it was, is, or will be a dog. We can formalize that as:

$\forall x \, (\exists t \, (- \text{ to be a dog } -_{\text{time}}) \, (x, t) \rightarrow$
$\quad \forall w \, ((- \text{ to exist } -_{\text{time}}) \, (x, w) \rightarrow (- \text{ to be a dog } -_{\text{time}}) \, (x, w)))$

Note that no tenses are needed in this formalization.

These examples suggest the following definitions.

Perpetuating attributes in time An informal predicate is a *perpetuating attribute* if for any object, if it is true of that object at some time, then it is true of that object at all subsequent times that the object exists. The general form of an assertion that a simple unary predicate P (converted to an infinitive I_P) is a perpetuating attribute is:

$\forall x \, \forall t \, (I_P(-, -_{\text{time}}) \, (x, t) \rightarrow$
$\quad \forall w \, ((- \text{ to exist } -_{\text{time}}) \, (x, w) \wedge (t <_{\text{time}} w) \rightarrow I_P(-, -_{\text{time}}) \, (x, w)))$

Essential attributes in time An informal predicate is an *essential attribute in time* if for any object, if it is true of that object at some time, then it is true of that object at all times that the object exists. The general form of an assertion that a simple unary predicate P (converted to an infinitive I_P) is an essential attribute is:

$\forall x \, \exists t \, ((I_P(-, -_{\text{time}}) \, (x, t) \rightarrow$
$\quad \forall w \, ((- \text{ to exist } -_{\text{time}}) \, (x, w) \rightarrow I_P(-, -_{\text{time}}) \, (x, w)))$

We extend this guide to atomic predicates with internal structure by replacing I_P with A, where A is any atomic predicate using I_P.

These examples show that we can talk of possibilities in the future or the past in our logic. Birta could have been a bird; Ralph could have been a dog; Birta could

---

[27] It is an even more magical belief that a prince could become a dog, stepping up in the ladder of being.

someday be a princess. With branching time, a much fuller way to incorporate possibilities in time can be done: in the future Birta could be euthanized or she could be run over by a truck, both possibilities being equally "real".[28] See Appendix B here.

Example 6 Electrons have spin.

Analysis We might debate whether electrons are things, but the formalization is straightforward in classical predicate logic:

$\forall x \, (\, (\text{— to be an electron —}_{\text{time}}) \, (x) \rightarrow (\text{— to have spin —}_{\text{time}}) \, (x) \,)$

Taking account of time in the formalization is straightforward, too, for the example is certainly meant as omnitemporal:

$\forall x \, \forall t \, (\, (\text{— to be an electron —}_{\text{time}}) \, (x, t)$
$\rightarrow (\text{— to be have spin —}_{\text{time}}) \, (x, t) \,)$

But now the question of whether electrons are things is more pressing. What identity conditions do we have for electrons in time? Friedrich Waismann says in "The Decline and Fall of Causality":

> Observing an electron is, unavoidably, interfering with it. In the act of observation it is pushed by a photon, and this must alter its velocity. The situation is sometimes described by saying that such a minute object, if observed, i.e. interfered with, is taking a zigzag course, being tossed about under the impact of the photons like a boat in a heaving sea. In reality, it is far worse: for we cannot even speak of "the same" particle. Suppose we observe an atomic object and an instant later a similar one near-by, then we can't even be sure that it is "the same". Owing to the interaction between the object and the process of observing, which cannot be controlled, it is not possible to follow its course continuously. Two observations, even if following one another very shortly, should rather be regarded as disconnected events, and it is not possible to combine them unambiguously into a single comprehensive picture. Nor is there any way of telling what "happens" between one observation and the next. In other words, any picture of what is "really" going on contains gaps which cannot be filled in. That is why any attempt at tracing the path of an atomic particle is doomed to fail. As a consequence, the question as to whether a particle, really and truly, is the same is not only undecidable but devoid of meaning. p. 231

Example 7 An object can't be both P and not-P.

Analysis This is an object-based version of the law of excluded middle. In classical predicate logic, we formalize this as a scheme, where P stands for any atomic predicate:

$\forall x \, \neg \, (\, P(x) \land \neg \, P(x) \,)$

---

[28] Note that all the issues about identity across possible worlds already occur with identity across times. If in a model "— is a dog —$_{\text{time}}$" is not an essential attribute in time, then we cannot use "is a dog" as part of the criteria of identity for the universe of things in the model.

For predicates (properties) in time such as "is a puppy", this law is false: my dog Birta was a puppy and she was also a mature dog, just not at the same time. The version of the "law" that takes account of time is: "There is no object that can be both P and not-P at the same time":

$$\neg \exists x \exists t \, (I_P (x, t) \wedge \neg I_P (x, t))$$

This is a tautology.

29 Relativizing Quantifiers

All

We noted in Volume 0 that in classical predicate logic we have a choice to read "all" as "each and every one" or "each and every one and there is at least one". We chose the first, and so in classical predicate logic we formalize:

(1) All dogs are mammals.

$\forall x\,((\text{— is a dog})\,(x) \rightarrow (\text{— is a mammal})\,(x))$

How do we formalize (1) taking account of time? We could take the tense to indicate the present and formalize it as:

$\forall x\,((\text{— to be a dog —}_{\text{time}})\,(x, \text{now}) \rightarrow (\text{— to be a mammal —}_{\text{time}})\,(x, \text{now}))$

Or we could read "is" as omnitemporal: all dogs at any time are mammals:

$\forall x\,\forall t\,((\text{— to be a dog —}_{\text{time}})\,(x, t) \rightarrow (\text{— to be a mammal —}_{\text{time}})\,(x, t))$

To distinguish between the present-tense reading of (1) and the omnitemporal reading is an unavoidable analysis we have to make before formalization. We can't assume that "All A are B" is always omnitemporal, since "All police officers are men" was true a hundred years ago but is not now.

For relative uses of "all" in past-tense propositions, consider:

(2) All dodos were birds.

We have to decide whether this is about some time in the past or all times in the past:

(3) $\exists t\,((t <_{\text{time}} \text{now}) \wedge$
    $\forall x\,((\text{— to be a dodo —}_{\text{time}})\,(x, t) \rightarrow (\text{— to be a bird —}_{\text{time}})\,(x, t)))$

(4) $\forall t\,((t <_{\text{time}} \text{now}) \rightarrow$
    $\forall x\,((\text{— to be a dodo —}_{\text{time}})\,(x, t) \rightarrow (\text{— to be a bird —}_{\text{time}})\,(x, t)))$

A reading of (2) as omnitemporal is ruled out by the past tense, though we know that it would be equivalent to (4) since there are no longer dodos. But that is analysis, not formalization: both (2) and the formalizations could be true if there were dodos now that are all marsupials. Another reading of (2) is that there was some interval of time in the past when all dodos were birds. By the downward and upward closure of truth in time, (3) will serve for that.

For relative uses of "all" in future-tense propositions, consider:

(5) All dogs will be greyhounds.

This might be true if at some time in the future all other dogs had been killed by cats because they were too slow. We can formalize (5) as:

$\exists t\,((\text{now} <_{\text{time}} t)\,\wedge$
$\quad \forall x\,((-\text{ to be a dog }-_{\text{time}})\,(x, t) \to (-\text{ to be a greyhound }-_{\text{time}})\,(x, t)\,))$

Or we can read (5) as meaning that there is some time in the future after which at all times dogs will be greyhounds:

$\exists t\,((\text{now} <_{\text{time}} t)\,\wedge\,(\forall w\,(t \leq_{\text{time}} w) \to$
$\quad(\forall x((-\text{ to be a dog }-_{\text{time}})\,(x, w) \to (-\text{ to be a greyhound }-_{\text{time}})\,(x, w)))$

The choice of "$(t \leq_{\text{time}} w)$" rather than "$(t <_{\text{time}} w)$" reflects a reading of "after" that includes the time at which all dogs start to be greyhounds.

Taking these examples as archetypes, we can set out conventions for formalizing relative uses of "all"

Formalizing relative uses of "all"

Let α and β be common or collective nouns that we can convert into predicates, which in turn we can convert into infinitives I_α and I_β. Depending on context or by agreement in specific cases, the following choices are good formalizations.

All α are β.

   (about the present)
   $\forall x\,(I_\alpha(-,-_{\text{time}})\,(x, \text{now}) \to I_\beta(-,-_{\text{time}})\,(x, \text{now}))$

   (omnitemporal)
   $\forall x\,\forall t\,(I_\alpha(-,-_{\text{time}})\,(x, t) \to I_\beta(-,-_{\text{time}})\,(x, t))$

All α were β.

   (about some time in the past)
   $\exists t\,((t <_{\text{time}} \text{now}) \wedge \forall x\,(I_\alpha(-,-_{\text{time}})\,(x, t) \to I_\beta(-,-_{\text{time}})\,(x, t)))$

   (about all times in the past)
   $\forall t\,((t <_{\text{time}} \text{now}) \to (\forall x\,(I_\alpha(-,-_{\text{time}})\,(x, t) \to I_\beta(-,-_{\text{time}})\,(x, t))))$

All α will be β.

   (about some time in the future)
   $\exists t\,((\text{now} <_{\text{time}} t) \wedge \forall x\,(I_\alpha(-,-_{\text{time}})\,(x, t) \to I_\beta(-,-_{\text{time}})\,(x, t)))$

   (about all times after some time in the future)
   $\exists t\,((\text{now} <_{\text{time}} t) \wedge \forall w\,((t \leq_{\text{time}} w) \to$
   $\quad(\forall x\,(I_\alpha(-,-_{\text{time}})\,(x, w) \to I_\beta(-,-_{\text{time}})\,(x, w))))$

Some

For relative uses of "some" consider:

(6)   Some dogs are terriers.

In classical predicate logic we formalize (6) reading "some" as "at least one", leaving

144 Chapter 29

as an option to take account of the plural according to context:

$\exists x\,((-\text{ to be a dog})\,(x) \wedge (-\text{ to be a terrier})\,(x))$

Taking time into account, (6) seems to be about the present, so we can formalize it as:

$\exists x\,((-\text{ to be a dog}-_{\text{time}})\,(x, \text{now}) \wedge (-\text{ to be a terrier}-_{\text{time}})\,(x, \text{now}))$

Similarly, we can formalize each of the following:

Some dogs were terriers.

$\exists t\,((t <_{\text{time}} \text{now}) \wedge$
$\qquad \exists x\,((-\text{ to be a dog}-_{\text{time}})\,(x, t) \wedge (-\text{ to be a terrier}-_{\text{time}})\,(x, t)))$

Some dogs will be terriers.

$\exists t\,((\text{now} <_{\text{time}} t) \wedge$
$\qquad \exists x\,((-\text{ to be a dog}-_{\text{time}})\,(x, t) \wedge (-\text{ to be a terrier}-_{\text{time}})\,(x, t)))$

Taking these examples as archetypal, we can adopt conventions for formalizing relative uses of "some".

Formalizing relative uses of "some"

Let α and β be common or collective nouns that we can convert into predicates, which in turn we can convert into infinitives I_α and I_β. Then the following are good formalizations.

Some α are β.

$\exists x\,(I_\alpha(-, -_{\text{time}})\,(x, \text{now}) \wedge I_\beta(-, -_{\text{time}})\,(x, \text{now}))$

Some α were β.

$\exists t\,((t <_{\text{time}} \text{now}) \wedge \exists x\,(I_\alpha(-, -_{\text{time}})\,(x, t) \wedge I_\beta(-, -_{\text{time}})\,(x, t)))$

Some α will be β.

$\exists t\,((\text{now} <_{\text{time}} t) \wedge \exists x\,(I_\alpha(-, -_{\text{time}})\,(x, t) \wedge I_\beta(-, -_{\text{time}})\,(x, t)))$

One example will illustrate how differently we treat relativizing in classical predicate logic and in our temporal predicate logic.

Example 1 Birta hates every cat.

Analysis The example is true about my dog Birta, I believe. But that doesn't mean she hates every cat that ever lived, lives, or will live. She can't hate something she's not aware of. Rather, it means something like:

Birta hates every cat she has encountered or will encounter at all times after she encounters it.

We can formalize this as:

$\forall x \, \exists t \, (\, (\, (- \text{ to be a cat } -_{\text{time}}) \, (x, t) \, \wedge$
$((- \text{ to encounter } -_{\text{time}})/\text{obj}(x)) \, (\text{Birta}, t) \,)$
$\to \, (\forall w \, ((t <_{\text{time}} w) \to \, ((- \text{ to hate } -_{\text{time}})/\text{obj}(x)) \, (\text{Birta}, w)))\,)$

You might think this shows what a lot of work it is to formalize in temporal logic what we can so easily formalize in classical predicate logic:

(*) $\forall x \, (\, (- \text{ is a cat}) \, (x) \to (- \text{ hates } -) \, (\text{Birta}, x) \,)$

But the discussion shows that (*) is not a good formalization. The only way we could rescue it would be to restrict the universe of the realization. But restricting what cats are in the universe to allow for (*) to be good might not be compatible with what we need to put into the universe for formalizations of other propositions and would push a lot of the work of formalizing into the choice of an appropriate universe. Many "straightforward" formalizations in classical predicate logic can be seen to be wrong once we consider what the ordinary language propositions mean. Allowing for only timeless formalizations forces us to put on blinders; after working in classical predicate logic long enough we are not even aware how those blinders narrow our view.

30 Examples of Formalizing: Meaning Axioms

Example 1 Birta is a dog if and only if she is a domestic canine.
Analysis In classical predicate logic, we formalize this as:

(— to be a dog) (Birta) ↔ ((— to be a canine)/domestic) (Birta)

It's not a tautology, but it follows from a meaning axiom:

$\forall x$ ((— to be a dog) (x) ↔ ((— to be a canine)/domestic) (x))

Meaning axioms in classical predicate logic are thought of as atemporal. But here we have to take account of time:

$\forall t \, \forall x$ ((— to be a dog —$_{\text{time}}$) (x, t) ↔
    ((— to be a canine —$_{\text{time}}$)/domestic) (x, t))

Instead of atemporal, the meaning axiom is omnitemporal.

Example 2 Birta is a puppy.
Analysis We saw in Chapter 14 that we cannot formalize in classical predicate logic the meaning axiom relating "— is a puppy" to "— is a dog":

Something is a puppy iff it is a dog that is not mature.

Now we can use:

$\forall t \, \forall x$ ((— to be a puppy —$_{\text{time}}$) (x, t)
    ↔ ((— to be a dog —$_{\text{time}}$)/(~mature)) (x, t))

Example 3 Bill is a bachelor.
Analysis Understanding "a bachelor" to be a man who is not married and never was married, we can formalize the example as:

(— to be a bachelor) (Bill, now)

relative to the meaning axiom:

$\forall x \, \forall t$ ((— to be a bachelor) (x, t) ↔ (((— to be a man)/(~married)) (x, t)
    ∧ ($\forall w$ ($w <_{\text{time}} t$) → ¬ (— to be a man)/married) (x, w))))

Example 4 Caesar was assassinated.

$\exists t$ (Max (— to be assassinated —$_{\text{time}}$) (Caesar, t) ∧ ($t <_{\text{time}}$ now))

Analysis This formalization leaves open the possibility that Caesar was assassinated more than once. We can eliminate that possibility with a meaning axiom:

$\forall x$ ($\exists t$ (— to be assassinated —$_{\text{time}}$) (x, t) →
    $\exists ! w$ Max (— to be assassinated —$_{\text{time}}$) (x, w))

31 Examples of Formalizing: How Many Times

Example 1 *Spot barked just once.*

$\exists! t$ Max $(-$ to bark $-_{\text{time}})$ (Spot, t)

Analysis I'll leave to you to formalize "Spot barked just two times", "Spot barked just three times", ...

Example 2 Zoe: *Every time Dick feeds Spot, Spot barks.*

Analysis It would seem simple to formalize this:

$\forall t\,(\,((-$ to feed $-_{\text{time}})/\text{obj}(\text{Spot}))\,(\text{Dick}, t)\, \rightarrow\, (-$ to bark $-_{\text{time}})\,(\text{Spot}, t)\,)$

But this is not right: for it to be true Spot has to bark at the same time as Dick feeds him. Yet Zoe knows that typically Dick will get Spot's food, carry it to the back yard, and then, when Spot sees him or smells the food, Spot barks. If we count the time in which Dick feeds Spot to include all the time from when Dick first gets the food until Spot begins eating it, then we can formalize the example as:

$\forall t\,(\,\text{Max}\,((-$ to feed $-_{\text{time}})/\text{obj}(\text{Spot}))\,(\text{Dick}, t) \rightarrow$
$\exists w\,(\,W_{\text{time}}(w, t) \wedge (-$ to bark $-_{\text{time}})\,(\text{Spot}, w))\,)$

Example 3 *Sometimes when Dick feeds Spot, Spot barks.*

$\exists t\,(\,\text{Max}((-$ to feed $-_{\text{time}})/\text{obj}(\text{Spot}))\,(\text{Dick}, t) \rightarrow$
$\exists w\,(\,W_{\text{time}}(w, t) \wedge (-$ to bark $-_{\text{time}})\,(\text{Spot}, w))\,)$

Analysis The formalization is good for the more ample understanding of "Dick feeds Spot" described in the last example, reading "sometimes" as "at least one time".

Example 4 *Every time one billiard ball hits another, the second one moves.*

Analysis A straightforward formalization of this is:

(a) $\quad \forall t\, \forall x\, \forall y\, (\,(\,((-$ to be a ball $-_{\text{time}})/\text{billiard})\,(x, t)$
$\wedge\,((-$ to be a ball $-_{\text{time}})/\text{billiard})\,(y, t)$
$\wedge\,((-$ to hit $-_{\text{time}})/\text{obj}(y))\,(x, t)\,)$
$\rightarrow (-$ to move $-_{\text{time}})\,(y, t)\,)$

But this is false: the second ball does not move at the very moment the first ball touches it but only some very small time after that. So was your high school physics teacher lying to you when she said this was true? Physicists normally ignore the difference in time between when the balls touch and the second ball begins to move, since that time is so small in comparison to any other measurement they'd want to make in that situation.[29] When they are reasoning about atoms hitting, they'd be more precise about when the second atom begins to move. Here it is not what we

---

[29] See "Models and Theories" for what we ignore in devising and applying theories in science.

148   Chapter 31

mean by "hits" or "moves" that determines whether (a) is a good formalization but how small we allow intervals of time in the universe of our model. If we allow only times corresponding to fixed time markers with decimal expansions to a tenth of a second, then (a) is a good formalization.

Example 5   Every time Dick yells near Spot, Spot jumps.

Analysis   Zoe knows this is true. Every time Dick yells, Spot jumps a little later — not exactly at the moment when Dick yells. But we can't construe "Dick yells" to include a time after he stops yelling. Rather, if Zoe thinks the example is true she should be more precise and say:

Every time Dick yells near Spot, a little time after that Spot jumps.

We have no way to say "a little time after", even introducing into our semi-formal language vocabulary for measuring times that we'll see in Chapter 33.

Example 6   When Spot barks, he barks loudly.

$\forall t$ (Max (— to bark —$_{time}$) (Spot, t) →
      $\exists w$ ((— to bark —$_{time}$)/loudly) (Spot, w) $\wedge$ $W_{time}(w, t)$))

Analysis   This is to understand "when" as "whenever". It allows that at some times Spot barks a little and then starts to bark loudly.

Example 7   If Spot barks, then he barks loudly.

Formalization as for the previous example.

Analysis   This is to construe the conditional as about not just the present but all times.

Example 8   There never was a unicorn.

$\forall t \neg \exists x$ ((— to be a unicorn —$_{time}$) (x, t) $\wedge$ $(t <_{time}$ now))

Analysis   I've formalized "never" as "always not". The formalization is equivalent to reading "never" as "there is no time":

$\neg \exists t \exists x$ ((— to be a unicorn —$_{time}$) (x, t) $\wedge$ $(t <_{time}$ now))

Example 9   Puff never barks.

$\forall t \neg$ (— to bark —$_{time}$) (Puff, t)

Analysis   We don't need to worry about restricting this to only times when Puff exists, for at times when he doesn't exist it's true, too. The formalization is equivalent to reading "never" as "there is no time":

$\neg \exists t$ (— to bark —$_{time}$) (Puff, t)

Example 10   No one ever trained Puff.

$\neg \exists x \exists t$ ((— to be a person —$_{time}$) (x, t) $\wedge$ ((— to train —$_{time}$)/obj(Puff)) (x, t))

Analysis Alternatively we could use:

$\forall t \neg \exists x\, ((\text{— to be a person —}_{time})\,(x, t)$
$\qquad \land\, ((\text{— to train —}_{time})/\text{obj}(\text{Puff}))\,(x, t)\,)$

This is equivalent to the formalization above.

Example 11 *Some day Flo will be a woman.*

$\exists t\, ((\text{— to be a woman —}_{time})\,(\text{Flo}, t) \land (\text{now} <_{time} t)\,)$
$\qquad \land\, (\sim(\text{— to be a woman —}_{time}))\,(\text{Flo, now})$

Analysis Here "someday" means a future time, not necessarily a day. In ordinary speech the proposition also means that Flo is not now a woman, which I've included in the formalization.

Example 12 *Nearly every time that Dick feeds Spot, Spot barks.*

Not formalizable.

Analysis The example is true. But we can't formalize it for the same reason we can't formalize "Nearly every dog barks": the informal quantifier "nearly every" is outside the scope of our methods.

Example 13 *Spot hardly ever barks at Tom.*

Not formalizable.

Analysis The example is true. But we can't formalize it. The quantifier "hardly ever" means "almost never", which though clear enough in ordinary speech, is not a quantification we can formalize in our logic.

Example 14 *Most of the time Spot is happy.*

Not formalizable.

Analysis We can read "most of the time" as "most times", but that doesn't help because the quantifier "most" is outside the scope of our methods.

Example 15 *Much of the time Spot is happy.*

Not formalizable.

Analysis The phase "much of the time" suggests a conception of time as a mass. Perhaps it is equivalent to "most of the time", but even so it would be outside the scope of our methods.

32 Examples of Formalizing: The Internal Structure of Atomic Predicates

In order to simplify the discussions in this chapter, I will not use "Max" in the formalizations, leaving to you to formulate fuller analyses.

Example 1 *Spot barked loudly.*
    Therefore, *Spot barked.*

Analysis The informal inference is valid, a characteristic inference for the restrictor "loudly". We can formalize it as the valid semi-formal inference:

$$\frac{\exists t \, (((- \text{ to bark } -_{\text{time}})/\text{loudly}) (\text{Spot}, t) \land (t <_{\text{time}} \text{now}))}{\exists t \, ((- \text{ to bark } -_{\text{time}}) (\text{Spot}, t) \land (t <_{\text{time}} \text{now}))}$$

But it seems to me that the example should be read as concluding that Spot barked at the time he barked loudly. After all, that's what ensures that the semi-formal inference is valid. With that cross-referencing of a time variable we have to formalize the two apparently atomic propositions within the scope of the same quantifier. That means we have a single proposition, not two, and hence not an inference. The hypothetical nature of the apparent inference is captured by using a conditional. If the inference is valid, the conditional should be a tautology.

Should we use the existential quantifier to govern the conditional, since we interpret the premise to mean that Spot barked at some time? We are evaluating an inference, so both premise and conclusion must be viewed as schemes: it is not possible for the premise to be true and conclusion false no matter how we make the schemes into proposition by stipulating a time for both. So we use the universal quantifier in formalizing the inference:

$$\forall t \, (((- \text{ to bark } -_{\text{time}})/\text{loudly}) (\text{Spot}, t) \land (t <_{\text{time}} \text{now}))$$
$$\rightarrow ((- \text{ to bark } -_{\text{time}}) (\text{Spot}, t) \land (t <_{\text{time}} \text{now})))$$

A fundamental criterion of formalization is that an informally valid inference should be formalized as a formally valid inference. But when we use words that normally indicate an inference such as "so", "therefore", and "hence" to link two or more propositions that have some (implicit) cross-referencing of variables, we have no choice but to formalize the apparent inference as a conditional, governed by a universal quantifier. If the apparent inference is valid, the semi-formal conditional should be a tautology, which in this case it is.

Example 2 *Spot barked loudly at 1 p.m. May 3, 1998.*
    $((- \text{ to bark } -_{\text{time}})/\text{loudly}) (\text{Spot}, 1 \text{ p.m. May 3 1998})$

Analysis To make this discussion simpler, I've not included the clause to show that the time is in the past tense.

Here "loudly" restricts the predicate "(— to bark —$_{time}$)". So the formalization is true iff Spot barked loudly in comparison to all things at all times for which "(— to bark —$_{time}$)" applies. That is, Spot barked loudly in comparison to all things that did bark, are barking, or will bark. *All our modifiers are used omnitemporally in this way.*

But now consider:

Spot barked loudly at 1 p.m. May 3, 1998 in comparison to
those things that were barking then.

Here "loudly" is meant to restrict not "(— to bark —$_{time}$)" but that predicate applied to things at the time 1 p.m. May 3, 1998. We can extend our system to formalize this by changing the definition of an atomic predicate to be one in which the time-marked blank is filled with a time term and the ordinary blanks are left unfilled. Then we can formalize the example as:

((— to bark 1 p.m. May 3 1998)/loudly) (Spot)

In this way we can use restrictors to make comparisons to objects at a time rather than to objects at all times so *modifiers need not be omnitemporal.*

We could also allow for an ordinary blank to be filled and not the time blank:

((Spot to bark —$_{time}$)/loudly) (1 p.m. May 3 1998)

We could use this to formalize:

Spot barked loudly at 1 p.m. May 3, 1998 in comparison to
how Spot barked at all times.

I'll leave to others to modify the systems here to allow for such formalizations.

Example 3 Slowly, Tom escorted Zoe and Dick pushed Manuel.

$\exists t$ ((((— to escort —$_{time}$)/obj(Zoe))/slowly) (Tom, t)
$\wedge$ (((— to push —$_{time}$)/obj(Manuel))/slowly) (Dick, t) $\wedge$ ($t <_{time}$ now))

Analysis Though it looks as if "slowly" is modifying a proposition, we know that propositions are not fast or slow. Rather, by being placed at the beginning we understand that "slowly" is meant to apply to the two parts that follow for the same time (Manuel is in a wheelchair). We couldn't formalize this in classical predicate logic with modifiers because of this time issue (Volume 1, Chapter 30, Example 12).

Example 4 Dick was quickly irritated and chased Spot.

$\exists t \exists w$ (((— to be irritated —$_{time}$)/quickly) (Dick, t) $\wedge$ ($t <_{time}$ now)
$\wedge$ ((— to chase —$_{time}$)/obj(Spot)) (Dick, w)
$\wedge$ ($w <_{time}$ now) $\wedge$ ($t <_{time} w$))

Analysis This is to read "and" as "and then". (Compare this to the analysis in Example 13 of Chapter 33 of Volume 1 not taking time into account.)

152   Chapter 32

Example 5 Dick will sing and dance.

$\exists t \, (\, ((-$ to sing $-_{time}) + (-$ to dance $-_{time})) \, (\text{Dick}, t) \wedge (\text{now} <_{time} t) \,)$

Dick will sing and not dance.

$\exists t \, (\, ((-$ to sing $-_{time}) + \sim (-$ to dance $-_{time})) \, (\text{Dick}, t) \wedge (\text{now} <_{time} t) \,)$

Analysis I understand the second example to mean that Dick will not sing and dance at the same time. The formalization is odd, though, because in our usual talk we would understand these examples to be about a more-or-less specific time rather than about some indefinite future time.

Example 6 Dick will not sing and dance.

$\exists t \, (\sim (\, (-$ to sing $-_{time}) + (-$ to dance $-_{time})) \, (\text{Dick}, t) \wedge (\text{now} <_{time} t) \,)$

Analysis Here the negation applies to a conjoined predicate. From the formalization it follows that Dick exists at any time that makes the unquantified part of the formalization true. This formalization is odd for the same reason as the last example.

Example 7 Tom and Dick sang. So Tom sang and Dick sang.

Analysis The example is a characteristic inference for conjunctions of terms in classical predicate logic. Taking time into consideration it's valid, too.

$\underline{\exists t \, ((-$ to sing $-_{time}) \, (\text{Tom} \wedge \text{Dick}, t) \wedge (t <_{time} \text{now}) \,)}$
$\exists t \, ((-$ to sing $-_{time}) \, (\text{Tom}, t) \wedge (t <_{time} \text{now}) \,)$
$\quad \wedge \, \exists t \, ((-$ to sing $-_{time}) \, (\text{Dick}, t) \wedge (t <_{time} \text{now}) \,)$

But we understand the conclusion as meaning that Dick sang and Tom sang at the time that they sang together. So we should formalize the apparent inference as a conditional as in Example 2:

$\exists t \, (\, ((-$ to sing $-_{time}) \, (\text{Tom} \wedge \text{Dick}, t) \wedge (t <_{time} \text{now}) \,$
$\rightarrow (-$ to sing $-_{time}) \, (\text{Tom}, t) \wedge (-$ to sing $-_{time}) \, (\text{Dick}, t) \,)$

This is a tautology. Note that an ordinary atomic predicate is positive for existence for each term in a conjunction of terms.

Example 8 I never hit an old man.
  My brother is an old man.
  Therefore, I never hit my brother.

Analysis The formalization of this in classical predicate logic is valid:

$\neg \exists x \, (\, ((-$ hit$)/\text{obj}(x)) \, (\text{Arf}) \wedge ((-$ is a man$)/\text{old}) \, (x) \,)$
$\underline{(-$ is a man$)/\text{old}) \, (\text{Bob})}$
$\neg \, ((-$ hit$)/\text{obj}(\text{Bob})) \, (\text{Arf})$

But the example is invalid: the premises are true and conclusion false. Taking time into account, we can show that:

$$\neg \exists t \, \exists x \, (\, ((- \text{ to hit } -_{\text{time}})/\text{obj}(x)) \, (\text{Arf}, t) \wedge (t <_{\text{time}} \text{now})$$
$$\wedge \, ((- \text{ to be a man } -_{\text{time}})/\text{old}) \, (x, t) \,)$$
$$\underline{(- \text{ to be a man } -_{\text{time}})/\text{old}) \, (\text{Bob}, \text{now})}$$
$$\neg \exists t \, (\, ((- \text{ to hit } -_{\text{time}})/\text{obj}(\text{Bob})) \, (\text{Arf}, t \wedge (t <_{\text{time}} \text{now}) \,)$$

As a rule, *a description of an object referred to in a variable restrictor should be at the same time as that of the main predication* unless context or phrasing indicates otherwise, as in "I never hit someone who is an old man now".

Example 9 A little boy will become a famous man.

$$\exists x \, \exists t \, \exists w \, (\, ((- \text{ to be a boy})/\text{little}) \, (x, \text{now})$$
$$\wedge \, ((- \text{ to be a man})/\text{famous}) \, (x, w) \wedge (\text{now} <_{\text{time}} w) \,)$$
$$\vee \, (\, ((- \text{ to be a boy})/\text{little}) \, (x, t) \wedge (\text{now} <_{\text{time}} t)$$
$$\wedge \, ((- \text{ to be a man})/\text{famous}) \, (x, w) \wedge (t <_{\text{time}} w) \,) \,)$$

Analysis This example would be true if there is not now such a little boy but there will be one in the future. We need the conjunction "$(t <_{\text{time}} w)$" to account for "become", for if something becomes, it is not already.

Example 10 Antichrist will be an orator.

$$\exists t \, (\, (- \text{ to be an orator } -_{\text{time}}) \, (\text{Antichrist}, t) \wedge (\text{now} <_{\text{time}} t) \,)$$

Analysis Here is what Peter Øhrstrøm and Per F. V. Hasle say in *Temporal Logic* about this and the previous example, which were much discussed in the middle ages.

> One of the crucial problems motivating the work on "ampliatio" was the problem regarding the naïve conception of tensed statements. According to that conception, a proposition of the type "A will be B" is equivalent to the claim of the existence of a future in which "A is B" [is true], and similarly a proposition of the type "A has been B" is regarded as true if and only if the proposition "A is B" was true at some past time. But this naïve conception cannot be upheld in all cases. Consider for instance the statement
>
> > "The little boy will become a famous man".
>
> This proposition can certainly be true, even though the statement "the little boy is a famous man" cannot be fulfilled at any time. The solution was to interpret the statement as being equivalent to:
>
> > "For a given person x, x is now a little boy
> > and x will become a famous man".
>
> "The little boy" thus refers to something in the present although the verb is referring to the future. But even this more refined treatment cannot encompass all cases, as we can see from the sentence "Antichrist will be an orator". Crucial to this example is the theological observation that Antichrist does not yet exist. The statement could consequently not be paraphrased in the same way as the statement about the little boy, but was understood as being equivalent to:

154   Chapter 32

"For some person *x*: it is true that *x* will be Antichrist, and *x* will be an orator".

In an analogous manner the proposition, "Something white was black", might be true because the following statement is true: "Something which has been white was black", or it might be true because of the truth of "Something which is white was black". pp. 39–40

Example 11   Spot almost barked at 4 p.m. June 2, 2002.
Therefore, *Spot did not bark at 4 p.m. June 2, 2002.*

$$\frac{((-\text{ to bark }-_{\text{time}})/\text{almost}) (\text{Spot}, 4 \text{ p.m. June 2 2002})}{(\sim(-\text{ to bark }-_{\text{time}})) (\text{Spot}, 4 \text{ p.m. June 2 2002})}$$

Analysis   The informal inference is valid, and the formalization is valid by *Neg-time*. We need predicate negation in the conclusion because from the informal conclusion we can conclude that Spot existed at that time.

Example 12   Spot almost barked.
Therefore, *Spot did not bark.*

Analysis   In classical predicate logic the informal inference is valid, a characteristic inference for the negator "almost".

But when we consider time, it seems that this is not valid. The premise means that at some time, Spot almost barked, while the conclusion can be and I think most naturally is read as meaning that Spot never barked. What is valid is:

There was a time when Spot almost barked.
So there was a time when Spot didn't bark.

As with Example 2, to formalize this we need to relate the times in the premise and conclusion, which we can by using a conditional:

$\exists t\, (\, (((-\text{ to bark }-_{\text{time}})/\text{almost}) (\text{Spot}, t) \wedge\ (t <_{\text{time}} \text{now}))$
    $\rightarrow\, \sim(-\text{ to bark }-_{\text{time}}) (\text{Spot}, t)\,)$

This is a tautology.

Example 13   Spot barked at almost 4 p.m. June 2, 2002.
Not formalizable.

Analysis   The example means that there is some time that is almost 4 p.m. June 2, 2002 at which Spot barked. But to say that a time is almost 4 p.m. June 2, 2002 is not to use "almost" as in the last example but as "very near in time". Even if we have some way to measure time, it would be difficult to formalize uses of "almost" with times.

Example 14   Tom will telephone in a day.
Analysis   It is tempting to read "in" here as a variable restrictor:

Tom will telephone in a week.
Tom will telephone in a month.
Tom will telephone in a few minutes.

But "in a day" is giving a duration from the present, for which we have to be able to measure times.

33 Measuring Time?

In our daily lives we measure time using a standard clock and fixed time markers. To do that here we would have to incorporate in the language a collection of fixed time markers and enough arithmetic to calculate differences in lengths of times and summing lengths of time. Yet people measured times long before there were standard fixed time markers.

Example 1 Lala: *I will return before there are three full moons.*

$\exists t\,((-\text{ to return }-_{\text{time}})\,(\text{Lala}, t) \wedge (\text{now} < t) \wedge$
$\quad \neg\,\exists t_1\,\exists t_2\,\exists t_3\,(\text{Max}\,(-\text{ is full }-_{\text{time}})\,(\text{moon}, t_1) \wedge$
$\quad \text{Max}\,(-\text{ is full }-_{\text{time}})\,(\text{moon}, t_2) \wedge \text{Max}\,(-\text{ is full }-_{\text{time}})\,(\text{moon}, t_3)$
$\quad \wedge\,(\text{now} <_{\text{time}} t_1 <_{\text{time}} t_2 <_{\text{time}} t_3 <_{\text{time}} t)\,))$

Analysis We can count and do a lot of arithmetic in predicate logic by counting variables and quantifiers (see Example 19 of Chapter 9 of Volume 0).

This measuring involves no comparison. It involves no standard, only counting full moons. It does, however, need some assumptions about full moons, which I'll leave to you.

Example 2 Spot barked longer than Dick yelled.

Analysis To formalize this, we could add a predicate for comparing times:

$(-_{\text{time}} \text{ is shorter than } -_{\text{time}})$

Adding some axioms for it, we could formalize the example:

$\exists t_1\,\exists t_2\,(\,\text{Max}\,(-\text{ to bark }-_{\text{time}})\,(\text{Spot}, t_1) \wedge (t_1 <_{\text{time}} \text{now})$
$\quad \wedge\,\text{Max}\,(-\text{ to yell }-_{\text{time}})\,(\text{Dick}, t_2) \wedge (t_2 <_{\text{time}} \text{now})$
$\quad \wedge\,(-_{\text{time}} \text{ is shorter than } -_{\text{time}})\,(t_1, t_2)\,)$

This is to read the example as about some particular time when Spot barked and Dick yelled, not as summing up over several such times.

Example 3 Lala will be away less time than when Ugo was sick.

Analysis Ugo was sick with smallpox, and everyone knows that's what's meant when they talk about the time when he was sick. Over the years they have come to use that as a standard length of time. We can formalize the example using the length-of-time-comparison predicate suggested in the last example:

$\exists t_1\,\exists t_2\,(\,\text{Max}\,(-\text{ to be away }-_{\text{time}})\,(\text{Lala}, t_1) \wedge (\text{now} <_{\text{time}} t_1)$
$\quad \wedge\,\text{Max}\,(-\text{ to be sick }-_{\text{time}})\,(\text{Ugo}, t_2)) \wedge (t_2 <_{\text{time}} \text{now})$
$\quad \wedge\,(-_{\text{time}} \text{ is shorter than } -_{\text{time}})\,(t_1, t_2)\,)$

Example 4 The time when Lala was away was longer than the time between when Ugo was sick and Mizza killed a bear.

Analysis To use the length-of-time-comparison predicate in a formalization of this example we need that the interval of times between when Ugo was sick and Mizza killed a bear is a time, though there's no suggestion of it being used as a standard. Generally, to make comparisons like this we'd need to assume that the interval between any two maximal intervals of predication is a time. I'll leave to you to develop that and a general method of formalization that would cover this example.

Example 5 *Tom will telephone in three minutes.*

Analysis Let's assume we have a single time in our model that we'll use for the standard of a minute, say, 11:06 a.m. Mountain Daylight Time April 30, 2018, which we'll call "minute". Then using the length-of-time-comparison predicate, we first define:

$(-_{time}$ is the same length as $-_{time})$ (t, w) $\equiv_{Def}$
  $\neg$ $(-_{time}$ is shorter than $-_{time})$ (t, w) $\wedge$
  $\neg$ $(-_{time}$ is shorter than $-_{time})$ (w, t)

Then we can formalize the example without assuming that an interval of time is a time:

$\exists t$ (($(-$ to telephone $-_{time})$ (Tom, t) $\wedge$ (now $<_{time} t$)
  $\wedge$ $\neg \exists t_1 \exists t_2 \exists t_3$ (($t_1 <_{time} t_2 <_{time} t_3$) $\wedge$
  $\wedge$ $(-_{time}$ is the same length as $-_{time})$ $(t_1,$ minute$)$
  $\wedge$ $(-_{time}$ is the same length as $-_{time})$ $(t_2,$ minute$)$
  $\wedge$ $(-_{time}$ is the same length as $-_{time})$ $(t_3,$ minute$)$
  $\wedge$ (now $\leq_{time} t_1 <_{time} t_2 <_{time} t_3 \leq_{time} t$)))

It's not enough to say that there are not three minutes in that interval, since minutes can overlap. This works unless there is a minute in that interval that's not in the universe of times. We can't say in the semi-formal language that there are enough times. All we can do is say that we'll have "all minutes" in the universe of times of our model—assuming that makes sense.

Example 6 *Tom will telephone in a week.*

Analysis We usually count weeks according to a calendar. But we could take one interval of time in a model to use as a standard for the length of a week and formalize this example in the manner we did the previous one.

Example 7 *Suzy arrived shortly after Tom arrived.*

Analysis Suzy agreed to meet Tom shortly after 8 p.m. at the restaurant. Suzy shows up at 8:32 p.m. Tom is irritated and tells Suzy she's late. Suzy says she did arrive shortly after 8 p.m; after all, she didn't arrive at 9 p.m. It isn't that "shortly" is vague; it's that Tom and Suzy have subjective interpretations of that word. We often measure time subjectively: time flies when you're having fun; time drags when you're waiting in the dentist's office. Perhaps there is a way to formalize such measures of time, indexing each time measure according to the person whose standard is invoked. But I won't try.

Tom says that in the future, when he says "shortly after" he means within 10 minutes. Suzy says, O.K. They've replaced their subjective interpretations with an objective one. They both know how to use that. With that stipulation you can formalize the example.

Example 8 Spot barked shortly before Dick yelled.

Analysis Should we agree with Tom and Suzy and take "shortly" to mean within 10 minutes? What we mean by "shortly after" for meeting someone at a restaurant isn't what we mean by "shortly after" for yelling after a dog barks, which I reckon is no more than one minute. That is, there is at most one minute between the maximum interval of time of Spot barking and the maximum interval of time of Dick yelling. With that interpretation, we can formalize the example.

$$\exists t_1 \exists t_2 \, (\text{Max} \, (- \text{ to bark} -_{\text{time}}) \, (\text{Spot}, t_1) \land (t_1 <_{\text{time}} \text{now})$$
$$\land \, \text{Max} \, (- \text{ to yell} -_{\text{time}}) \, (\text{Dick}, t_2) \land (t_2 <_{\text{time}} \text{now}) \land (t_2 <_{\text{time}} t_1)$$
$$\land \, \neg \exists t_3 \, (-_{\text{time}} \text{ is the same length as } -_{\text{time}}) \, (t_3, \text{minute})$$
$$\land \, (\text{now} <_{\text{time}} t_1 <_{\text{time}} t_3 <_{\text{time}} t_2) \,)$$

This works unless there is a minute between when Spot barked and Dick yelled that's not in the universe of times. (Compare Example 5 above.)

Evaluating "Spot barked shortly before Dick yelled" seems so easy for us in our daily life. We do it effortlessly. But the assumptions we use in that evaluation are substantial and not easily formalized.

Example 9 Every time Dick yells near Spot, a little time after Spot jumps.

Analysis A little time after Dick yells isn't the shortly after of ten minutes that Tom and Suzy agreed on or even the shortly after of one minute for when Dick yells after Spot barks, but probably only 2 seconds. We'd have to have lots of standards to invoke for "a little time", "shortly after", "soon after", and other informal measures, each dependent on the particular predications being compared.

Example 10 Dick spent more time training Spot than Suzy spent training Puff.

Analysis To formalize this we would need to have a way to sum up times.

Example 11 December 21st or December 22nd is the shortest day of the year.

Analysis I can't see how to formalize this. We could add fixed time markers, and a predicate to pick out calendar days, and a predicate to pick out calendar years, but "December 21" is meant to pick out a day in each calendar year. Instead of treating it as a name, we might use a predicate "$-_{\text{time}}$ is December 21", but then we would have to add meaning axioms relating that to fixed time markers.

Example 12 Every instant of time has the same length.

Analysis This is easy to formalize with the length-of-time-comparison predicate:

$\forall t_1 \, \forall t_2 \, (\text{instant}(t_1) \wedge \text{instant}(t_2) \rightarrow$
$(-_{\text{time}} \text{ is the same length as } -_{\text{time}})(t_1, t_2))$

It can be true in a model. But it isn't consonant with the view of time as made up of dimensionless points of time. I don't see how to formalize that conception. At best, we could add to this example that every instant of time is shorter than any time that is not an instant:

$\forall t_1 \, \forall t_2 \, ((\text{instant}(t_1) \wedge \neg(\text{instant}(t_2))) \rightarrow$
$(-_{\text{time}} \text{ is shorter than } -_{\text{time}})(t_1, t_2))$

Aside: *Standards for measuring times*

There are several ways time is measured objectively in our society. One is in relation to the earth's movement relative to the sun. We can divide the time between summer solstices into parts, defining a second that way. Or we can measure time in relation to the earth's movement relative to a distant star, which is called "sidereal time". Or we can use the international standard that a second is 9,192,631,770 periods of the radiation corresponding to the transition between the two hyperfine levels of the ground state of the cesium 133 atom at 0 degrees Kelvin. People adopt these different measures for different purposes.

34 Other Tenses

In English we modify our talk of the division of past-present-future using other tenses. For the most part those are peculiar to English. Still, it's worth seeing how we can formalize some uses of those, partly as a guide to formalizing reasoning in English and partly for developing a method which might be applicable to formalizing the use of tenses in other languages.

The analyses I'll make here are based on the grammar taught to students of English as a second or foreign language as well as my own intuitions.[30] These are abstractions from how we ordinarily talk, and, as with our formalization of "and", there are many exceptions.

The past perfect and the future perfect
Consider the two propositions:

> Dick ate.
> Dick had eaten.

The truth-conditions for these are the same: the predicate "— to eat $—_{time}$" applies to Dick at some time (interval) in the past. The past perfect by itself as opposed to the past does not indicate anything different about the description.

It is in compound sentences that the two tenses are put to different uses. We rarely use the past perfect by itself. It is hard to imagine a context where we would say only "Dick had eaten" except in reply to a question or comment that already establishes a time. What is typical are sentences like:

(1) Dick had eaten when Zoe arrived.

The use of the past perfect with "when" indicates that the time of "Dick had eaten" is completely before the time of "Zoe arrived".

$$\exists t_1 \exists t_2 \,(\, \text{Max}(-\text{ to eat }-_{time})(\text{Dick}, t_1) \wedge (t_1 <_{time} \text{now})$$
$$\wedge \text{ Max}(-\text{ to arrive }-_{time})(\text{Zoe}, t_2) \wedge (t_2 <_{time} \text{now})$$
$$\wedge (t_1 <_{time} t_2) \wedge \neg O(t_1, t_2)\,)$$

Generally, the past perfect is used to indicate that the time of the predication is completely before the time of another whose time is also in the past, which we can formalize using maximal intervals of predication.

---

[30] See particularly Betty Schrampfer Azar's *Understanding and Using English Grammar*.

In linguistics the word "tense" is often used for those parts or constructions of a language that have to do with indicating time relative to the time of speaking. The word "aspect" is used for those parts or constructions of a language that have to do with indicating time relative to the act itself, for example whether the act is completed, habitual, or ongoing. I use the word "tense" to cover both these notions of tense and aspect, for together they are part of how we reason taking account of time. Robert I. Binnick examines the complexities of these issues in *Time and the Verb: A Guide to Tense and Aspect*.

Contrast this with:

(2) Dick ate when Zoe arrived.

We'd understand this to mean that Dick ate after or beginning at the same time as when Zoe arrived. That is, the time when Zoe arrives overlaps from before the time that Dick ate. To formalize that we can define:

$$O_{\leq}(w_1, w_2) \equiv_{Def} \exists t\, (W_{time}(t, w_1) \wedge W_{time}(t, w_2) \wedge$$
$$\exists t'\, (W_{time}(t', w_1) \wedge (\forall w\, (W_{time}(w, w_2) \rightarrow t' \leq_{time} w))))$$

Then we can formalize (2) as:

(3) $\exists t_1 \exists t_2\, (\,\text{Max}(-\text{ to eat }-_{time})(\text{Dick}, t_1) \wedge (t_1 <_{time} \text{now})$
$\wedge\, \text{Max}(-\text{ to arrive }-_{time})(\text{Zoe}, t_2) \wedge (t_2 <_{time} \text{now}) \wedge O_{\leq}(t_2, t_1)\,)$

A similar analysis applies to the future perfect.

The past continuous and the future continuous[31]

Consider the two propositions:

Dick ate.
Dick was eating.

We formalize these the same, not distinguishing their truth-conditions in terms of how long an interval they're meant to be about. It is in compound sentences that the two tenses are put to different uses. Consider:

Dick was eating when Zoe arrived.

We cannot use (3) to formalize this because the point of using the past continuous with "when" is to indicate that the time of "Dick was eating" started before and continued up to and including the time when Zoe arrived. We use instead:

$\exists t_1 \exists t_2\, (\,\text{Max}(-\text{ to eat }-_{time})(\text{Dick}, t_1) \wedge (t_1 <_{time} \text{now})$
$\wedge\, \text{Max}(-\text{ to arrive }-_{time})(\text{Zoe}, t_2) \wedge (t_2 <_{time} \text{now}) \wedge O_{<}(t_1, t_2)\,)$

A similar analysis applies to the future continuous.

The past perfect continuous and the future perfect continuous

The past perfect continuous is normally used in contexts where a relative time can be established, as in:

Dick had been eating when Zoe arrived.

For this to be true, the time of "Dick ate" has to overlap from before the time of "Zoe arrived" with just a single instant. I'll leave the formalization of that to you.

Formalizing propositions in which the future perfect continuous is used follows similarly, as with "Dick will have been eating when Zoe arrives". But normally we

---

[31] The label "progressive" is often used in place of "continuous".

use the past or future perfect continuous only with some indication of a specified duration. More typical are:

> Dick had been eating for ten minutes when Zoe arrived.
>
> Dick will have been eating for nearly a half hour when Zoe arrives.

To formalize these we'd need to use ways to measure times.

The present perfect continuous
Consider:

> Dick has been studying.

The time of this predication is relative not to the time of another predication but to now: it means that "Dick studied" is true of an extended time directly up to and including now. We can formalize this with:

$$\exists t \, (\, \text{Max} \, (- \text{ to study } -_{\text{time}}) \, (\text{Dick}, t) \, \wedge \, O_{\leq}(t, \text{now}) \,)$$

The simple present
Consider:

> Dick: Do you see that bird?
> Zoe: Yes, I see it.

We can formalize "I see it" as we do "Zoe is seeing it". When someone uses the simple present to talk of now, we can formalize as we do the present continuous.

Sometimes in telling a story the simple present is used to describe a now that is in the past of the storyteller, as in:

> A sausage falls from the grill. Spot barks. Dick yells. Spot eats it.
> He wags his tail. It was the best of times; it was the wurst of times.

Then we have a choice whether to formalize "Spot barks" as in the present of a now designated as the past of the subjective present of the storyteller, as we would "Spot is barking", or as in the past, as we would with "Spot barked".

We've also seen that the simple present can be used for the future:

> Dick will cook dinner when Zoe arrives.

Here "Zoe arrives" should be formalized as "Zoe will arrive".

Another use of the simple present we've seen is to indicate the omnitemporal:

> All dogs are mammals.

$$\forall x \, \forall t \, (\, (- \text{ to be a dog } -_{\text{time}}) \, (x, t) \, \rightarrow \, (- \text{ to be a mammal } -_{\text{time}}) \, (x, t) \,)$$

Another use of the simple present is for what is called "the habitual", which we'll look at in the next chapter.

35 The Habitual

Consider:

(1)   Dick drinks.

We understand by this that Dick habitually drinks (alcohol), not just now, where what we mean by "habitually" is perhaps as vague as what we mean by "loudly".
   Perhaps we could use "habitually" as a predicate restrictor and formalize (1):

   $((-\text{ to drink }-_{\text{time}})/\text{habitually})$ (Dick, now)

But from this we can conclude "$(-\text{ to drink }-_{\text{time}})$ (Dick, now), which we can't deduce from (1): though Dick might drink quite often, he might not be drinking now.
   But "habitually" is not right either. It might not be a habit of Dick to drink in the sense of Dick doing it on a regular basis, yet (1) is true. Rather, we understand (1) as something like:

   Dick has the ability and preference to drink (alcohol) and
   does drink from time to time.

Some would say that instead of "preference" we should say "disposition". But either way, we can't formalize "Dick drinks from time to time".
   Suppose that Rolando is going to stay with Dick and Zoe, and Dick wants to know whether he'll have to drive him everywhere. Zoe assures Dick:

(2)   Rolando drives.

Dick understands this to mean that Rolando has the ability to drive and perhaps more, such as that he has a driver's license, though Rolando might not have driven for several years.
   Examples like these are sometimes said to be in the *present habitual* tense. That is a bad label, for both (1) and (2) are in the simple present tense. They are about now: Dick now has the ability and disposition to drink; Rolando now has the ability to drive.
   It is not a matter of tense; it is a matter of meaning. In the habitual sense, "Dick drank" could be true or false but not equivalent to the use of "Dick drank" to mean that he was drinking at some time in the past. In the habitual sense, "Rolando will drive" can be true or false but not equivalent to the use of "Rolando will drive" to mean that at some time in the future he will be driving. Just as we mark "to have" as "to have$_p$" to indicate that we mean the sense of possession, we can label "to drink" and "to drive" to indicate this habitual sense, formalizing (1) and (2) as:

   $(-\text{ to drink}_{\text{habitual}} -_{\text{time}})$ (Dick, now)

   $(-\text{ to drive}_{\text{habitual}} -_{\text{time}})$ (Rolando, now)

These predicates can then be used with other tenses. Zoe might tell Suzy "Dick didn't drink, though he drinks now", which we can formalize as:

$\exists t \, (\sim (- \text{ to drink}_{\text{habitual}} -_{\text{time}}) (\text{Dick}, t) \land (t <_{\text{time}} \text{now})$
$\land \, (- \text{ to drink}_{\text{habitual}} -_{\text{time}}) (\text{Dick}, \text{now}) \,)$

The predicate "$(- \text{ to drink}_{\text{habitual}} -_{\text{time}})$" is atomic and hence positive for existence. So we have the formalization:

I drink.
Therefore, I am.

$$\frac{(- \text{ to drink}_{\text{habitual}} -_{\text{time}}) (\text{Arf}, \text{now})}{(- \text{ to exist} -_{\text{time}}) (\text{Arf}, \text{now})}$$

This is a valid inference.

Summary of Quantifying over Times

Classical predicate logic takes no account of time: saying that a predicate applies to an object is a timeless assertion. But almost all our reasoning is about objects in time, and what is true of such objects varies with times.

To extend classical predicate logic to take account of time, we have to be clear about how we can make reference to a thing in time. We can understand reference to things that existed and no longer exist and to things that will exist but don't now exist in terms of the method of pointing. We know what it means to point to a thing of that kind, and that method, rather than the actual pointing, is all we need.

We want to talk of times. So we need that times are or can be conceived of as things, specifiable and re-identifiable to quantify over. But our conception of time in our ordinary talk and experience is as a mass: every part of time is time, and there are no smallest times. Nonetheless, we can view parts of that mass of time as individual times that we can pick out, just as we pick out parts of water or mud with containing descriptions.

Our goal is not to investigate what time is, but how to formalize our talk of time in our reasoning. But we need some assumptions about time or times for that. On the view of time as a mass, one time can be within another. On the view of time as made up of instants, longer times are collections of those and there are times within times. So we assume a relation on times of one being within another. We also assume that times are ordered. Combining that with times being within other times, we take times to be intervals. But if a time is within another time, then neither can be before the other, and if two times are overlapping, neither is before the other. So we don't assume that the ordering of times is total.

We assume that the individuals in a universe of a model persist in time stably enough to be used as values of variables and that picking out an individual as a value is independent of any particular time. So the existential quantifier, which is used to assert that there is a thing that can be taken to be the value of a variable, does not take account of when a thing exists: it is supratemporal. In the syntax, we use time variables and time names to take values from a universe of times in a model, and we adopt an order-predicate for times and a within-predicate for times.

Predications about an individual or individuals now say what is true of that object or objects not only in time but at a time. Hence, each predicate needs a blank to be filled with a term for a time. The exception is the identity predicate, for that is used to assert that two terms refer to the same object. The use of a time blank and the ordering of times allow us to dispense with markers for past, present, and future by taking the infinitive form of an ordinary language predicate. For example, "— to be a dog —$_{time}$" is now a predicate that applies or does not apply to Birta at a time and applies or does not apply to my desk at a time. If true of Birta at some time, it is true of her at all times within that time. And if true of her at all times within a time, it is true of her at that larger time. That is, we assume that truth is closed downwards and upwards in time.

To talk of when things exist, we add to the formal language an existence predicate "— to exist —$_{time}$" which allows us to say that Birta exists at this time but not at that. It, too, is closed downwards and upwards for truth. We assume further that there are no gaps in time when a thing exists.

This is enough for a formal system, **QT** for quantifying over things in time. In order to be able to formalize more, we extend it to a logic **QT+internal** in the same way we extend classical predicate to allow for atomic predicates to have internal structure.

The examples from English we wish to formalize involve tenses based on a past-present-future division of time. So to formalize examples from English, we assume that the ordering is quasi-linear: any two times can be compared as before or after unless they overlap.

For quasi-linear orderings of time we assume further that for each true predication there is a maximal interval of time in which the predicate applies to that (those) object(s), and that interval is itself a time. We assume further that for each object that there is a maximal interval of time in which it exists, which is a time too. Further, we assume that when two times overlap, the overlap, which is an interval, is itself a time. And we assume that each predicate we use in a semi-formal language is positive for existence: if it applies to an object or objects at a time, then that object or objects exist at that time. These assumptions added to **QT+internal** give the theory **QTL**.

Using this logic to formalize propositions and inferences in English, we developed conventions for formalizing. We examined long-standing issues such as the difference between accidental properties and essential attributes. Each is worthy of much more analysis beyond these initial investigations.

SPACE in PREDICATE LOGIC

36 Things in Time and in Space

Physical things
Dogs and rocks and tables exist in time and they exist in space.

Some say there are things that are not in time or space. Numbers, qualities, God, they say, are not of any time or place. Let's put aside reasoning about things conceived in this way, as we put aside such reasoning in our work on things in time.

Some say there are things that exist in time but not in space. Ideas, and symphonies, and souls, they say are of particular times but they are not of any place. There is no place you can point to where Beethoven's Fifth Symphony is, yet there was a time before it existed. There is no place I can point to where my idea of justice is. Yet, at least on some conceptions, an idea is not forever: it is related to, dependent in some way on the existence of a body, but it is not in the body or at the body. I cannot be clearer because I do not understand these views. So let's put aside reasoning about things conceived in this way.

Still, we can reason about things in time without worrying about where they are. That's what we've done so far in this volume. It's what we do in using English: each sentence is tensed but need not be marked for location. There are no standard location markers, much less ones required of every sentence.

No one, so far as I know, claims that there are things at locations but not at times. This seems to be a big difference between our conceptions of things in time and things in space, at least for those who speak a language such as English. Yes, we can reason about things in locations without worrying about when they exist. But that seems so strange I can't figure out how to do it.

So here let's focus on how to take account of space in addition to time in our reasoning, building on what we already have in our temporal predicate logic. The things, the individuals we will be concerned with will be things in time and space: *physical things*.

Mud, water, and gold also exist in time and space, but not the way individual things do. As we cannot reason about masses in predicate logic, we will not be concerned with masses here.

There are many similarities between our conceptions of space and time that guide us in our experience, in our speech, and in our reasoning.[32] We can draw on those as we develop a logic for reasoning about things in space as well as time, though we'll also find major differences.

Distant in space compared to distant in time
Just as we had to consider how we can talk about things distant in time, we have to consider how we can talk about things distant in space.

---

[32] Richard Taylor, in "Spatial and Temporal Analogies and the Concept of Identity" discusses many of those.

Someone asserts that there is a moon about a planet that revolves around a star so far away we cannot possibly have evidence for whether such a moon exists. Most of us believe that the proposition is true or false, here and now. Yet when someone asserts "Tom and Suzy will have a baby within three years" many of us doubt that is true or false here and now. What is distant in space, we assume, is different from what is in the future.

To even phrase the issue that way begs the question. The issue is whether there *is* distant in space and whether there *is* in the future.

Right now I think of Fred Kroon in New Zealand. He is not here, and I have no access to him where he is. Similarly, I have no access to him last year, nor next year. Anything that is not within my physical sense-field is not accessible to me, either through being distant in space or distant in time. But I could call him now and talk to him on the phone, whereas I cannot speak to him in the past or in the future. This seems a big difference.

The access I have to Fred Kroon when I am talking to him on the phone is access through inference: I have good reason to believe it is his voice and he is talking to me.[33] We use our reasoning abilities to have some limited access to things distant, and to things past, and to things future. It might seem that there is a difference in the kinds of evidence available to us in reasoning about things distant in space compared to distant in time. For distant in space and distant in the past we use the same kind of evidence: memory, knowledge of how the world is constructed, physical clues. For distant in the future don't we use the same? We use physical evidence and knowledge of how the world is constructed: there will be a mountain there tomorrow—as we look at a mountain now. We use memory: I remember that there is a road to my house and I reason about driving into town next week. You might say that the road or the mountain could be destroyed tomorrow and hence my access to the future is at best imperfect. But all the clues I have about things distant in space and distant in the past could lead me astray, too. The only difference in kinds of evidence I see is our sense of memory, of lived past. But we also have memory of places we have been.

Still, there is a difference: I am here and Fred Kroon is there, yet there is no Fred Kroon in the future. There is no place in the future from which he could be talking to me. But that is no argument for the difference between the far distant and the future; it is only a particular instance of the grand metaphysical assumption that the far distant *is* in a way that the future is not.

Our doubts about the the nature of the future led us to say that we can reason with future-tense sentences by treating them as if they are true or false, not assuming they actually are true or false. We build models in which they are assumed to be true and models in which they are assumed to be false, and compare those. We have no need for grand metaphysical assumptions about the future in order to reason this way.

---

[33] More than once I have thought I was talking to a friend only to discover that it was a recording.

Similarly, we can avoid making a grand metaphysical assumption that there *is* in the far distant and that sentences about the far distant are indeed true or false. We need only treat those as if they are true or false in order to reason with them, not necessarily accepting that they are true or false.

In reasoning about the future our worries about what *is* in the future led us to consider how we can reason about things that do not exist now. We resolved that by reflecting on our understanding of naming: we know what it means to point to a thing of that kind, and that method, rather than the actual pointing, is all we need. That discussion in Chapter 15 can apply equally to reasoning about things that do not exist here.

37 Locations

Space, perhaps even more than time, is to us a mass: every part of space is space, and there are no smallest parts of space. Yet we talk of places as if they are distinct and can be counted.

(1)   Suzy left her cell phone some place in Tom's apartment.
There's no place where Puff and Harry have been together.
There are three places where Spot dug in the yard.
Everywhere that Mary went her lamb was sure to go.

At least one, none, three, all places—this looks like talk about individuals and times that we formalize in predicate logic. But to use predicate logic to talk about places or locations we need to figure out how we can pick out parts of space.

We pick out parts of the mass of water as individuals with containing descriptions: a glass of water, a lake of water. Or we pick out parts physically: that water there, the water in my hand. Similarly, we pick out bits of mud: a puddle of mud, the mud here on my shoe. But there are no natural descriptions for parts of space.

Time is a mass: every part of time is time, and there are no smallest bits of time. Yet we imagine that time comes in distinguishable bits because we have fixed time markers that seem to name times: May 16, 2018 or 3:18 p.m. June 11, 2011. But we have no comparable names for locations that direct us to pay attention to specific parts of space.

Why not agree on a co-ordinate system for specifying locations? Mathematical co-ordinate systems for space are based on the view of space as constituted of indivisible, dimensionless points. It would be a big project to devise co-ordinates without that assumption. And then we would have to incorporate a great deal of mathematics into our formal systems to relate locations described with co-ordinates. Even then we'd have no reason to think we could devise a mathematical description of the space within the shed in my corral much less just any region of space in which we are interested. To say that there is a mathematical description of the boundaries of the region of space in which your body is at the precise moment you are reading this is only a theoretical possibility. It has no part in our reasoning, for we could not give such a description.

We point to locations. But locations aren't like things: we can't understand the pointing in terms of what kind you're pointing to. All locations are the same kind. We pick out locations with descriptions or markers or moving our hands around. In our descriptions, in our picking out locations, we are as much creating a location as when when we talk about the "quantity" of water 1 cm in diameter that lies 3 cm below the surface of the water in my bathtub and 10 cm northwest from the drain. It's not there, not a quantity, not something we can quantify over until we have picked it out.

A universe of locations, all the locations we're talking about, is not given in advance—unless we've agreed on just a few as the only ones. There is no collection of all parts of space just as there is no collection of all bits of water in my bathtub that we can quantify over. What we have with masses—time as well as space—are only ways of picking out. We have ways that we agree for picking out locations; then there is a location as we pick it out.

This is no different from what we do when specifying a universe of individuals for a model of predicate logic. We don't specify each individual separately, unless we decide in advance that we're talking about only a small number of objects. When we say that the universe includes all dogs, we haven't specified the things, only which kind of things, and that's equivalent to how we identify things of that kind. "All dogs" is pretty vague. But we all agree (more or less) on how we pick out a thing as a dog and not as an elephant or rock or microbe or table.

A universe of a model is not a collection of things, as if each thing were separate and distinct and fixed across time and space, ready for us to investigate. That "collection" idea is just the idea of a predicate or kind, which is what we actually use via methods of picking out. There are large questions about how we know how to pick out, how we agree on how we pick out, how we understand picking out, how we designate at all. But some resolution based on some ways of picking out is needed and is what we have to assume in order to have a model of predicate logic. What we do not have to assume is that by saying "the universe contains all dogs" we have specified each and every dog individually as part of the universe, fixed in advance. That is so close to a platonic view of things and collections that it removes the notion of a universe of things from our ability to use predicate logic. At best, it's an abstraction, ignoring how we might pick out this dog or that dog, so as to make the mechanics of the logic simpler. But the abstraction is not the reality. Or perhaps you think it is a reality; if so it's a quite unusable one. "$\forall x \, ((- \text{ is a mammal})(x))$" is true in a model with a universe that "contains" all dogs means that whatever thing we pick out as being a dog, it is a mammal. This is the basis of the semantics of predicate logic using assignments of references.[34]

But with locations there is no abstraction we can make to arrive at all locations within my ranch, no more than we can talk of all bits of water in my bathtub. We can't even imagine what those are. We only have more or less clear ways of specifying parts of space within my ranch.

To reason about locations, to formalize propositions such as those at (1), we'll have to assume that we can pick out parts of space under some general description like "the places within Arf's ranch" along with some more or less clear way of specifying places. The ways of specifying places will vary from model to model: we'll likely pick out places within my ranch differently from how we pick out places on the surface of a piece of granite.

---

[34] But see the discussion of "pure" reference on pp. 86–87 of Volume 1.

To use our ways of picking out locations, we need to add to the language of our temporal predicate logic new *location (place) variables*: $l_0, l_1, l_2, \ldots$. These are meant to take values from a universe of places **P** in a model that is distinct from the universe of physical things **U** and is distinct from the universe **T** of times. That is, whatever we pick out as a location is not a physical thing and is not a time; whatever we pick out as a time is not a location and is not a physical thing; whatever we pick out as a physical thing is not a time and is not a location. We'll also need new *name symbols* $e_0, e_1, e_2, \ldots$ that we can realize as location names. The variables and name symbols together are *location terms*, for which I'll use the meta-variables $l, l', p, p', p_0, p_1, p_2, \ldots$; for the semantics, I'll use **p, p′, p′′, p₁, p₂,** ... to stand for locations in **P**. And we'll add an equality predicate for locations,

Before we go further, we have to decide how reference to physical things, to times, and to locations relate to one another.

Aside: A psychologist on space
For comparison, here are some definitions that John J. Riesen gives in "The Generation and Early Development of Spatial Inferences".

> In this chapter the terms are used in the following ways: "Location" refers to the spot at which an object rests. "Orientation" refers to an observer's location and facing direction. "Place" refers to the immediate vicinity in which an observer could easily read or play ball, or where an observer could view the layout around but without talking. p. 40

These definitions are too vague (what kind of ball: football? a dog chasing a ball?, baseball?). They are inadequate (something can be oriented to another object without an observer). They rest on unexamined assumptions that affect whatever conclusions about psychology he draws (a bird in flight has no location, nor does a bush). The distinctions and precision we make in this text are not just pedantic.

38   References to Things, Times, and Locations Are Independent

In classical predicate logic we reason about individual things. How we pick out those things, how we refer to an individual thing has been a central concern in this series of volumes. We've always assumed that each individual thing can be picked out as the focus of our attention independently of any other individual thing. Indeed, that is part of what we mean by saying that an object is an individual thing. And we refer to individual things in a timeless fashion (Chapter 15).

In our temporal predicate logic, reference to a thing is reference to a thing that can cross many times, a thing that exists through all its "changes". Individual things are conceived not as timeless but as supratemporal, and reference is supratemporal. Talk of individual things is how we impose—or perhaps recognize—stability in the flux of our experience.

In our temporal predicate logic we treat times as things. We quantify over them and pick them out. Any time can be the focus of our attention independently of any other time or any individual thing, or at least that's what we agreed in Chapter 15.

In our work in this and previous volumes we did not even consider locations in discussing reference to an individual thing or to a time. Just as an individual thing may exist at many times, it may exist at many locations. That is part of its stability. A physical thing is independent of where it is as much as when it is, though we might pick out the physical thing with a description that involves a time and a location. Physical things and reference to them are supralocational as well as supratemporal.

There may be many locations at a particular time. But we can focus our attention and refer to that time as reference for a variable or name independently of any location. Reference to times is supralocational. Similarly, a location exists across many times. Reference to locations is supratemporal.

Yet physicists tell us this is wrong. Times are not independent of locations, and locations are not independent of times. There is not time and space but only space-time. Nonetheless, physicists use four variables in their equations: one for time and three for location. In any case, we are concerned here with how our usual conceptions of time and space are involved in judging our reasoning; we are not concerned with characterizing time, space, or space-time. Perhaps our logic will have to be modified to use in reasoning about the very large, the very small, and the very fast. But it won't be just reference that will be in question for that kind of reasoning. The larger issue will be whether it is correct within the physicist's view to think of the world as made up of individual things.

Though a location might be distinguishable only in relation to other locations, we can refer to it without consideration of those. Reference to a location does not depend on other locations. This is not to say that space can be broken into independent parts. It is only to say that we can focus on a particular location as reference for a variable or name independently of consideration of any other location.

Might reference to a location depend essentially on a thing? I might talk of an individual thing in picking out a location, such as when I say that "Dogshine" is the ranch where my home is now. But the location would be the same whether I used that description or another involving a different object, taking "Dogshine" to be the ranch where my donkey Bon Bon lived last year. A location is independent of individual things; it is the same regardless of what is there. Or at least that is what I'll assume here.[35] Reference to locations is independent of any individual thing.

In sum, we'll take the following assumption as the basis for developing a predicate logic of space and time.

Physical Things, the World, and Propositions The world is made up at least in part of *physical things*: individual things that exist in time and space.

Times are (can be conceived of as) individual things.

Locations are (can be conceived of as) individual things.

Reference to a physical thing does not depend on any other physical thing, time, or location.

Reference to a time-as-thing does not depend on any other time-as-thing, physical thing, or location.

Reference to a location-as-thing does not depend on any other location-as-thing, physical thing, or time.

The only propositions in which we are interested in are those that are about physical things and/or times-as-things and/or locations-as-things.

To adopt this approach to formalizing reasoning about physical things is to stand back from time and space and things in time and space to do our metalogic as if we could survey times and places and physical things. We can constrain how physical things and times and places relate in establishing the truth of predications, but we do that outside of the times and locations we are talking about.

In temporal predicate logic we allow a predicate to be used in a semi-formal language only if it is extensional: it doesn't matter how we refer to an object or time in evaluating whether a predication using it is true. Let's now require that it doesn't matter how we refer to a location.

---

[35] See Appendix E here for the view that locations are specifiable only in terms of things.

39 Assumptions about Locations

The within-relation
Basic to our conception of space is that one location can be within another. This is so whether we think of locations as parts of the mass of space, each having no smallest part, or consider space to be constituted of indivisible points and larger locations collections of those. So we will assume that in every model there is a relation $W_{location}$ on the universe of places P that is meant to be the relation of one location being within another.

It's simplest to assume that each location is within itself. And two distinct locations cannot be within each other, for otherwise the relation would be circular.

My computer is within my office, and my office (which is a separate building) is within my ranch—thinking of small volumes around those. So it follows—you don't have to go and look—that my computer is within my ranch. The relation is transitive.

The part-whole relation on locations In every model, there is a binary relation $W_{location}$ on the universe of places P that satisfies:

$W_{location}(p, p)$

If $W_{location}(p_1, p_2)$ and $W_{location}(p_2, p_1)$, then $p_1 = p_2$.

If $W_{location}(p_1, p_2)$ and $W_{location}(p_2, p_3)$ then $W_{location}(p_1, p_3)$.

These three assumptions about the within-relation on locations are the same as we made for the within-relation on times, which I'll now write as "W_{time}".

To talk about this relation, we'll add "$W_{location}(-, -)$" to the formal language, allowing only location terms to fill the blanks.

We agreed that every time is determined by its parts, for if not there would have to be some unknown or at least unacknowledged property of times that would distinguish them. Similarly, let's agree that every part of space, every location, is determined by what locations are within it.

Parts determine locations
$p = p'$ iff (for all p_1, $W_{location}(p_1, p)$ iff $W_{location}(p_1, p')$)

Overlaps
Two locations can overlap:

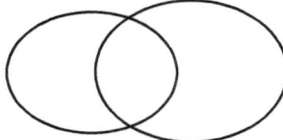

If we can describe each of these locations, we can describe the part of space that is the overlap as the locations that are in both the original locations, since a part of space is determined by the locations in it. So it seems right to assume that whatever methods of picking out locations we use for a model will include picking out overlaps, too. But if we make that assumption, we can't require that the method for picking out a location allows for smallest locations to be volumes of 2 cm in diameter, for though the two locations in the picture above are each greater than 2 cm across, the overlap is less.

We'll consider later how to formulate semantic and syntactic definitions and conditions for picking out locations and overlaps.

Connected regions
Suppose someone says that Spot is sitting in this location:

(1)

That's just silly. You've drawn two locations, we'd say, not one. A location should be all one piece, connected, just as an interval of time is one piece.

Intervals of time are defined in terms of the order relation on times. There is no comparable basic relation on locations, nothing that stands at the heart of our conception of space to help us define a connected region.

Can't we use points, lines, and a between-relation to require that any two locations within a location can be joined within the location itself? Those are fundamental notions in axiomatic geometry.[36] But points, lines, and a between-relation in geometry are given for a conception of space as made up of dimensionless points. Here a location can be a volume or an area, for which it's not even clear what "between" means. Is p_2 between p_1 and p_3?

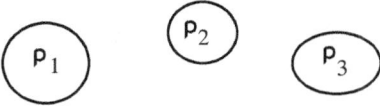

Nor can we use the mathematical theory of topology because for that to apply, P itself must be a location, the empty set has to be a location, the union of any number of locations has to be a location, and the intersection of any two locations must be a location. Each of those assumptions except perhaps the last seems inappropriate to apply to all models

---

[36] See my *Classical Mathematical Logic*.

Perhaps we could require that if two locations p_1 and p_2 are strictly within p, then there is some location p_3 that overlaps the two and is strictly within p.

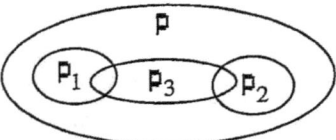

But this picture assumes that p_3 is connected, whereas it, too, could have two parts:

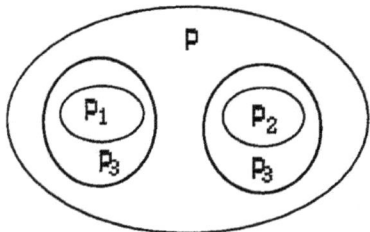

In any case, there's nothing wrong with calling the two parts of (1) *a* location, so long as we have that each of the two parts is a location. And surely each part in (1) is, since we can pick them out. If we have the two parts together as a "bit" of space to quantify over in a model, we should have each part separately. But that has to remain an informal constraint on our ways of picking out locations.

So we'll assume only that there is a within-relation on the universe of places and that locations are determined by what locations are within them. We distinguish the universe of locations from the universe of times by how we take account of space in evaluating the truth of propositions about physical things.

40 True in a Location

Where is Spot sitting?

(1)

Of course he's sitting there (pointing).
 What are you pointing to? Locations aren't like things: we can't understand the pointing in terms of what kind you're pointing to. All locations are the same kind.
 We can pick out a time when Spot was running by agreeing on some beginning and ending, even though there's no reason to think that there are specific points in time when Spot began to run and ended running. The choice of that time is not arbitrary, for we would certainly agree that it doesn't include when Spot first lifted his head and sniffed the air while he was lying down. But it is not (clearly) fixed either. Agreements are all we have.
 There are no natural boundaries for where Spot is sitting nothing even as uncertain as beginnings and endings. We cannot talk about *the* location where Spot is sitting. The choice of what counts as where Spot is sitting is not clearly fixed.
 We pick out locations where Spot is sitting with descriptions or markers or moving our hands around him. The locations we describe include more or less "space" around Spot. The choices of those places are not arbitrary, for we would not accept that Spot is sitting at a location two meters in diameter around the bus stop down the street.
 So one of us "draws" a location, a volume, around him:

(You'll have to imagine a volume is depicted.) Another disagrees and picks out a different location:

180   Chapter 40

The overlap, the common part of the regions might seem to be a good candidate for us to agree on where Spot is sitting, which we could refine more and more.

Overlaps leading to smaller regions in which Spot is sitting are an attempt to come to some minimal location in which Spot is sitting. There is none, for Spot breathes in and out, he fidgets, he sheds hair, he wags his tail. The time would have to be very small for there to be a minimal location, so small we couldn't even perceive it. We have the opposite: if Spot is sitting in Dick and Zoe's yard, he is sitting in the block where they live, and he is sitting in their town. If Spot is sitting in one part of space, he's sitting in any part of space that includes that part. Instead of truth being closed downward as for time, truth is closed outward for locations. We talk of a predicate being true of an object *at* a time, but we talk of a predicate being true of an object at a time *in* a location.

Can we put any conditions on what counts as a place where Spot is sitting, or where he is running, or where the rock was falling last midnight? Consider:

Would we count this as a location where Spot is sitting? I think not. That's not because someone might not recognize that it's Spot there. It's because Spot has to be in the location—all of him, not just some part. Later we'll have a way to say that, but for now this is an informal condition.

Would we count this as a location in which Spot is sitting?

(2)

This is meant to represent a location that is the smallest possible volume of space that we can imagine Spot is in at the time of (1). This would not distinguish Spot sitting from Spot floating or levitating. Without some ground below him and sky above we don't have up and down, which is needed for him to be sitting or for a rock to be falling.

All I can say, and all I think we have any intuition about, is that the location in which Spot is sitting has to include all of him and some part of the world that "justifies" our saying he is sitting there. That's some help, but not much.

Here is another example, a photograph of a horse running taken by Eadweard Muybridge in the late 19th Century:

The exposure time, if I understand correctly, was 2000th of a second. We've already agreed that if the horse was running from the minute before to the minute after this picture, as it surely was, then it was running in every time within that interval. So yes, the horse is running in this small time—and we recognize that in the picture. But is the horse running in this location? Does there have to be more space in front of and behind the horse for us to say it's running in that location? Here is another photograph by Muybridge:

Here I think we're right to say the horse isn't running in this location because part of it, part of its hoof, is not in the location.

What about Spot being a dog? Won't (2), the smallest volume we can imagine around Spot at the time of (1), do for where he was a dog at that time? After all, isn't he a dog wherever he is? That's to take "to be a dog" as a property, an attribute of him at a time, as opposed to "to sit" or "to run". Dividing predicates into classifications versus process, as we've seen before, is fraught with problems. It will be enough that we'll have tools to examine formally the idea that a predicate is an attribute of an object iff it's true of the object in any location in which the object is.

I offer these examples and comments only to stimulate discussion in the hope that we can come to some clearer idea about what we count as a location in which a predication is true. For now, we'll have to rely on agreements—or disagreements.

To codify these observations we need to be able to talk about where a predication is true. Previously we had "($-$ to sit $-_{time}$)". Now we will have in addition (not replacing it) the predicate "($-$ to sit $-_{time}$, $-_{location}$)" where the last blank can be filled only with a location term. If "(— to sit $-_{time}$, $-_{location}$)" is true of a physical thing o at time t in location p, then "(— to sit $-_{time}$)" is true of o at t. We'll build on our logic of quantifying over times as things, **QT**.

As a convention, when we write $A(x_1, \ldots, x_n, t)$ and $A(x_1, \ldots, x_n, t, l)$ together these are meant to be the same ordinary atomic wff except that in the first the infinitive has a time-marked blank, while in the latter the infinitive has both a time-marked blank and a location-marked blank. Now we can make explicit the principles we saw in this discussion.

Outward closure of truth for locations
If A is an ordinary atomic temporal predicate, and $A(x_1, \ldots, x_n, t, l)$ is true of $o_1, \ldots, o_n$ at time t in place p, then for any p′ such that $W_{location}(p, p')$, $A(x_1, \ldots, x_n, t, l)$ is true of $o_1, \ldots, o_n$ at time t in p′.

True at a time and location, then true at that time
If $A(x_1, \ldots, x_n, t, l)$ is true of $o_1, \ldots, o_n$ at time t in place p, then $A(x_1, \ldots, x_n, t)$ is true of $o_1, \ldots, o_n$ at time t.

From the first we can conclude the following.

Inward closure of falsity for locations
If $A(x_1, \ldots, x_n, t, l)$ is false of $o_1, \ldots, o_n$ at time t in p, and $W_{location}(p', p)$, then $A(x_1, \ldots, x_n, t, l)$ is false of $o_1, \ldots, o_n$ at time t in p′.

Now suppose that Spot was barking from 5:01 to 5:03 p.m. in Dick and Zoe's yard. Then he was barking at 5:02 p.m.—not just at some place, but within their yard. And if he was barking at all times from 5:01 to 5:03 p.m. in Dick and Zoe's yard, then he was barking from 5:01 to 5:03 p.m. in Dick and Zoe's yard. It's not just that he was barking somewhere during that time. It's that he was barking within that location.

Downward and upward closure of truth in time at a location
$A(x_1, \ldots, x_n, t, l)$ is true of $o_1, \ldots, o_n$ at time t in place p iff
$A(x_1, \ldots, x_n, t, l)$ is true of $o_1, \ldots, o_n$ at t′ in p for all t′ such that $W_{time}(t', t)$.

Aside: No location markers
In English we mark every sentence for time but not for location. If the marking is specific, say 9:17 a.m. July 26, 2018, we don't feel we have to give a location. I think that's because of an unstated assumption that by specifying the time we have also specified the location. "Spot was barking at 9:17 a.m. July 26, 2018" if true also tells us the location: where he was barking at that time. Indeed, we often don't have a better way to specify the location. But, as we have seen, we really haven't specified a location but at best a range of locations.

41 Locational Predicates

If Spot is sitting, then he's sitting in some places and not sitting in others. If Spot is barking, then he's barking in some places and not in others. If this rock was falling last midnight, then it was falling in some places at midnight and not falling in others. It seems that if a predicate is true of an object at a time, then it's true in some places and not in others. And if the predicate is true of an object at a time in a place, then the object exists there.

But consider:

Bob and Dick are brothers.

This is true. But in what location is it true? Dick is in New Zealand, and Bob is in California. So is it true within the location of New Zealand and California together, as if that were one location? Or is it true only of a location big enough to include New Zealand and California?

Opinion divides.[37] Some say that it's true everywhere. But that seems odd, for it means Bob and Dick are brothers on the moon. Others say it isn't true in any location. Others say that it's nonsense to ask where it's true. "Where are Bob and Dick brothers?" makes as little sense as asking what color my idea of justice is.

Consider, too:

John F. Kennedy was President of the U.S.A. in 1963.

In what location is this true? Where was he President? In the boundaries of the U.S.A.? But he was President of the U.S.A. when he visited Berlin that year. Is being President of the U.S.A. an attribute of him, so that it is true in any location in which he exists? But wasn't he President of the U.S.A. in Mongolia, though he never visited there? Perhaps we should say that John F. Kennedy was President in all places. But was he President on a distant star? Perhaps we've misconstrued the claim and the predicate should be "— to be President of —", which is a relation, or "— to be President / of (U.S.A.)". But the U.S.A. is a social construct, not a physical thing, so a name for it can't fill a blank. Any predicate that involves a social construct seems to be inherently relational, though we have no way to represent that relation in our logic. Perhaps, then, we should say that it's wrong to think of John F. Kennedy as President in a location.

Yet some relations are true about objects in some locations and not others, for example,

— to be next to —, —$_{time}$

This is true of Spot's dish and Spot's doghouse in the location depicted in (1) on p. 179, but it's not true of Spot's dish and his doghouse in a small location around

---

[37] Among the few people I've asked.

the bus stop down the street at that time. Whenever it's true of two objects, it's true of those objects in some locations and not in others. With our new vocabulary we can formalize these distinctions.

Locational, omnilocational, and alocational predicates
An ordinary atomic predicate $A(x_1, \ldots, x_n, t)$ is:

locational $\equiv_{\text{Def}}$ if $A(x_1, \ldots, x_n, t)$ is true of $\mathsf{o}_1, \ldots, \mathsf{o}_n$ at t, then there is some location p such that $A(x_1, \ldots, x_n, t, l)$ is true of $\mathsf{o}_1, \ldots, \mathsf{o}_n$ at t in location p, and there is some p' such that $A(x_1, \ldots, x_n, t, l)$ is not true of $\mathsf{o}_1, \ldots, \mathsf{o}_n$ at t in location p'.

omnilocational $\equiv_{\text{Def}}$ if $A(x_1, \ldots, x_n, t)$ is true of $\mathsf{o}_1, \ldots, \mathsf{o}_n$ at t, then for every location p, $A(x_1, \ldots, x_n, t, l)$ is true of o at t in location p.

alocational $\equiv_{\text{Def}}$ for some $\mathsf{o}_1, \ldots, \mathsf{o}_n$ at t, $A(x_1, \ldots, x_n, t)$ is true of $\mathsf{o}_1, \ldots, \mathsf{o}_n$ at t, but there are no objects $\mathsf{b}_1, \ldots, \mathsf{b}_n$, time t, and location p such that $A(x_1, \ldots, x_n, t, l)$ is true of $\mathsf{b}_1, \ldots, \mathsf{b}_n$ at t in location p.

With a locational predicate, such as "— to sit —$_{\text{time}}$", locations distinguish where a predication for a thing and time is true and where it is false. Locations can't make that distinction if there's no place in which the predication is false.

We can accommodate the view that it's nonsense to say that "— and — to be brothers —$_{\text{time}}$" is true of a thing at a time in a place by treating nonsense as false, as we usually do in formal logic. Then this predicate would be classified as alocational. We could more strictly reflect the idea that it's nonsense by not allowing for a time-and-location marked version of the predicate. But then we'd have to have two kinds of predicate variables reflecting that division: each has a time-marked version, but only those of the first kind would also have a time-and-space marked version. That would elevate a semantic distinction into a formal distinction, and it would complicate the theory considerably in comparison to assimilating nonsense to false.[38]

For particular predicates we can make explicit the classification by adding a meaning axiom. For example, we could add:

"(— to sit —$_{\text{time}}$)" is locational:
$$\forall x \, \forall t \, (\, (\text{— to sit —}_{\text{time}}) \, (x, t)$$
$$\rightarrow \exists l \, (\text{— to sit —}_{\text{time}, \, \text{—location}}) \, (x, t, l) \,)$$

---

[38] We might say that a predicate is *semi-locational* if it's true of some physical things at times in some locations, but also true of some physical things at times in no location. We might say that a predicate is *partly omnilocational* if for some physical things and times it's true in some locations and not others, but for some physical things and times it's true in every location. But I have no examples of these.

"(— and — to be brothers —$_{time}$)" is omnilocational:

$\forall x \, \forall t \, (\, (— \text{ and } — \text{ to be brothers } —_{time}, —_{location}) \, (x, y, t)$
$\rightarrow \forall l \, (— \text{ and } — \text{ to be brothers } —_{time}, —_{location}) \, (x, y, t, l) \,)$

"(— and — to be brothers —$_{time}$)" is alocational:

$\forall x \, \forall t \, \forall l \, \neg \, (\, (— \text{ and } — \text{ to be brothers } —_{time}, —_{location}) \, (x, y, t, l) \,)$

We can make these abbreviations for all predicates.

locational (A) $\equiv_{Def}$
$\quad \forall \ldots \forall t \, (\, A(x_1, \ldots, x_n, t) \rightarrow$
$\quad\quad \exists l \, A(x_1, \ldots, x_n, t, l) \wedge \exists l \, \neg \, A(x_1, \ldots, x_n, t, l) \,)$

omnilocational (A) $\equiv_{Def}$
$\quad \forall \ldots \forall t \, (\, A(x_1, \ldots, x_n, t) \rightarrow \forall l \, A(x_1, \ldots, x_n, t, l) \,)$

alocational (A) $\equiv_{Def}$
$\quad \forall \ldots \forall t \, (\, A(x_1, \ldots, x_n, t) \rightarrow \forall l \, \neg \, A(x_1, \ldots, x_n, t, l) \,)$

Our examples have been of simple atomic predicates. But these classifications apply equally to atomic predicates with internal structure. An atomic predicate might not be locational even though the simple predicate in it is locational. For example

(— to be a dog —$_{time}$)   is locational

(— to be a dog —$_{time}$)/of (Dick)   is either omnilocational or alocational

To examine these ideas further, we need a way to say that a thing exists in a location.

42 Existence in Space and Time

We talk about when things are with "(— to exist —$_{time}$)". Now we want to talk about where things are. So let's adopt a new predicate:

(— to exist —$_{time}$, —$_{location}$)

We're reasoning about only physical things in time and space. So if a thing exists at a time, it exists in a location. And if a thing exists in a time and location, we can say it exists at that time without mentioning the location.

Exists at a time, then exists at some location at that time
If (— to exist —$_{time}$) (x, t) is true of o at time t, then there is some location p such that (— to exist —$_{time}$, —$_{location}$) is true of o at time t in location p.

Exists at a time and location, then exists at that time
If (— to exist —$_{time}$, —$_{location}$) (x, t, l) is true of o at t in p, then (— to exist —$_{time}$) (x, t) is true of o at t.

Since "—$_{location}$" means "within the location", existence predications are closed outward with respect to locations.

Outward closure of existence in space
If (— to exist —$_{time}$, —$_{location}$) (x, t, l) is true of o at t in p, and $W_{location}(p, p')$, then (— to exist —$_{time}$, —$_{location}$) (x, t, l) is true of o at t in p'.

As with ordinary predicates, the existence predicate is closed downward and upward for truth in time at a location.

Downward and upward closure of existence in time at a location
(— to exist —$_{time}$, —$_{location}$) (x, t, l) is true of o at time t within place p
iff (— to exist —$_{time}$, —$_{location}$) (x, t, l) is true of o within p for all t´ such that $W_{time}(t', t)$.

Spot existed in front of his doghouse at 4:18 p.m. July 1, 2000 and he existed at the bus stop near Dick and Zoe's yard at 5:12 p.m. that day. During that time he wandered from his yard to Tom's home, then to the fire hydrant a block away, then to the bus stop. The principle of the continuity of existence in time plus the principle that what exists at a time exists at some location at that time guarantee that at each time during that interval he existed somewhere. We may believe that during that time Spot existed at locations that connect his doghouse and the bus stop. But to adopt a semantic condition that those places lie along a continuous path from the one location to the other would be as difficult as requiring that locations be connected regions.

We've agreed that the ordinary predicates we use in a semi-formal language are positive for existence in time:

If $A(x_1, \ldots, x_n, t)$ is true of $\mathsf{o}_1, \ldots, \mathsf{o}_n, \mathsf{t}$, then for each i,
$(-\text{ to exist }-_{\text{time}})(x_i, t)$ is true of o_i, t.

And we said that if Spot is sitting in a location, he's in that location; if he is a dog in a location, then he's in that location; if he is chasing a rabbit in a location, he's in that location. It would seem, then, that if a predicate is true of a thing at a time in a place, then the thing has to exist in that place. But if we take "— to be President of the U.S.A. $-_{\text{time}}$" to be omnilocational, then John F. Kennedy would have to exist in every location in 1993. It's only locational predicates that distinguish where things are.

Locational predicates are positive for existence in time and location
If A is locational and $A(x_1, \ldots, x_n, t, l)$ is true of $\mathsf{o}_1, \ldots, \mathsf{o}_n$ at time t in location p, then for each i, $(-\text{ to exist }-_{\text{time}}, -_{\text{location}})(x, t, l)$ is true of o_i at t within p.

So some predicates distinguish where things are. Do things distinguish locations?

43 Where Things Are

Dick and Zoe and Suzy are at the park.

>Suzy: Where is the band going to play tomorrow?
>Zoe: Over there where Dick is.

Zoe is directing Suzy's attention to Dick and some region around him that's big enough for a band to set up yet is within the park. Later Suzy asks:

>Suzy: Where is the lawn chair?
>Zoe: Over there where Dick is.

Suzy understands Zoe to be directing her attention to a small region around Dick, perhaps no more than 3 meters in diameter.

We pick out places with objects. But "there, where Dick is" doesn't designate a specific place. It directs our attention to places around Dick. We understand what places to look at by context, including our knowledge of how large a region is needed for some predication to be true and what other things are meant to be in the region.

Yet it seems to be a part of our conception of physical things that they do establish locations. We've all heard:

> No two things can be in the same place at the same time.

We may believe it, but it's false. My heart is within any place where my body is, but my heart and my body are two distinct things. Perhaps we can adjust the saying:

> No two things, neither of which is a part of the other, can be in the same place at the same time.

But this is wrong, too. In 2012, my dog Birta was in her doghouse and my dog Chocolate was in the same doghouse. Of course they were there at different times within 2012. For this dictum to be true, the time has to be small enough that the two objects are completely separated by location at that time.

And we've all heard:

> Nothing can be in two places at the same time.

But in 2012 Birta my dog was in lots of different places. For this to be true, it must be about very small times.

It might seem that at small enough times a physical object does determine a unique region of space: the volume that exactly encloses it, a minimal location for it. But my dog Birta changes constantly. Change is the one constant in her and all our lives. We ignore the changes to say that there is one thing, one individual we call "Birta", just as we say there is one rock to be the value of this variable, or one house, though both change. Birta is a thing that is supratemporal. And Birta exists, over time, in lots of locations. We ignore the location of Birta in viewing her as one thing; we don't say that there is a different thing at the different locations. But to assume a minimal location for her at either 10:01 a.m. January 6, 2006 or some smaller time is different. It is not to find a unity of her but to ignore what we know to be true: there is no time small enough for us to classify her as stable in a minimal location. It isn't that we ignore all the changes in her during an interval of time to establish a location. It's that we somehow find a unity of location of her, perhaps an average of her many locations during that time, pretending to a stability of size we know very well is not there. Yet such a conception of stability in location seems to be part of our notion of a physical thing.

This tension between our believing that physical things determine locations and knowing that each physical thing is constantly changing is inherent in building our world and logic on the assumption that there are individual things. We look for some resolution. We could add to our logic that each physical object picks out a unique minimal location at small enough times. That could be viewed as an abstraction, as we abstract in any scientific theory, ignoring much of our experience that would show that the assumption we make is strictly speaking false. Or it could be a constraint on what we take as locations and times: with few enough locations and/or only large enough times, at every time for each thing there could be a minimal location at a small enough time within that. But such an assumption does not seem fundamental enough to adopt for all our models. We can add it for specific models. And if there are few locations, a minimal location for Spot in the picture (1) on p. 179 might also be a minimal location for his dish. Or there might be many minimal locations where Spot is sitting at the time of that picture.

Still, it seems essential to how we conceive of physical things that they can be distinguished by location. We can formulate that assumption without using minimal locations. It's enough that if a and b are different, then at any time t at which both exist, there is some location p in which a exists and b doesn't, or there is a location p in which b exists and a doesn't. We need both possibilities, since at any time and location at which my body exists, my heart exists there, too—I hope.

Locations distinguish physical things
If $a \neq b$, then at every time t at which both exist, there is some place p such that
$(-\text{ to exist }-_{\text{time}}, -_{\text{location}}) (x, t, l)$ is true of a, t, p and is not true of b, t, p
or $(-\text{ to exist }-_{\text{time}}, -_{\text{location}}) (x, t, l)$ is true of b, t, p and is not true of a, t, p.

This assumption multiplies how many locations we must have in a model: for any time there must be enough locations to distinguish all the things that exist at that time. That's why we can't use it as a definition of identity, since without this assumption there might not be enough locations.[39]

Now we can formulate a logic for reasoning about things in space and time.

---

[39] Compare the predicate logic criterion of identity discussed in Volume 0.

44 A Formal Language for Reasoning about Things in Time and Space

Vocabulary

    infinitive symbols I_j^n for $j \geq 0$ and $n \geq 1$, where n indicates the arity

    time order predicate $- <_{\text{time}} -$

    time part predicate $W_{\text{time}}(-,-)$

    location part predicate $W_{\text{location}}(-,-)$

    temporal existence predicate $(-\text{ to exist }-_{\text{time}})$

    temporal-and-spatial existence predicate $(-\text{ to exist }-_{\text{time}}, -_{\text{location}})$

    equality predicate $(- \equiv -)$

    time equality predicate $(- \equiv_{\text{time}} -)$

    location equality predicate $(- \equiv_{\text{location}} -)$

    predicate modifier symbols

        restrictor symbols $R_0, R_1, \ldots$

        variable restrictor symbols R_n^i for $n \geq 0$ and $i \geq 1$

        logical variable restrictor $\text{obj}(-)$

        negator symbols $N_0, N_1, \ldots$

        logical negator $\sim$

    individual name symbols $c_0, c_1, \ldots$ ⎫
    individual variables $x_0, x_1, \ldots$ ⎬ *individual terms*

    time name symbols $b_0, b_1, \ldots$ ⎫
    time variables $t_0, t_1, \ldots$ ⎬ *time terms*

    location name symbols $e_0, e_1, \ldots$ ⎫
    location variables $l_0, l_1, \ldots$ ⎬ *location terms*

    internal logical connectives

        term conjoiner $\wedge$

        predicate conjoiner $+$

        modifier conjoiner $\&$

        predicate disjoiner $\cup$

    external logical connectives

        propositional connectives $\neg, \rightarrow, \wedge, \vee$

        quantifiers $\forall, \exists$

punctuation
> *parentheses* ( )
> *comma* ,
> (*unmarked*) *blank* —
> *time-marked blank* —$_{time}$
> *location-marked blank* —$_{location}$
> *slash* /

Individual terms and conjunctions of individual terms
  i. Each individual variable and each individual name symbol is an individual term of degree 1.
  ii. If d is an individual term or a formal conjunction of individual terms of degree n, and d' is an individual term or formal conjunction of individual terms of degree m, then $(d \wedge d')$ is a *conjunction of individual terms* of degree $n + m$.
  iii. A concatenation of symbols is a formal conjunction of individual terms iff it is a conjunction of individual terms of degree n for some $n > 1$.

Formal modifiers
  i. Every predicate restrictor symbol is a formal restrictor of degree 1.

  If R is a k-ary variable predicate restrictor symbol and $d_1, \ldots, d_k$ are individual terms or conjunctions of individual terms, then $R(d_1, \ldots, d_k)$ is a formal restrictor of degree 1.

  If d is a individual term or conjunction of individual terms, then obj (d) is a formal restrictor of degree 1.

  Every predicate negator symbol is a formal negator of degree 1.

  The logical negator ~ is a formal negator of degree 1.

  ii. If R is a formal restrictor of degree n, then (~R) is a formal restrictor of degree $n + 1$.
  iii. If R is a formal restrictor of degree n, and N is a formal negator, then (R / N) is a formal restrictor of degree $n + 1$.
  iv. If N and N´ are formal negators, then (N / N´) is a formal restrictor of degree 2.
  v. If R and R´ are formal restrictors of degree n and degree m respectively, then (R & R´) is a formal restrictor of degree $n + m$. It is a *conjunction of restrictors*.
  vi. A concatenation of symbols is a formal modifier iff it is a formal restrictor or formal negator of degree n for some $n \geq 1$.

  Formal modifiers of degree > 1 are *complex*.

Formal ordinary atomic predicates

i. If I is a *k*-ary infinitive symbol, then $I(-,\ldots,-,-_{time})$ with *k* unmarked blanks is a *formal ordinary atomic predicate* of degree 0. It is a *simple* formal atomic predicate.

If I is a *k*-ary infinitive symbol, then $I(-,\ldots,-,-_{time},-_{location})$ with *k* unmarked blanks is a *formal ordinary atomic predicate* of degree 0. It is a *simple* formal atomic predicate.

ii. If $Q(-,\ldots,-,-_{time})$ is a *k*-ary formal ordinary atomic predicate of degree *n*, then $(Q/\sim)(-,\ldots,-,-_{time})$ is a *k*-ary formal ordinary atomic predicate of degree $n+1$. It is a *pure negated predicate*.

If $Q(-,\ldots,-,-_{time},-_{location})$ is a *k*-ary formal ordinary atomic predicate of degree *n*, then $(Q/\sim)(-,\ldots,-,-_{time},-_{location})$ is a *k*-ary formal ordinary atomic predicate of degree $n+1$. It is a *pure negated predicate*.

iii. If $Q(-,\ldots,-,-_{time})$ is a *k*-ary formal ordinary atomic predicate of degree *n*, and M is a formal modifier other than $\sim$, and Q is not a pure negated predicate, then $(Q/M)(-,\ldots,-,-_{time})$ is a *k*-ary formal ordinary atomic predicate of degree $n+1$.

If $Q(-,\ldots,-,-_{time},-_{location})$ is a *k*-ary formal ordinary atomic predicate of degree *n*, and M is a formal modifier other than $\sim$, and Q is not a pure negated predicate, then $(Q/M)(-,\ldots,-,-_{time},-_{location})$ is a *k*-ary formal ordinary atomic predicate of degree $n+1$.

iv. If $Q_1(-,-_{time})$ and $Q_2(-,-_{time})$ are unary formal ordinary atomic predicates of degrees *n* and *m*, respectively, and at least one of them is not a pure negated predicate, then $(Q_1 + Q_2)(-,-_{time})$ is a unary formal ordinary atomic predicate of degree $n+m+1$. It is an *internally conjoined* formal atomic predicate.

If $Q_1(-,-_{time},-_{location})$ and $Q_2(-,-_{time},-_{location})$ are unary formal ordinary atomic predicates of degrees *n* and *m*, respectively, and at least one of them is not a pure negated predicate, then $(Q_1 + Q_2)(-,-_{time},-_{location})$ is a unary formal ordinary atomic predicate of degree $n+m+1$. It is an *internally conjoined* formal atomic predicate.

v. If $Q_1(-,-_{time})$ and $Q_2(-,-_{time})$ are unary formal ordinary atomic predicates of degrees *n* and *m*, respectively, and neither is a pure negated predicate, then $(Q_1 \cup Q_2)(-,-_{time})$ is a unary formal ordinary atomic predicate of degree $n+m+1$. It is an *internally disjoined* formal atomic predicate.

If $Q_1(-, -_{\text{time}}, -_{\text{location}})$ and $Q_2(-, -_{\text{time}}, -_{\text{location}})$ are unary formal ordinary atomic predicates of degrees n and m, respectively, and neither is a pure negated predicate, then $(Q_1 \cup Q_2)(-, -_{\text{time}}, -_{\text{location}})$ is a unary formal ordinary atomic predicate of degree $n + m + 1$. It is an *internally disjoined* formal atomic predicate.

vi. A concatenation of symbols is a formal atomic predicate iff it is a formal atomic predicate of degree n for some n.

Formal atomic predicates of degree > 0 are *modified formal atomic predicates*.

Note that we do not conjoin or disjoin a pair of predicates one of which is only time-marked and the other is time-and-space marked.

Wffs

i. If $A(-, \ldots, -, -_{\text{time}})$ is an n-ary formal ordinary atomic predicate, and $u_1, \ldots, u_n$ are individual terms, and t is a time term, then the following is an *ordinary* wff of *length* 1:

$$(A(-, \ldots, -, -_{\text{time}})(u_1, \ldots, u_n, t))$$

The terms $u_1, \ldots, u_n$ *fill* the unmarked blanks in that order, and term t *fills* the time-marked blank.

If $A(-, \ldots, -, -_{\text{time}}, -_{\text{location}})$ is an n-ary formal ordinary atomic predicate, $u_1, \ldots, u_n$ are individual terms, and t is a time term, and l is a location term, then the following is an *ordinary* wff of *length* 1:

$$(A(-, \ldots, -, -_{\text{time}}, -_{\text{location}})(u_1, \ldots, u_n, t, l))$$

The terms $u_1, \ldots, u_n$ *fill* the unmarked blanks in that order, and term t *fills* time-marked blank, and term l *fills* the location-marked blank.

ii. If t and w are time terms, each of the following is a wff of length 1:

$$((- <_{\text{time}} -)(t, w))$$
$$(W_{\text{time}}(-, -)(t, w))$$
$$((- \equiv_{\text{time}} -)(t, w))$$

iii. If l and p are location terms, each of the following is a wff of length 1:

$$(W_{\text{location}}(-, -)(l, p))$$
$$((- \equiv_{\text{location}} -)(l, p))$$

iv. If u and v are individual terms, the following is a wff of length 1:

$$((- \equiv -)(u, v))$$

v. If u is an individual term and t is a time term, the following is a wff of length 1:

$$((- \text{ to exist } -_{\text{time}})(u, t))$$

vi. If u is an individual term, t is a time term, and l is a location term, then the following is a wff of length 1:

$$((- \text{ to exist } -_{\text{time}}, -_{\text{location}})(u, t, l))$$

Each occurrence of each variable in a wff of length 1 is *free*.

vii. If A is a wff of length n, then $(\neg A)$ is a wff of length $n + 1$.
An occurrence of a variable in $(\neg A)$ is free iff it is free in A.

viii. If A and B are wffs, and the maximum of the lengths of A and B is n, then each of $(A \to B)$ and $(A \wedge B)$ and $(A \vee B)$ is a wff of length $n + 1$.
An occurrence of a variable in $(A \to B)$ is free iff the corresponding occurrence of the variable in A or in B is free, and similarly for $(A \wedge B)$ and $(A \vee B)$.

ix. If A is a wff of length n and some occurrence of an individual variable x is free in A, then each of $(\forall x A)$ and $(\exists x A)$ is a wff of length $n + 1$.
An occurrence of a variable in either $(\forall x A)$ or $(\exists x A)$ is free iff the variable is not x and the corresponding occurrence in A is free.

x. If A is a wff of length n and some occurrence of a time variable t is free in A, then each of $(\forall t A)$ and $(\exists t A)$ is a wff of length $n + 1$.
An occurrence of a variable in either $(\forall t A)$ or $(\exists t A)$ is free iff the variable is not t and the corresponding occurrence in A is free.

xi. If A is a wff of length n and some occurrence of a location variable l is free in A, then each of $(\forall l A)$ and $(\exists l A)$ is a wff of length $n + 1$.
An occurrence of a variable in either $(\forall l A)$ or $(\exists l A)$ is free iff the variable is not l and the corresponding occurrence in A is free.

A concatenation of symbols is a *wff* iff it is a wff of length n or some $n \geq 1$.

A wff of length 1 is *atomic*; all other wffs are *compound*.

A wff is *closed* if there is no occurrence of a variable free in it; otherwise it is *open*.

In $(\forall x A)$ the initial $\forall x$ has *scope* A and *binds* each free occurrence of x in A, and that occurrence is *bound* by that quantifier, and similarly for $(\exists x A)$.

In $(\forall t A)$ the initial $\forall t$ has *scope* A and *binds* each free occurrence of t in A, and that occurrence is *bound* by that quantifier, and similarly for $(\exists t A)$.

In $(\forall l A)$ the initial $\forall t$ has *scope* A and *binds* each free occurrence of l in A, and that occurrence is *bound* by that quantifier, and similarly for $(\exists l A)$.

I'll leave to you to prove the unique readability of wffs (compare the comments on p. 141 of Volume 1).

Substituting for variables

An individual term or conjunction of terms *u* is *free for an occurrence of an individual variable x* in A (that is, free to replace it) iff both:
- The occurrence of *x* is free.
- If *y* is a variable that occurs in *u*, the occurrence of *u* does not lie within the scope of some occurrence of $\forall y$ or $\exists y$.

An individual term or conjunction of terms *u* is *free for x* in A iff it is free for every free occurrence of *x* in A, and in that case we write A(*u* | *x*) to mean the wff that results from replacing every free occurrence of *x* in A with *u*.

A time term *w* is *free for an occurrence of a time variable t* in A (that is, free to replace it) iff both:
- The occurrence of *t* is free.
- If *w* is a variable, the occurrence does not lie within the scope of some occurrence of $\forall w$ or $\exists w$.

A time term *w* is *free for t* in A iff it is free for every free occurrence of *t* in A, and then we write A(*w* | *t*) to mean the wff that results from replacing every free occurrence of *t* in A with *w*.

A location term *p* is *free for an occurrence of a location variable l* in A (that is, free to replace it) iff both:
- The occurrence of *l* is free.
- If *p* is a variable, the occurrence does not lie within the scope of some occurrence of $\forall p$ or $\exists p$.

A location term *p* is *free for l* in A iff it is free for every free occurrence of *l* in A, and then we write A(*p* | *l*) to mean the wff that results from replacing every free occurrence of *l* in A with *p*.

The universal closure of a wff

Let $x_{n_1}, \ldots, x_{n_r}$ be a list of all the individual variables that occur free in A such that $n_1 < \cdots < n_r$. Let $t_{m_1}, \ldots, t_{m_s}$ be a list of all the time variables that occur free in A such that $m_1 < \cdots < m_s$. Let $l_{i_1}, \ldots, l_{i_v}$ be a list of all the time variables that occur free in A such that $i_1 < \cdots < i_v$. The *universal closure* of A is:

$$\forall \ldots A \equiv_{\text{Def}} \forall x_{n_1} \cdots \forall x_{n_n} \forall t_{m_1} \cdots \forall t_{m_s} \forall l_{i_1} \cdots \forall l_{i_v} A$$

Abbreviations

A, B, C, A_0, A_1, B_0, B_1, C_0, C_1, D_0, D_1, ... stand for wffs.

$(u \equiv v)$ stands for $((- \equiv -)(u,v))$

$(t \equiv_{\text{time}} w)$ stands for $((- \equiv_{\text{time}} -)(t,w))$

$(l \equiv_{\text{location}} p)$ stands for $((- \equiv_{\text{location}} -)(l,p))$

Chapter 44

$(u \not\equiv v)$ stands for $\neg((-\equiv-)(u,v))$

$(t \not\equiv_{time} w)$ stands for $\neg((-\equiv_{time}-)(t,w))$

$(l \not\equiv_{location} p)$ stands for $\neg((-\equiv_{location}-)(l,p))$

$(t <_{time} w)$ stands for $((-<_{time}-)(t,w))$

$(W_{time}(t,w))$ stands for $(W_{time}(-,-)(t,w))$

$(t \leq_{time} w)$ stands for $(t <_{time} w) \vee (t =_{time} w)$

$(w_1 \leq_{time} w_2 \leq_{time} w_3)$ stands for $(w_1 \leq_{time} w_2) \wedge (w_2 \leq_{time} w_3)$
                               and similarly for $<$

$(W_{location}(t,w))$ stands for $(W_{location}(-,-)(t,w))$

$\sim Q$ stands for $Q/\sim$

$O(w,t)$ stands for $\exists t((W_{time}(t,w) \wedge W_{time}(t,w))$

$O_<(w_1,w_2)$ stands for $\exists t((W_{time}(t,w_1) \wedge W_{time}(t,w_2))$
                      $\wedge \; \exists t'(W_{time}(t',w_1) \wedge$
                           $\forall w (W_{time}(w,w_1) \rightarrow t' <_{time} w)))$

We use the same conventions for deleting parentheses as in classical predicate logic, adding that $\forall x, \exists x, \forall t, \exists t, \forall l, \exists l$ bind equally strongly.

Realizations and semi-formal languages

An ordinary language name, or predicate, or restrictor, or variable restrictor, or negator is *simple* iff it contains no part we could formalize as a name, predicate, propositional connective, variable, quantifier, internal conjunction, internal disjunction, predicate, predicate modifier, or a combination of those. A *simple infinitive* is a simple predicate with the verb in it converted to an infinitive; it has the same arity as the predicate from which it is derived.

A *realization* of the formal language is an assignment of:

- Simple names to none, some, or all of the individual, time, or location name symbols.
- Simple infinitives to none, some, or all of the infinitive symbols.
- Simple non-variable restrictors to none, some, or all of the non-variable restrictor symbols.
- Simple variable restrictors to none, some, or all of the variable restrictor symbols.
- Simple negators to none, some, or all of the negator symbols.

The *realization of a formal wff* is the formula we get when we replace the formal symbols with the parts of ordinary language that are assigned to them; it is a *semi-formal wff*. The *semi-formal language* for a realization is the collection of realizations of formal wffs. Parts of the semi-formal language inherit the terminology applied to parts of the formal language with the word "formal" deleted.

The simple names, predicates, and modifiers of the realization are *categorematic*. The rest of the vocabulary is *logical* or *syncategorematic*. Categorematic parts of the formal language joined by punctuation or logical vocabulary are categorematic.

The *pure language of space and time* is the language in which no symbol is realized. The *pure language of space* is the pure language of space and time without time terms and without the predicates W_{time}, $<_{time}$, and $\equiv_{time}$.

45 Semantics for a Formal Theory for Reasoning about Things in Time and Space

Universes

For a realization we adopt three universes:

A universe U which is a non-empty collection of objects called *physical things*.

A universe T which is a non-empty collection of things called *times*.

A universe P which is a non-empty collection of things called *locations*.

No thing is in more than one of U, T, and P.

Relations on the universes

There is a binary relation $<$ on T.

There is a binary relation W_{time} on T.

There is a binary relation $W_{location}$ on P.

$O(w_1, w_2) \equiv_{Def}$ there is some t such that $W_{time}(t, w_1)$ and $W_{time}(t, w_2)$.

These relations satisfy the following conditions:

W_{time} *is a part-whole relation*

$\quad W_{time}(t, t)$

$\quad$ If $W_{time}(t, w)$ and $W_{time}(w, t)$, then $t = w$.

$\quad$ If $W_{time}(t_1, t_2)$ and $W_{time}(t_2, t_3)$, then $W_{time}(t_1, t_3)$.

Parts determine times

$\quad t = t'$ iff (for all w, $W(w, t)$ iff $W(w, t')$)

$<$ *is an ordering*

$\quad$ Not $t < t$.

$\quad$ If $t_1 < w$ and $w < t_2$, then $t_1 < t_2$.

Parts and wholes are unrelated in the ordering

$\quad$ If $W_{time}(t_1, t_2)$, then neither $t_1 < t_2$ nor $t_2 < t_1$.

Parts of times are related to other times in the ordering as the whole is related

$\quad$ If $W_{time}(t_1, t_2)$ and $t_2 < t_3$, then $t_1 < t_3$.

$\quad$ If $W_{time}(t_1, t_2)$ and $t_3 < t_2$, then $t_3 < t_1$.

Overlapping times are not related in the ordering

$\quad$ If $O(w_1, w_2)$, then neither $w_1 < w_2$ nor $w_2 < w_1$.

Times are intervals (sequentially connected)
    If $W_{time}(t_1, w)$ and $W_{time}(t_2, w)$, then $O(t_1, t_2)$ or $t_1 < t_2$ or $t_2 < t_2$.

    If $W_{time}(t_1, w)$ and $W_{time}(t_2, w)$ and $t_1 < t_2$, then
        for every t_3 such that $t_1 < t_3 < t_2$, $W_{time}(t_3, w)$.

$W_{location}$ *is a part-whole relation*
    $W_{location}(t, t)$
    If $W_{location}(t, w)$ and $W_{location}(w, t)$, then $t = w$.
    If $W_{location}(t_1, t_2)$ and $W_{location}(t_2, t_3)$ then $W_{location}(t_1, t_3)$.

Parts determine locations
    $p = p'$ iff (for all p_1, $W_{location}(p_1, p)$ iff $W_{location}(p_1, p')$)

Assignments of references

An assignment of references σ assigns:

    To each individual variable x an object σ(x) from U.

    To each individual name c an object σ(c) from U such that for every assignment of references τ, σ(c) = τ(c).

    To each time variable t a time σ(t) from T.

    To each time name b a time σ(b) from T such that for every assignment of references τ, σ(b) = τ(b).

    To each location variable l a location σ(l) from P.

    To each location name e a location σ(e) from P such that for every assignment of references τ, σ(e) = τ(e).

Completeness of the collection of assignments of references
There is at least one assignment of references.

For every assignment of references σ, and every individual variable x, and every object in U, either σ assigns that object to x or there is an assignment τ that differs from σ only in that it assigns that object to x.

    We write $\tau \sim_x \sigma$ to mean that τ differs from σ at most in what it assigns x.

For every assignment of references σ, and every time variable t, and every time in T, either σ assigns that time to t or there is an assignment τ that differs from σ only in that it assigns that time to t.

    We write $\tau \sim_t \sigma$ to mean that τ differs from σ at most in what it assigns t.

For every assignment of references σ, and every location variable l, and every location in P, either σ assigns that location to l or there is an assignment τ that differs from σ only in that it assigns that location to l.

    We write $\tau \sim_l \sigma$ to mean that τ differs from σ at most in what it assigns l.

Satisfaction of atomic wffs

$A(u_1, \ldots, u_n, t)$ stands for an ordinary atomic wff that contains no location-marked blank, and $A(u_1, \ldots, u_n, t, l)$ is the same wff with a location-marked blank added.

For each assignment of references σ, there is a valuation ν_σ that assigns to each atomic wff a truth-value T or F.

We write $\nu_\sigma(A(u_1, \ldots, u_n, t)) = T$ or $\nu_\sigma \vDash A(u_1, \ldots, u_n, t)$ to mean that A is *true of* or *applies to the objects* $\sigma(u_1), \ldots, \sigma(u_n)$ in that order *at time* $\sigma(t)$.

We write $\nu_\sigma(A(u_1, \ldots, u_n, t, l)) = T$ or $\nu_\sigma \vDash A(u_1, \ldots, u_n, t, l)$ to mean that $A(u_1, \ldots, u_n, t, l)$ is *true of* or *applies to the objects* $\sigma(u_1), \ldots, \sigma(u_n)$ in that order *at time* $\sigma(t)$ *in location* $\sigma(l)$.

The valuations of the atomic wffs satisfy the following conditions.

$\nu_\sigma \vDash (w <_{time} t)$ iff $\sigma(w) < \sigma(t)$.

$\nu_\sigma \vDash W_{time}(w, t)$ iff $W_{time}(\sigma(w), \sigma(t))$.

$\nu_\sigma \vDash W_{location}(l, p)$ iff $W_{location}(\sigma(w), \sigma(t))$.

$\nu_\sigma \vDash u \equiv v$ iff $\sigma(u)$ is the same physical thing as $\sigma(v)$.

$\nu_\sigma \vDash w \equiv_{time} t$ iff $\sigma(w)$ is the same time as $\sigma(t)$.

$\nu_\sigma \vDash l \equiv_{location} p$ iff $\sigma(l)$ is the same location as $\sigma(p)$.

The extensionality condition

Let $u_1, \ldots, u_n$ be the individual terms that appear in A (if any), w_1, w_2 be the time terms in A (if any), and p_1, p_2 the location terms in A (if any). For any σ and τ and individual terms $v_1, \ldots, v_n$ and time terms w'_1, w'_2, and location terms p'_1, p'_2, if for all i, $\sigma(u_i) = \tau(v_i)$, and $\sigma(w_1) = \tau(w'_1)$, and $\sigma(w_2) = \tau(w'_2)$, and $\sigma(p_1) = \tau(p'_1)$, and $\sigma(p_2) = \tau(p'_2)$, then $\nu_\sigma \vDash A$ iff

$\nu_\tau \vDash A(v_1/u_1, \ldots, v_n/u_n, w'_1/w_1, w'_2/w_2, p'_1/p_1, p'_2/p_2)$

Downward and upward closure of truth in time

$\nu_\sigma \vDash A(x_1, \ldots, x_n, t)$ iff

for all τ such that $\tau \sim_w \sigma$, if $\nu_\tau \vDash W_{time}(w, t)$, then $\nu_\tau \vDash A(w/t)$.

Things exist in time

For every σ there is a $\tau \sim_t \sigma$ such that $\nu_\tau \vDash (-\text{ to exist } -_{time})(x, t)$.

Downward and upward closure of existence in time

$\nu_\sigma \vDash (-\text{ to exist } -)(x, t)$ iff

for all $\tau \sim_w \sigma$, if $\nu_\tau \vDash W_{time}(w, t)$, then $\nu_\tau \vDash (-\text{ to exist } -)(x, w)$.

Ordinary atomic predicates are positive for existence in time

If $\nu_\sigma \vDash A(x_1, \ldots, x_n, t)$, then for each i,

$\nu_\sigma \vDash (-\text{ to exist } -_{time})(x_i, t)$.

Continuity in time of when a thing exists
If $\upsilon_\sigma \vDash (-\text{ to exist }-)(x, t_1)$ and $\upsilon_\sigma \vDash (-\text{ to exist }-)(x, t_2)$, then
  i. there is some τ such that $\tau \sim_{t_3} \sigma$ and $\upsilon_\tau \vDash (t_3 \leq_{\text{time}} t_1)$
     and $\upsilon_\tau \vDash (t_3 \leq_{\text{time}} t_2)$, and $\upsilon_\tau \vDash (-\text{ to exist }-)(x, t_3)$, and
  ii. if τ is such that $\tau \sim_{t_3} \sigma$ and $\upsilon_\tau \vDash (t_1 <_{\text{time}} t_3 <_{\text{time}} t_2)$,
     then $\upsilon_\tau \vDash (-\text{ to exist }-)(x, t_3)$.

Outward closure of truth for locations
If $\upsilon_\sigma \vDash A(x_1, \ldots, x_n, t, l)$, then for all τ such that $\tau \sim_p \sigma$,
  if $\upsilon_\tau \vDash W_{\text{location}}(l, p)$, then $\upsilon_\tau \vDash A(x_1, \ldots, x_n, t, p/l)$.

True at a time and location, then true at that time
If $\upsilon_\sigma \vDash A(x_1, \ldots, x_n, t, l)$, then $\upsilon_\sigma \vDash A(x_1, \ldots, x_n, t)$.

Downward and upward closure of truth in time at a location
$\upsilon_\sigma \vDash A(x_1, \ldots, x_n, t, l)$ iff for all τ such that $\tau \sim_w \sigma$,
  if $\upsilon_\tau \vDash W_{\text{time}}(w, t)$, then $\upsilon_\tau \vDash A(x_1, \ldots, x_n, w/t, l)$.

Exists at a time, then exists at some location at that time
If $\upsilon_\sigma \vDash (-\text{ to exist }-_{\text{time}})(x, t)$, then there is some $\tau \sim_l \sigma$
  such that $\upsilon_\tau \vDash (-\text{ to exist }-_{\text{time}}, -_{\text{location}})(x, t, l)$.

Exists at a time and location, then exists at that time
If $\upsilon_\sigma \vDash (-\text{ to exist }-_{\text{time}}, -_{\text{location}})(x, t, l)$,
  then $\upsilon_\sigma \vDash (-\text{ to exist }-_{\text{time}})(x, t)$.

Outward closure of existence in space
If $\upsilon_\sigma \vDash (-\text{ to exist }-_{\text{time}}, -_{\text{location}})(x, t, l)$, then for all $\tau \sim_p \sigma$,
  if $\upsilon_\tau \vDash W_{\text{location}}(l, p)$, then $\upsilon_\tau \vDash (-\text{ to exist }-_{\text{time}}, -_{\text{location}})(x, t, l)$.

Downward and upward closure of existence in time at a location
$\upsilon_\sigma \vDash (-\text{ to exist }-_{\text{time}}, -_{\text{location}})(x, t, l)$ iff for all $\tau \sim_t \sigma$,
  if $\upsilon_\tau \vDash W_{\text{time}}(w, t)$, then $\upsilon_\tau \vDash (-\text{ to exist }-_{\text{time}}, -_{\text{location}})(x, t, l)$.

An ordinary atomic predicate $A(x_1, \ldots, x_n, t)$ is *locational* $\equiv_{\text{Def}}$ if A is true of $\mathsf{o}_1, \ldots, \mathsf{o}_n$ at t, then there is some location p such that $A(x_1, \ldots, x_n, t, l)$ is true of $\mathsf{o}_1, \ldots, \mathsf{o}_n$ at t in location p, and there is some p′ such that $A(x_1, \ldots, x_n, t, l)$ is not true of $\mathsf{o}_1, \ldots, \mathsf{o}_n$ at t in location p′.

Locational predicates are positive for existence in time and location
If A is locational and $\upsilon_\sigma \vDash A(x_1, \ldots, x_n, t, l)$, then for each i,
$\upsilon_\sigma \vDash (-\text{ to exist }-_{\text{time}}, -_{\text{location}})(x_i, t, l)$.

Locations distinguish physical things

If $\upsilon_\sigma \vDash (x \not\equiv y)$ and

$\upsilon_\tau \vDash (-\text{ to exist }-_{\text{time}})(x, t)$ and $\upsilon_\tau \vDash (-\text{ to exist }-_{\text{time}})(y, t)$,

then there is some assignment of references γ such that $\gamma \sim_l \tau$ and either:

$\upsilon_\gamma \vDash (-\text{ to exist }-_{\text{time}}, -_{\text{location}})(x, t, l)$
and $\upsilon_\gamma \vDash \neg (-\text{ to exist }-_{\text{time}}, -_{\text{location}})(y, t, l)$,

or

$\upsilon_\gamma \vDash \neg (-\text{ to exist }-_{\text{time}}, -_{\text{location}})(x, t, l)$
and $\upsilon_\gamma \vDash (-\text{ to exist }-_{\text{time}}, -_{\text{location}})(y, t, l)$.

The valuations of atomic wffs also satisfy the following conditions governing the internal structure of atomic predicates, where R, R', R_1, R_2, R_3 are restrictors or variable restrictors; N, N' are negators; $d, d', d_1, \ldots, d_n$ are individual terms or conjunctions of individual terms; $\vec{d}$ is a sequence of individual terms or sequences of conjunctions of individual terms of the appropriate length for the arity of the predicate or restrictor; P is a simple atomic predicate; $Q, Q_1, Q_2, \ldots$ are atomic predicates; $A, B, C, A_0, A_1, \ldots$ are atomic wffs; and A(X) means the part of language X appears in A.

RES If $\upsilon_\sigma \vDash (Q/R)(\vec{d}, t)$, then $\upsilon_\sigma \vDash Q(\vec{d}, t)$.
If $\upsilon_\sigma \vDash (Q/R)(\vec{d}, t, l)$, then $\upsilon_\sigma \vDash Q(\vec{d}, t, l)$.

Neg If $\upsilon_\sigma \vDash (Q/N)(\vec{d}, t)$, then $\upsilon_\sigma \nvDash Q(\vec{d}, t)$.
If $\upsilon_\sigma \vDash (Q/N)(\vec{d}, t, l)$, then $\upsilon_\sigma \nvDash Q(\vec{d}, t, l)$.

RN   If $\upsilon_\sigma \vDash (Q/(R/N))(\vec{d}, t)$, then $\upsilon_\sigma \nvDash (Q/R)(\vec{d}, t)$.
If $\upsilon_\sigma \vDash (Q/(R/N))(\vec{d}, t, l)$, then $\upsilon_\sigma \nvDash (Q/R)(\vec{d}, t, l)$.

NN' If $\upsilon_\sigma \vDash (Q/(N/N'))(\vec{d}, t)$, then $\upsilon_\sigma \nvDash (Q/N)(\vec{d}, t)$.
If $\upsilon_\sigma \vDash (Q/(N/N'))(\vec{d}, t, l)$, then $\upsilon_\sigma \nvDash (Q/N)(\vec{d}, t, l)$.

Pure negated restrictors

If $\upsilon_\sigma \vDash (Q/\sim R)(\vec{d}, t)$, then $\upsilon_\sigma \vDash Q(\vec{d}, t)$ and $\upsilon_\sigma \nvDash (Q/R)(\vec{d}, t)$.

If $\upsilon_\sigma \vDash (Q/\sim R)(\vec{d}, t, l)$, then $\upsilon_\sigma \vDash Q(\vec{d}, t, l)$ and $\upsilon_\sigma \nvDash (Q/R)(\vec{d}, t, l)$.

Negators of restrictors and pure negated restrictors

If $\upsilon_\sigma \vDash (Q/(R/N))(\vec{d}, t)$, then $\upsilon_\sigma \vDash (Q/\sim R)(\vec{d}, t)$.

If $\upsilon_\sigma \vDash (Q/(R/N))(\vec{d}, t, l)$, then $\upsilon_\sigma \vDash (Q/\sim R)(\vec{d}, t, l)$.

Commutativity of $\wedge$

If $\upsilon_\sigma \vDash A(d \wedge d', t)$, then $\vDash A(d' \wedge d, t)$
where $d \wedge d'$ appears in A and $d' \wedge d$ replaces some but not necessarily all of those appearances.

If $\vee_\sigma \vDash A(d \wedge d', t, l)$, then $\vDash A(d' \wedge d, t, l)$
>where $d \wedge d'$ appears in A and $d' \wedge d$ replaces some but not necessarily all of those appearances.

Associativity of $\wedge$

If $\vee_\sigma \vDash A(d_1 \wedge d_2) \wedge d_3, t)$, then $\vee_\sigma \vDash A(d_1 \wedge (d_2 \wedge d_3), t))$
>where $(d_1 \wedge d_2) \wedge d_3$ appears in A and $d_1 \wedge (d_2 \wedge d_3)$ replaces some but not necessarily all of those appearances.

If $\vee_\sigma \vDash A(d_1 \wedge d_2) \wedge d_3, t, l)$, then $\vee_\sigma \vDash A(d_1 \wedge (d_2 \wedge d_3), t, l))$
>where $(d_1 \wedge d_2) \wedge d_3$ appears in A and $d_1 \wedge (d_2 \wedge d_3)$ replaces some but not necessarily all of those appearances.

$\wedge$ *implies* $\wedge$

If $\vee_\sigma \vDash P(d, t)$ and u is a term that appears in d, then $\vee_\sigma \vDash P(u, t)$.

If $\vee_\sigma \vDash P(d, t, l)$ and u is a term that appears in d, then $\vee_\sigma \vDash P(u, t, l)$.

Duplicated reference in conjoined terms

If d is a conjunction of terms in which both x and y appear and $\vee_\sigma \vDash x \equiv y$,
>then $\vee_\sigma \nvDash A(d, t)$ and $\vee_\sigma \nvDash A(d, t, l)$

Commutativity of $+$

If $\vee_\sigma \vDash A(Q_1 + Q_2)$, then $\vee_\sigma \vDash A(Q_2 + Q_1)$
>where $(Q_1 + Q_2)$ appears in A, and $(Q_2 + Q_1)$ replaces some but not necessarily all of those appearances.

Associativity of $+$

If $\vee_\sigma \vDash A((Q_1 + Q_2) + Q_3)$, then $\vee_\sigma \vDash A(Q_1 + (Q_2 + Q_3))$
>where $(Q_1 + Q_2) + Q_3$ appears in A, and $Q_1 + (Q_2 + Q_3)$ replaces some but not necessarily all of those appearances.

Restrictors and $+$

If $\vee_\sigma \vDash (Q_1 + (Q_2/R)) (d, t)$, then $\vee_\sigma \vDash (Q_1 + Q_2) (d, t)$.

If $\vee_\sigma \vDash (Q_1 + (Q_2/R)) (d, t)$, then $\vee_\sigma \vDash (Q_1 + Q_2) (d, t, l)$.

$+$ *implies* $\wedge$

If $\vee_\sigma \vDash (Q_1 + Q_2) (d, t)$, then $\vee_\sigma \vDash Q_1(d, t)$ and $\vee_\sigma \vDash Q_2(d, t)$.

If $\vee_\sigma \vDash (Q_1 + Q_2) (d, t, l)$, then $\vee_\sigma \vDash Q_1(d, t, l)$ and $\vee_\sigma \vDash Q_2(d, t, l)$.

Commutativity of &

If $\vee_\sigma \vDash A(R_1 \, \& \, R_2)$ iff $\vee_\sigma \vDash A(R_2 \, \& \, R_1)$
>where $R_1 \, \& \, R_2$ appears in A and $R_2 \, \& \, R_1$ replaces some but not necessarily all of those appearances.

Associativity of &

If $\nu_\sigma \vDash A((R_1 \& R_2) \& R_3)$, then $\nu_\sigma \vDash A(R_1 \& (R_2 \& R_3))$
 where $(R_1 \& R_2) \& R_3$ appears in A and $R_1 \& (R_2 \& R_3)$ replaces some but not necessarily all of those appearances.

& is equivalent to time-indexed $\wedge$

$\nu_\sigma \vDash (Q/(R \& R'))(\vec{d}, t)$ iff $\nu_\sigma \vDash (Q/R)(\vec{d}, t)$ and $\nu_\sigma \vDash (Q/R')(\vec{d}, t)$

$\nu_\sigma \vDash (Q/(R \& R'))(\vec{d}, t, l)$ iff $\nu_\sigma \vDash (Q/R)(\vec{d}, t, l)$ and $\nu_\sigma \vDash (Q/R')(\vec{d}, t, l)$

Negators and the pure negator in joined restrictors

If $\nu_\sigma \vDash (A/(R_1 \& (R_2/N)))(d, t)$, then $\nu_\sigma \vDash (A/(R_1 \& \sim R_2))(d, t)$

If $\nu_\sigma \vDash (A/(R_1 \& (R_2/N)))(d, t, l)$, then $\nu_\sigma \vDash (A/(R_1 \& \sim R_2))(d, t, l)$

Commutativity of $\cup$

If $\nu_\sigma \vDash A(Q_1 \cup Q_2)$, then $\nu_\sigma \vDash A(Q_2 \cup Q_1)$
 where $(Q_1 \cup Q_2)$ appears in A, and $(Q_2 \cup Q_1)$ replaces some but not necessarily all of those appearances.

Associativity of $\cup$

If $\nu_\sigma \vDash A((Q_1 \cup Q_2) \cup Q_3)$, then $\nu_\sigma \vDash A(Q_1 \cup (Q_2 \cup Q_3))$
 where $(Q_1 \cup Q_2) \cup Q_3$ appears in A, and $Q_1 \cup (Q_2 \cup Q_3)$ replaces some but not necessarily all of those appearances.

Equivalence of $\cup$ and $\vee$

$\nu_\sigma \vDash (Q_1 \cup Q_2)(d, t)$ iff $\nu_\sigma \vDash Q_1(d, t)$ or $\nu_\sigma \vDash Q_2(d, t)$.

$\nu_\sigma \vDash (Q_1 \cup Q_2)(d, t, l)$ iff $\nu_\sigma \vDash Q_1(d, t, l)$ or $\nu_\sigma \vDash Q_2(d, t, l)$.

$\cup$ implies $\vee$ with restrictors

If $\nu_\sigma \vDash ((Q_1 \cup Q_2)/R)(d, t)$,
 then $\nu_\sigma \vDash (Q_1/R))(d, t)$ or $\nu_\sigma \vDash (Q_2/R))(d, t)$.

If $\nu_\sigma \vDash ((Q_1 \cup Q_2)/R)(d, t, l)$,
 then $\nu_\sigma \vDash (Q_1/R))(d, t, l)$ or $\nu_\sigma \vDash (Q_2/R))(d, t, l)$.

Restrictors and $\cup$

If $\nu_\sigma \vDash ((Q_1 \cup Q_2)/R)(d, t)$ and $\nu_\sigma \vDash Q_1(d, t)$,
 then $\nu_\sigma \vDash (Q_1/R)(d, t)$.

If $\nu_\sigma \vDash ((Q_1 \cup Q_2)/R)(d, t, l)$ and $\nu_\sigma \vDash Q_1(d, t, l)$,
 then $\nu_\sigma \vDash (Q_1/R)(d, t, l)$.

The pure negator and negation

$\nu_\sigma \vDash (Q/\sim R)(\vec{d}, t)$ iff $\nu_\sigma \vDash Q(\vec{d}, t)$ and $\nu_\sigma \nvDash (Q/R)(\vec{d}, t)$

$\upsilon_\sigma \vDash (Q/\sim R)(\vec{d}, t, l)$ iff $\upsilon_\sigma \vDash Q(\vec{d}, t, l)$ and $\upsilon_\sigma \nvDash (Q/R)(\vec{d}, t, l)$

$\upsilon_\sigma \vDash \sim Q(\vec{d}, t)$ iff $\upsilon_\sigma \nvDash Q(\vec{d}, t)$ and $\upsilon_\sigma \vDash (- \text{ to exist } -_{\text{time}})(\vec{d}, t)$

$\upsilon_\sigma \vDash \sim Q(\vec{d}, t, l)$ iff $\upsilon_\sigma \nvDash Q(\vec{d}, t, l)$ and $\upsilon_\sigma \vDash (- \text{ to exist } -_{\text{time}})(\vec{d}, t, l)$

If $\upsilon_\sigma \vDash (Q/N)(\vec{d}, t)$, then $\upsilon_\sigma \vDash \sim Q(\vec{d}, t)$.

If $\upsilon_\sigma \vDash (Q/N)(\vec{d}, t, l)$, then $\upsilon_\sigma \vDash \sim Q(\vec{d}, t, l)$

Satisfaction of compound wffs

The valuations for all assignments of wffs are extended to all wffs simultaneously:

$\upsilon_\sigma(\neg A) = T$	iff	$\upsilon_\sigma(A) = F$
$\upsilon_\sigma(A \to B) = T$	iff	$\upsilon_\sigma(A) = F$ or $\upsilon_\sigma(B) = T$
$\upsilon_\sigma(A \wedge B) = T$	iff	$\upsilon_\sigma(A) = T$ and $\upsilon_\sigma(B) = T$
$\upsilon_\sigma(A \vee B) = T$	iff	$\upsilon_\sigma(A) = T$ or $\upsilon_\sigma(B) = T$
$\upsilon_\sigma(\exists x\, A) = T$	iff	for some τ such that $\tau \sim_x \sigma$, $\upsilon_\tau(A) = T$
$\upsilon_\sigma(\forall x\, A) = T$	iff	for every τ such that $\tau \sim_x \sigma$, $\upsilon_\tau(A) = T$
$\upsilon_\sigma(\exists t\, A) = T$	iff	for some τ such that $\tau \sim_t \sigma$, $\upsilon_\tau(A) = T$
$\upsilon_\sigma(\forall t\, A) = T$	iff	for every τ such that $\tau \sim_t \sigma$, $\upsilon_\tau(A) = T$
$\upsilon_\sigma(\exists l\, A) = T$	iff	for some τ such that $\tau \sim_l \sigma$, $\upsilon_\tau(A) = T$
$\upsilon_\sigma(\forall l\, A) = T$	iff	for every τ such that $\tau \sim_l \sigma$, $\upsilon_\tau(A) = T$

The *valuation* υ is defined on all closed wffs A by: $\upsilon(A) = T$ iff every σ, $\upsilon_\sigma(A) = T$.

Models

A *model* M is a realization, universe of individuals, universe of times, universe of locations, complete collection of assignments of references, valuations for atomic wffs satisfying the conditions above, extension of the valuations to all wffs by the inductive definition, and the valuation on all closed wffs.

A proposition B of the semi-formal language is *true in the model* iff $\upsilon(A) = T$, in which case we write $M \vDash B$. It is *false in the model* otherwise, and we write $M \nvDash B$.

A formal wff A is *valid* or a *tautology* iff in every model its realization is true; in that case we write $\vDash A$. The formal inference Γ therefore A is *valid*, written $\Gamma \vDash A$, means that there is no model in which the realizations of all the wffs in Γ are true and the realization of A is false. These definitions are extended to semi-formal wffs via formal wffs of which they are realizations.

Sufficiency of the collection of models

Given any realization, collection of things in time, collection of times, collection of locations, complete collection of assignments of references, and assignment of truth-values to the atomic predications satisfying the conditions above they together constitute a model.

Classical predicate logic with quantifying over times and locations
The formal language, definitions of models, tautology, and semantic consequence constitute *classical predicate logic with quantifying over times and locations*, **QTS**.

I'll leave to you to define a model of what is true at a particular time and location (compare the Aside on p. 105).

46 An Axiom System for QTS

We adopt all the axioms of **QT** from Chapter 21 except that A, B, C stand for any wffs of the language of **QTS**. Then we add the following axioms.

Axioms governing $\forall$

1. a. $\forall \ldots (\forall l\, (A \to B) \to (\forall l\, A \to \forall l\, B))$
    if l is free in both A and B
   b. $\forall \ldots (\forall l\, (A \to B) \to (\forall l\, A \to B))$
    if l is free in A and not free in B
   c. $\forall \ldots (\forall l\, (A \to B) \to (A \to \forall l\, B))$
    if l is free in B and not free in A

2. $\forall \ldots (\forall l\, \forall p\, A \to \forall p\, \forall l\, A)$

3. $\forall \ldots (\forall l\, A(l) \to A(p/l))$
    where p is free for l in A

4. $\forall \ldots (\forall x\, \forall l\, A \to \forall l\, \forall x\, A)$

5. $\forall \ldots (\forall l\, \forall x\, A \to \forall x\, \forall l\, A)$

6. $\forall \ldots (\forall t\, \forall l\, A \to \forall l\, \forall t\, A)$

7. $\forall \ldots (\forall l\, \forall t\, A \to \forall t\, \forall l\, A)$

Axioms governing the relation between $\forall$ and $\exists$

8. a. $\forall \ldots (\exists l\, A \to \neg \forall l\, \neg A)$
   b. $\forall \ldots (\neg \forall l\, \neg A \to \exists l\, A)$

Axioms for equality and extensionality for location terms

9. $\forall l\, (l \equiv_{location} l)$

10. $\forall \ldots \forall l\, \forall p\, (l \equiv_{location} p \to (A(l) \to A(p/l)))$
    where A is atomic and p replaces some
    but not necessarily all occurrences of l in A

Axioms for space

$W_{location}$ *is a part-whole relation*

$\forall l\, W_{location}(l, l)$

$\forall l\, \forall p\, (W_{location}(l, p) \land W_{location}(p, l) \to (l \equiv_{location} p))$

$\forall l_1\, \forall l_2\, \forall l_3\, (W_{location}(l_1, l_2) \land W_{location}(l_2, l_3) \to W_{location}(l_1, l_3))$

Parts determine locations

$\forall l_1\, \forall l_2\, ((l_1 \equiv_{location} l_2) \leftrightarrow \forall l_3\, (W_{location}(l_3, l_1) \leftrightarrow W_{location}(l_3, l_2)))$

Outward closure of truth for locations
$\forall \ldots \forall l \, (A(y_1, \ldots, y_n, t, l) \rightarrow (\forall p \, (W_{\text{location}}(l, p) \rightarrow A(y_1, \ldots, y_n, t, p/l)$

Downward and upward closure of truth in time in a location
$\forall \ldots (A(x_1, \ldots, x_n, t, l) \leftrightarrow \forall w \, W_{\text{time}}(w, t) \rightarrow A(x_1, \ldots, x_n, w/t, l)$

True at a time and location, then true at that time
$\forall \ldots (A(x_1, \ldots, x_n, t, l) \rightarrow A(x_1, \ldots, x_n, t))$

Exists at a time, then exists at some location at that time
$\forall x \, \forall t \, (\, (-\text{ to exist } -_{\text{time}}) \, (x, t) \rightarrow \exists l \, (-\text{ to exist } -_{\text{time}}, -_{\text{location}}) \, (x, t, l) \,)$

Exists at a time and location, then exists at that time
$\forall x \, \forall t \, \forall l \, (\, (-\text{ to exist } -_{\text{time}}, -_{\text{location}}) \, (x, t, l) \rightarrow$
$\qquad (-\text{ to exist } -_{\text{time}}, -_{\text{location}}) \, (x, t) \,)$

Outward closure of existence in space
$\forall x \, \forall t \, \forall l \, (\, (-\text{ to exist } -_{\text{time}}, -_{\text{location}}) \, (x, t, l) \rightarrow$
$\qquad \forall p \, (\, (W_{\text{location}}(l, p) \rightarrow (-\text{ to exist } -_{\text{time}}, -_{\text{location}}) \, (x, t, p) \,)$

Downward and upward closure of existence in time in a location
$\forall x \, \forall t \, \forall l \, (\, (-\text{ to exist } -_{\text{time}}, -_{\text{location}}) \, (x, t, l) \leftrightarrow$
$\qquad \forall w \, (W_{\text{time}}(w, t) \rightarrow (-\text{ to exist } -_{\text{time}}, -_{\text{location}}) \, (x, t, l) \,)$

Locational predicates are positive for existence in time and location
$\forall \ldots (\text{locational } (A(x_1, \ldots, x_n, t) \rightarrow$
$\qquad (A(x_1, \ldots, x_n, t, l) \rightarrow (-\text{ to exist } -_{\text{time}}, -_{\text{location}}) \, (x, t, l) \,) \,)$

Locations distinguish physical things
$\forall x \, \forall y \, \forall t \, (\, (x \not\equiv y \wedge (-\text{ to exist } -_{\text{time}}) \, (x, t) \wedge (-\text{ to exist } -_{\text{time}}) \, (y, t) \,) \rightarrow$
$\qquad (\exists l \, (-\text{ to exist } -_{\text{time}}, -_{\text{location}}) \, (x, t, l) \wedge$
$\qquad \neg (-\text{ to exist } -_{\text{time}}, -_{\text{location}}) \, (y, t, l) \,) \vee$
$\qquad (\exists l \, \neg (-\text{ to exist } -_{\text{time}}, -_{\text{location}}) \, (x, t, l) \wedge$
$\qquad (-\text{ to exist } -_{\text{time}}, -_{\text{location}}) \, (y, t, l) \,) \,)$

Axioms for the internal structure of atomic predicates
We use the same notational conventions for Q, R, etc. as for the semantics on p. 202.

RES $\quad \forall \ldots (\, (Q/R) \, (\vec{d}, t) \rightarrow Q(\vec{d}, t) \,)$

$\qquad \forall \ldots (\, (Q/R) \, (\vec{d}, t, l) \rightarrow Q(\vec{d}, t, l) \,)$

Neg $\quad \forall \ldots (\, (Q/N) \, (\vec{d}, t) \rightarrow Q(\vec{d}, t) \,)$

$\qquad \forall \ldots (\, (Q/N) \, (\vec{d}, t, l) \rightarrow Q(\vec{d}, t, l) \,)$

An Axiom System for **QTS** 209

RN $\forall \ldots ((Q/(R/N))(\vec{d},t) \rightarrow (Q/R)(\vec{d},t))$
 $\forall \ldots ((Q/(R/N))(\vec{d},t,l) \rightarrow (Q/R)(\vec{d},t,l))$

NN' $\forall \ldots ((Q/(N/N'))(\vec{d},t) \rightarrow (Q/N)(\vec{d},t))$
 $\forall \ldots ((Q/(N/N'))(\vec{d},t,l) \rightarrow (Q/N)(\vec{d},t,l))$

Pure negated restrictors
$\forall \ldots ((Q/\sim R)(\vec{d},t) \rightarrow (Q(\vec{d},t) \wedge \neg (Q/R)(\vec{d},t)))$
$\forall \ldots ((Q/\sim R)(\vec{d},t,l) \rightarrow (Q(\vec{d},t,l) \wedge \neg (Q/R)(\vec{d},t,l)))$

Negators of restrictors and pure negated restrictors
$\forall \ldots ((Q/(R/N))(\vec{d},t) \rightarrow (Q/\sim R)(\vec{d},t))$
$\forall \ldots ((Q/(R/N))(\vec{d},t,l) \rightarrow (Q/\sim R)(\vec{d},t,l))$

Commutativity of $\wedge$
$\forall \ldots (A(d \wedge d', t) \rightarrow A(d' \wedge d, t))$
 where $d \wedge d'$ appears in A and $d' \wedge d$ replaces some but not necessarily all of those appearances.
$\forall \ldots (A(d \wedge d', t, l) \wedge A(d' \wedge d, t, l))$
 where $d \wedge d'$ appears in A and $d' \wedge d$ replaces some but not necessarily all of those appearances.

Associativity of $\wedge$
$\forall \ldots (A((d_1 \wedge d_2) \wedge d_3, t) \rightarrow A(d_1 \wedge (d_2 \wedge d_3), t)))$
 where $(d_1 \wedge d_2) \wedge d_3$ appears in A and $d_1 \wedge (d_2 \wedge d_3)$ replaces some but not necessarily all of those appearances.
$\forall \ldots (A((d_1 \wedge d_2) \wedge d_3, t, l) \rightarrow A(d_1 \wedge (d_2 \wedge d_3), t, l)))$
 where $(d_1 \wedge d_2) \wedge d_3$ appears in A and $d_1 \wedge (d_2 \wedge d_3)$ replaces some but not necessarily all of those appearances.

$\wedge$ *implies* $\wedge$
$\forall \ldots (P(d,t) \rightarrow P(u,t))$ where u is a term that appears in d.
$\forall \ldots (P(d,t,l) \rightarrow P(u,t,l))$ where u is a term that appears in d.

Duplicated reference in conjoined terms
If d is a conjunction of terms in which both x and y appear
 $\forall \ldots (x \equiv y \rightarrow \neg A(d,t))$
 $\forall \ldots (x \equiv y \rightarrow \neg A(d,t,l))$

Commutativity of +
$\forall \ldots (A(Q_1 + Q_2) \rightarrow A(Q_2 + Q_1))$
 where $(Q_1 + Q_2)$ appears in A, and $(Q_2 + Q_1)$ replaces some but not necessarily all of those appearances.

Associativity of +
∀ ... (A((Q$_1$ + Q$_2$) + Q$_3$) → A(Q$_1$ + (Q$_2$ + Q$_3$)))
    where (Q$_1$ + Q$_2$) + Q$_3$ appears in A, and Q$_1$ + (Q$_2$ + Q$_3$) replaces some
    but not necessarily all of those appearances.

Restrictors and +
∀ ... ((Q$_1$ + (Q$_2$/R)) (d, t) → (Q$_1$ + Q$_2$) (d, t))
∀ ... ((Q$_1$ + (Q$_2$/R)) (d, t) → (Q$_1$ + Q$_2$) (d, t, l))

+ implies ∧
∀ ... ((Q$_1$ + Q$_2$) (d, t) → (Q$_1$(d, t) ∧ Q$_2$(d, t)))
∀ ... ((Q$_1$ + Q$_2$) (d, t, l) → (Q$_1$(d, t, l) ∧ Q$_2$(d, t, l)))

Commutativity of &
∀ ... (A(R$_1$ & R$_2$) ↔ A(R$_2$ & R$_1$))
    where R$_1$ & R$_2$ appears in A and R$_2$ & R$_1$ replaces some
    but not necessarily all of those appearances.

Associativity of &
∀ ... (A((R$_1$ & R$_2$) & R$_3$) → A(R$_1$ & (R$_2$ & R$_3$)))
    where (R$_1$ & R$_2$) & R$_3$ appears in A and R$_1$ & (R$_2$ & R$_3$) replaces some
    but not necessarily all of those appearances.

& is equivalent to time-indexed ∧
∀ ... ((Q/(R & R´)) ($\vec{d}$, t) ↔ ((Q/R) ($\vec{d}$, t) ∧ (Q/R´) ($\vec{d}$, t)))
∀ ... ((Q/(R & R´)) ($\vec{d}$, t, l) ↔ ((Q/R) ($\vec{d}$, t, l) ∧ (Q/R´) ($\vec{d}$, t, l)))

Negators and the pure negator in joined restrictors
∀ ... ((A/(R$_1$ & (R$_2$/N))) (d, t) → (A/(R$_1$ & ~R$_2$)) (d, t))
∀ ... ((A/(R$_1$ & (R$_2$/N))) (d, t, l) → (A/(R$_1$ & ~R$_2$)) (d, t, l))

Commutativity of ∪
∀ ... (A(Q$_1$ ∪ Q$_2$) → A(Q$_2$ ∪ Q$_1$))
    where (Q$_1$ ∪ Q$_2$) appears in A, and (Q$_2$ ∪ Q$_1$) replaces some
    but not necessarily all of those appearances.

Associativity of ∪
∀ ... (A((Q$_1$ ∪ Q$_2$) ∪ Q$_3$) → A(Q$_1$ ∪ (Q$_2$ ∪ Q$_3$)))
    where (Q$_1$ ∪ Q$_2$) ∪ Q$_3$ appears in A, and Q$_1$ ∪ (Q$_2$ ∪ Q$_3$) replaces some
    but not necessarily all of those appearances.

Equivalence of $\cup$ *and* $\vee$

$\forall \ldots ((Q_1 \cup Q_2)(d,t) \leftrightarrow (Q_1(d,t) \vee Q_2(d,t)))$

$\forall \ldots ((Q_1 \cup Q_2)(d,t,l) \leftrightarrow (Q_1(d,t,l) \vee Q_2(d,t,l)))$

$\cup$ *implies* $\vee$ *with restrictors*

$\forall \ldots (((Q_1 \cup Q_2)/R)(d,t) \rightarrow ((Q_1/R)(d,t) \vee (Q_2/R))(d,t)))$

$\forall \ldots (((Q_1 \cup Q_2)/R)(d,t,l) \rightarrow ((Q_1/R))(d,t,l) \vee (Q_2/R))(d,t,l)))$

Restrictors and $\cup$

$\forall \ldots (((Q_1 \cup Q_2)/R)(d,t) \wedge Q_1(d,t) \rightarrow (Q_1/R)(d,t))$

$\forall \ldots (((Q_1 \cup Q_2)/R)(d,t,l) \wedge Q_1(d,t,l) \rightarrow (Q_1/R)(d,t,l))$

The pure negator and negation

$\forall \ldots (Q/\sim R)(\vec{d},t) \leftrightarrow (Q(\vec{d},t) \wedge \neg (Q/R)(\vec{d},t)))$

$\forall \ldots (Q/\sim R)(\vec{d},t,l) \leftrightarrow (Q(\vec{d},t) \wedge \neg (Q/R)(\vec{d},t,l)))$

$\forall \ldots (\sim Q(\vec{d},t) \leftrightarrow (\neg Q(\vec{d},t) \wedge (- \text{ to exist } -_{\text{time}})(\vec{d},t)))$

$\forall \ldots (\sim Q(\vec{d},t,l) \leftrightarrow (\neg Q(\vec{d},t,l) \wedge (- \text{ to exist } -_{\text{time}})(\vec{d},t,l)))$

$\forall \ldots ((Q/N)(\vec{d},t) \rightarrow \sim Q(\vec{d},t))$

$\forall \ldots ((Q/N)(\vec{d},t,l) \rightarrow \sim Q(\vec{d},t,l))$

A strong completeness theorem follows as in the discussion of that for **QT** at the end of Chapter 21:

For any collection of wffs Γ and wff A, $\Gamma \vdash_{\mathbf{QTS}} A$ iff $\Gamma \vDash_{\mathbf{QTS}} A$.

47 Examples of Formalizing: The Nature of Space

In this chapter we'll see how we can incorporate or compare different conceptions of space within the pure language of time and space. Here and in other examples in this section, I'll use x for x_1, y for x_3, t for t_7, w for t_5, l for l_8, and p for l_{13}.

Example 1   There is more than one location.

$\quad \exists l \, \exists p \, (l \neq_{\text{location}} p)$

Analysis   In a model each physical thing exists at some time, and at that time it exists in a location. So if there are two physical things that exist at the same time, since they can be distinguished by location there must be two locations. So the example can be false in a model only if at no time do two physical things exist.

Example 2   There is a smallest unit of space.

$\quad \exists l \, \neg \exists p \, ((p \neq_{\text{location}} l) \wedge W_{\text{location}}(p, l))$

Analysis   There is a smallest unit of space iff there is some location such that no other location is contained in it. We can define:

$\quad \text{point}(l) \equiv_{\text{Def}} \neg \exists p \, ((p \neq_{\text{location}} l) \wedge W_{\text{time}}(p, l))$

This example is compatible with the view that space is composed of points and that any region is a collection of those points. But it is not compatible with the view of space as mass. Every part of a mass is a mass of the same kind, and masses have no smallest parts. However, this could be true in a model even on the understanding of space as mass because we could have a location without all of its parts in the universe of places.

Example 3   There is a place which contains no smallest unit of space.

$\quad \exists l \, \forall p \, (W_{\text{location}}(p, l) \rightarrow \neg \, \text{point}(p))$

Analysis   This is consistent with the assumption that there are smallest units of space, for it could be that those are contained only within particular locations. For this to be true in a model, there must be (potentially) infinitely many locations.

Example 4   There are no smallest units of space.

$\quad \neg \exists l \, \text{point}(l)$

Analysis   This is part of the conception of space as a mass, but is incompatible with the conception of space as made up of points. For this to be true in a model, there must be (potentially) infinitely many locations.

Example 5   There are only points in space.

$\quad \forall l \, \text{point}(l)$

Analysis In a model in which this is true, $W_{location}$ is the identity. I'll leave to you to consider the consequences of this for distinguishing physical things in a model.

Example 6 *Every location contains a point of space.*

$\forall l\, \exists p\, (W_{location}(p, l) \land \text{point}(p))$

Analysis This is true if space is composed of points, with some locations being collections of those points.

Example 7 *There are smallest units of space, and every location is a collection of those and nothing more.*

$\forall l\, \exists p\, (\text{point}(p) \land W_{location}(p, l)) \land$
$\forall l_1\, \forall l_2\, ((l_1 \equiv_{location} l_2) \leftrightarrow$
$\quad (\forall p\, (\text{point}(p) \rightarrow (W_{location}(p, l_1) \leftrightarrow W_{location}(p, l_2)))))$

Analysis If this is true in a model, "anything else" in a location besides points is irrelevant. This is the standard conception of space in mathematics.

Example 8 *Every location is contained in some other location.*

$\forall l\, \exists p\, (l \not\equiv_{location} p \land W_{location}(l, p))$

Analysis For this to be true in a model, the universe of places must be (potentially) infinite.

Example 9 *There are largest units of space.*

$\exists l\, \forall p\, (l \not\equiv_{location} p \rightarrow \neg W_{location}(l, p))$

Analysis The location of my ranch is within the location of Socorro County, which is within the location of New Mexico. Isn't there always a larger location?

We didn't take the comparable assumption for times as basic. Though it might seem more plausible here, it would rule out having a largest location—not a largest one "in reality", but a largest location we're talking about. It would also require that every universe of places be infinite.

Note that this could be true while there is no single largest unit of space that contains all locations.

Example 10 *There is no location that contains all other locations.*

$\neg \exists l\, \forall p\, W_{location}(p, l)$

Analysis To adopt this example as a condition on our models is not to deny that there is some totality of space. Indeed, the universe of locations P plays just that role in our models. But it is to deny that the totality of locations is itself a location.

Example 11 *Every two locations are within some location.*

$\forall l_1\, \forall l_2\, \exists p\, (W_{location}(l_1, p) \land W_{location}(l_2, p))$

Analysis Our conception of space seems to include that every two locations are within some location: the location of my office and the location of my corral are within the location of my ranch. We didn't take the comparable assumption for times as basic because two times could be in parallel timelines or on different branches of a branching-times universe. To adopt this assumption for locations would rule out the possibility of a universe of parallel spaces or branching spaces. My imagination is not rich enough to concoct such an example—it's hard enough to figure out what we might mean by branching times (see Appendix B). But we won't rule out such possibilities by taking this assumption as basic.

Example 12 *Every region of space is a location.*

Analysis Without some way of specifying regions, we can't formalize this.

Example 13 *Complements of locations are locations.*

Analysis We can define:

complement of $p \equiv_{Def}$ all q such that there is not a q'
with $W_{location}(q', p)$ and $W_{location}(q', q)$

Then we can formalize the example as:

$\forall p \, \exists l_1 \, (\forall l_2 \, (W_{location}(l_2, l_1)$
$\leftrightarrow \neg \exists l_3 \, W_{location}(l_3, p) \wedge W_{location}(l_3, l_1)))$

Example 14 *Space is three-dimensional.*

Analysis Nothing we've assumed or considered would ensure that the space of the universe of places is three-dimensional. We might want to have the universe of places two-dimensional when reasoning about what we can deduce from a map.

Euclidean geometry is based on a primitive notion of point. A point need not be a mysterious abstract thing, dimensionless, ethereal. A point is simply any part of space that has no dimension, no interior *that we recognize*. We can use Euclidean geometry without breaking our heads about how the abstract connects to the world
of experience, just as a carpenter uses Euclidean geometry by taking a dot made by a pencil to be a point and using a wooden triangle with sides 3 cm, 4 cm, and 5 cm to draw a right angle.[40]

In my *Classical Mathematical Logic*, formal systems for one-dimensional and two-dimensional Euclidean geometry are given along with axiomatizations. Those take as primitive a between-relation, identity of points, and a relation of congruence of distances. To adopt those here we would first need to assume that space is made up of points (Example 7). I have not seen a presentation in formal logic of three-dimensional Euclidean geometry.

---

[40] See "Mathematics as the Art of Abstraction" for a fuller discussion that includes how this view applies to mathematical theories generally.

Example 15 Space is dense.

Analysis   Density in space is defined for the conception of space as made up of points: between any two points in space there is a third point. It is a basic axiom of one-dimensional and of two-dimensional Euclidean geometry, as you can see in my *Classical Mathematical Logic*. But we've seen that there is no obvious way to define a between relation absent that assumption (p. 177).

Example 16   If nothing happens, there is no time.
  If nothing exists, there is no space.

Analysis   To ensure that these are true in a model, for every time and every place some ordinary atomic predication must be true of that time and place. So to formalize these we would have to use quantifiers over ordinary atomic predicates.

There are some who say that we can talk and perceive space only through objects and their relations. In Appendix E, I show that's wrong.

Example 17 Nothing exists everywhere.

$\neg \exists x \exists t \forall l \, (- \text{ to exist } -_{\text{time}}, -_{\text{location}}) \, (x, t, l)$

Analysis   I will not debate whether this is true "in reality". But we could have a model with a limited universe of places in which some thing does exist in all locations.

48 Informal Examples

Example 1 Zoe: Where was Spot barking?

Analysis Dick answers, "Over there by the gate". That's good enough. If Zoe insists, "I want to know precisely where Spot was barking", Dick will have a hard time answering, even though he saw and heard Spot barking. He can't pick out a unique location, only larger or smaller ones that will do. Though we often ask for *the* location where something happened, on the least reflection we realize that's not possible to specify.

Suppose Dick goes over and shows with his hands and steps off an area where Spot was barking. Zoe understands. But since truth is closed outwards for location, any location which includes that will count as where Spot was barking. So a small volume enclosing our solar system will count as where Spot was barking, too. Surely we'd like to rule that out as too big. But how too big? It is true that he was barking there. How large a location we'll accept as where Spot is barking depends on our interests, and those we cannot easily factor into our logic.

Example 2 Dick: Where is my cell phone? It's got to be somewhere.
 Zoe: Duh. *Everything is somewhere.*

Analysis I think most of us would agree with Zoe. That's because we think of "things" as physical things, objects in space and time. The formalization of what she said is a tautology in our system:

$$\forall x \, \forall t \, (\, (- \text{ to exist } -_{\text{time}}) \, (x, t) \rightarrow$$
$$\exists l \, (- \text{ to exist } -_{\text{time}}, -_{\text{location}}) \, (x, t, l) \,)$$

It seems to me that our first and strongest conception of "thing" is of physical things. Because of the central role that nouns play in our language, we extend the notion of thing to cover whatever we can talk about with a noun. So we talk as if there are things that exist in time but not space: symphonies and ideas. We talk as if there are abstract things not of time or space: numbers and qualities. We get into strange puzzles talking about those kind of "things", which we can rectify by asking on what basis the analogy is made to physical things.[41]

Example 3 Picking out the location of my ranch.

Analysis One way to pick out locations is with a co-ordinate system used in surveying. Surveyors employ survey stakes, sight distances, measure with tapes, and use GPS devices. That way of picking out locations, however, depends on establishing much of how we measure and much interpretation.

For example, suppose you want to buy my ranch Dogshine. You go to the

---

[41] See "Language and the World" in *Language and the World: Essays New and Old* for an extended discussion of this point.

county office and look at the survey of the ranch. That's a description. It gives the co-ordinates of how to find the boundaries and edges of the land. But you don't want to buy the land, the surface. You want to buy the ranch: the house, the garage, the corrals, the trees, the bushes, all of it. You want a fuller description. Often that's given by saying "and all that is on the land". But does that also include all that is under the land? That has to be specified; often rights to minerals or oil are withheld. And how far above the land is included? Up to the height of the tallest tree? Twelve meters above the surface? Sometimes we don't stop to think of locations as volumes. I talk of the place where the shed in my corral is, and if someone asks I say, yes, not just the piece of land it sits on but also the interior where the sheep and the donkey shelter from the rain. But if I want to build another shed in the place where the current one stands, then when I talk to the builder I'll mean the surface of the land where it is.

Example 5 Predications are extensional.

Analysis Since we've agreed that in our semi-formal languages every predicate is extensional, these three wffs are equivalent in a model:

$(-$ to bark $-_{time}, -_{location})$ (Birta, June 3 2011, Dogshine)

$(-$ to bark $-_{time}, -_{location})$ (Birta, t_7, l_{41})
> t_7 is assigned reference June 3, 2011
> l_{41} is assigned reference Dogshine

$(-$ to bark $-_{time}, -_{location})$ (x_{32}, Axmer, DD)
> x_{32} is assigned reference Birta,
> "Axmer" refers to June 3, 2011
> "DD" refers to Dogshine

Example 6 *Everywhere that Mary went the lamb was sure to go.*

Analysis It's not that in every location Mary is, the lamb is too, though perhaps at a somewhat later time. The lamb may be wandering and only rarely be in a small location around where Mary is or was. Nor is it that at every time the lamb is close to Mary, for Mary might walk from her home and the lamb not notice until a minute later and run furiously 50 meters to catch up to her. Even with a notion of close in time and close in space, I can't see how to formalize this. But it does me good to know that Mary and her lamb cared for each other so much.

49 Examples of Formalizing

Example 1 Birta is a dog.

Analysis We formalized this in Chapter 28, assuming that being a dog is an essential attribute:

$$\forall t\,((-\text{ to exist }-_{\text{time}})(\text{Birta}, t) \to (-\text{ to be a dog }-_{\text{time}})(\text{Birta}, t))$$

Taking account of space, and assuming that "$(-\text{ to be a dog }-_{\text{time}})$" is a locational predicate, we should then have as a consequence:

$$\forall t\,\forall l\,((-\text{ to exist }-_{\text{time}}, -_{\text{location}})(\text{Birta}, t, l)$$
$$\to (-\text{ to be a dog }-_{\text{time}}, -_{\text{location}})(\text{Birta}, t, l))$$

But nothing we've assumed so far in our logic guarantees this.

Example 2 If Birta is barking squealingly, then Birta is chasing a rabbit.

Analysis This is true: Birta never barked squealingly (in a high-pitched squeal) except when she was chasing a rabbit. It's an omnitemporal assertion.

(a) $\quad \forall t\,(((-\text{ to bark }-_{\text{time}})/\text{squealingly})(\text{Birta}, t) \to$
$\exists x\,(((-\text{ to chase }-_{\text{time}})/\text{obj}(x))(\text{Birta}, t) \wedge (-\text{to be a rabbit }-_{\text{time}})(x, t)))$

Both "$((-\text{ to bark }-_{\text{time}})/\text{squealingly})$" and "$((-\text{ to chase }-_{\text{time}})/\text{obj}(x))$" are locational. So it might seem from (a) we can conclude that whenever and wherever Birta is barking squealingly, she is chasing a rabbit.

(b) $\quad \forall t\,\forall l\,(((-\text{ to bark }-_{\text{time}}, -_{\text{location}})/\text{squealingly})(\text{Birta}, t, l) \to$
$\exists x\,((-\text{ to chase }-_{\text{time}}, -_{\text{location}})/\text{obj}(x))(\text{Birta}, t, l)$
$\wedge (-\text{to be a rabbit }-_{\text{time}})(x, t)))$

But (b) is false. During 5:01 p.m. Birta was barking squealingly and chasing a rabbit, but a very small location just enclosing Birta is big enough to count as a place in which she is barking, while a much larger location is needed for a place in which she is chasing a rabbit, perhaps including the rabbit.

The difference between this and the previous examples is that "to bark squealingly" and "to chase a rabbit" are not essential attributes of Birta: they are not true in any location where she is.

Essential attributes in time and space $P(-, -_{\text{time}})$ is an *essential attribute in time and space* iff for every σ and t, if $P(-, -_{\text{time}})$ is true of σ at t, then for every p in which σ exists at t, $P(-, -_{\text{time}}, -_{\text{location}})$ is true of σ at t in p.

$$\text{Attribute}\,(P(-, -_{\text{time}})) \equiv_{\text{Def}} \forall x\,\forall t\,(P(-, -_{\text{time}})(x, t) \to$$
$$\forall l\,((-\text{ to exist }-_{\text{time}}, -_{\text{location}})(x, t, l)$$
$$\to P(-, -_{\text{time}}, -_{\text{location}})(x, t, l)))$$

Examples of Formalizing 219

I'll let you define "perpetuating attribute in time and space" (compare Chapter 28).

If we add as a meaning axiom that "(— to be a dog —$_{time}$)" is an essential attribute in time and space, then we get the consequence we wanted in Example 1. In contrast, to say that Birta has the attribute of being a dog in any place regardless of whether she is in that place is to view Birta as timeless and locationless in having attributes. We don't say that Birta is a dog at a time when she doesn't exist, and we don't say that Birta is a dog in a place where she doesn't exist. We are reasoning about physical things.

Example 3 If Birta is a dog, then Birta is a mammal.

Analysis We formalized this in Chapter 28 as:

(a) $\forall t$ (((— to be a dog —$_{time}$) (Birta, t) → (— to be a mammal —$_{time}$) (Birta, t))

Assuming that both "(— to be a dog —$_{time}$)" and "(— to be a mammal —$_{time}$)" are locational predicates and essential attributes, we have as a consequence:

$\forall t\, \forall l$ (((— to be a dog —$_{time}$, —$_{location}$) (Birta, t, l)
→ (— to be a mammal —$_{time}$, —$_{location}$) (Birta, t, l))

Example 4 Barking is not an attribute of a thing.

⌝ Attribute ((— to bark —$_{time}$))

Analysis The formalization is true in any model that has enough times and places and the words have their usual meanings.

But someone might argue that barking is an attribute of a thing. That's to think of barking as in "Dogs bark", which is not an assertion about what a dog does at a time—it's certainly not true that all dogs bark at all times. It's about the ability of dogs: "All dogs have the ability to bark" (and that's false).

Example 5 Spot barked.

Analysis Considering only quasi-linear time models, we formalized this in Chapter 25 as:

(a) $\exists t$ (Max (— to bark —$_{time}$) (Spot, t) ∧ ($t <_{time}$ now))

But even assuming that "(— to bark —$_{time}$)" is locational, we can't conclude from (a) that there is some one location for the maximal interval:

(b) $\exists t\, \exists l$ (Max (— to bark —$_{time}$, —$_{location}$) (Spot, t, l) ∧ ($t <_{time}$ now))

Spot may have moved around a lot. To derive (b) we would need to assume that there is one location that contains all the locations in which Spot barked in that interval.

Example 6 Spot barked loudly at 1 p.m. May 3, 1998 in Kroon Park.

((— to bark —$_{time}$)/loudly) (Spot, 1 p.m. May 3 1998, Kroon Park)
∧ (1 p.m. May 3 1998 $<_{time}$ now)

Analysis Since "loudly" restricts the predicate "(— to bark —$_{time}$, —$_{location}$)", the formalization is true iff Spot barked loudly in comparison to all things at all times in all locations for which "(— to bark —$_{time}$, —$_{location}$)" applies. We need not, however, have only such omnitemporal and omnilocational readings of restrictors. As we saw in Example 2 of Chapter 32, if we fill the time-marked blank with the time-name before the restrictor is applied, we can compare how Spot barked to only things that barked at that time. If we also fill the location-marked blank before applying the restrictor, we can compare how Spot barked to only things that barked at that time and that place:

Spot barked loudly at 1 p.m. May 3, 1998 in Kroon Park
in comparison to those things that were barking then in that place.

((— to bark 1 p.m. May 3 1998, Kroon Park)/loudly) (Spot)

50 Location-Orienting Predicates

We orient ourselves in relation to physical things.

(1) Dick (on the phone): I'm in front of the grocery store.

As I show in Appendix E, this is not the only way we orient ourselves in space. But it's common and important enough for us to consider.

It seems simple to formalize (1), where "Smith's" is the name of the store:

$\quad$ (— to be in front of —, —$_{time}$) (Dick, Smith's, now)

The predicate "(— to be in front of —, —$_{time}$)" is surely locational. But there are two locations involved: where Dick is and where Smith's is. So we can't use:

$\quad$ (— to be in front of —, —$_{time}$, —$_{location}$)

It isn't just Dick and Smith's that are being related but locations where they are. We need instead:

$\quad$ (— to be in front of —, —$_{time}$, —$_{location}$, —$_{location}$)

Or better, to make it clearer which location is meant for which object:

(2) (—, —$_{location}$ to be in front of —, —$_{location}$, —$_{time}$)

This relates one object in a location to another object in a location at a time.

Smith's isn't something that moves around, but Spot is. So using (2), how can we formalize:

$\quad$ Spot is in front of where Dick was at 9 a.m. June 1, 2002.

We need instead:

$\quad$ (—, —$_{location}$, —$_{time}$ to be in front of —, —$_{location}$, —$_{time}$)

This relates two things via locations where they are at specific times.

It's not just where things are but what they're doing that can be in an orientation:

$\quad$ Spot is barking in front of Dick.

This is meant as "Spot is barking in front of where Dick is", for which we could use:

$\quad \exists l\, \exists p\, (\,($ — to bark —$_{time}$, —$_{location}$) (Spot, now, l) $\wedge$

$\quad\quad$ (—, —$_{location}$, —$_{time}$ to be in front of —, —$_{location}$, —$_{time}$)

$\quad\quad\quad$ (Spot, l, now, Dick, p, now))

The second reference to Spot is not needed; all we need to relate are the locations. It's not that Spot is in front of where Dick is, but the location where Spot is barking is in front of the location where Dick is.

That's clearer if we try to formalize:

(3) Spot is barking in front of where Dick is yelling.

For this to be true, Dick could be running back and forth yelling at Spot, who is also running and jumping, so long as Dick is facing in one direction during the entire time designated by "now". To formalize this we need to say that Spot is barking in some location, Dick is yelling in some location, and Dick gives the orientation for "in front of" at that time for the two locations. We get the locations via the predications for barking and yelling, so all we need to relate are those at a time, and Dick has to be mentioned to give the orientation. So let's use:

(4) ($-_{location}$ to be in front of $-$, $-_{location}$, $-_{time}$)

This relates one location to another location via an object at a time. Then we can formalize (3):

(5) $\exists l \, \exists p \, (\, (-$ to bark $-_{time}$, $-_{location})$ (Spot, now, l) $\wedge$
    $(-$ to yell $-_{time}$, $-_{location})$ (Dick, now, p) $\wedge$
    $(-_{location}$ to be in front of $-$, $-_{location}$, $-_{time})$ (l, Dick, p, now))

For this to be true, during the whole time designated by "now" Dick has to give a single orientation while he is yelling, moving around perhaps, but always facing one direction, or at least facing enough in that direction that we'd say there's a front and back to the place where he's yelling.

We can also formalize:

Dick is yelling in front of where Spot is barking.

And we can formalize:

Spot was barking at 3 p.m. June 3 2003 in front of
    where Dick was yelling at 2:58 p.m. June 3, 2003.

All we need to do is replace the two occurrences of "now" in (5) with these dates.
   Using (4) we can also formalize (1):

$\exists l \, \exists p \, (\, (-$ to exist $-_{time}$, $-_{location})$ (Dick, now, l) $\wedge$
    $(-_{location}$ to be in front of $-$, $-_{location}$, $-_{time})$ (l, Smith's, p, now))

We don't need to add that Smith's exist in the location we assign to p if we classify (4) as positive for existence for the object at the second location and time.

There are other phrases we use to orient in relation to things in space, such as "above" and "to the left of". Let's say a phrase is location orienting if it can be used to establish a relation between two locations due to a physical thing being in the second location at a time. Prefacing such a phrase by "to be", we can have predicates:

$(-_{\text{location}}$ to be above $-, -_{\text{location}}, -_{\text{time}})$

$(-_{\text{location}}$ to be to the left of $-, -_{\text{location}}, -_{\text{time}})$

We're using the object that is talked about in a location to give the orientation. This contrasts with the view that "to the left of" uses the orientation of the speaker.

We can extend our logic to include these *location-orienting predicates*, whose general form is:

$P(-_{\text{location}}, -, -_{\text{location}}, -_{\text{time}})$

The semantics for each are peculiar to it. For example (4) cannot be true of locations at a time if it is predicated of a ball because a ball has no front and back. The only general requirement we make is that each is positive for existence.

Location-orienting predication are positive for existence
If $P(-_{\text{location}}, -, -_{\text{location}}, -_{\text{time}})$ is true of p, p′, σ, t,
then $(-$ to exist $-_{\text{time}}, -_{\text{location}})$ is true of σ at t in p′.

Example 1 Spot is barking in front of Dick.

Analysis Why not view "in front of Dick" as a restrictor, an adverb that tells where? Since we also have "if front of Suzy", "in front of a ball", "in front of the house", it would be a variable restrictor. So for this example we'd have:

$\exists l\, (\,((- \text{ to bark } -_{\text{time}}, -_{\text{location}})/(\text{in front of (Dick)}) \,(\text{Spot, now}, l)\,)$

But the use of "in front of" in the example does not restrict the application of the predicate: it doesn't tell us to look at all barkings and restrict to those done in front of Dick. It restricts the location in an application of the predicate. Moreover, to formalize "Spot is in front of Dick" this way, we would have to use:

$\exists l\, (\,((- \text{ to exist } -_{\text{time}}, -_{\text{location}})/(\text{in front of (Dick)}) \,(\text{Spot, now}, l)\,)$

And we have good reason to reject modifiers of the existence predicate (p. 115).

Example 2 Spot is barking in front of where the band is playing.

Analysis A band playing in a location establishes an orientation of front and back. But it's not just the band. And it's not just the band playing. It's the orientation of the loudspeakers and the (perhaps hoped-for) audience. The members of the band could all turn around with their backs to the audience and continue playing, yet "in front of the band" would continue to be the same orientation. I don't see how to formalize this.

Example 3 Spot is barking in front of and to the left of Dick.

$\exists l\, \exists p\, (\,(- \text{ to bark } -_{\text{time}}, -_{\text{location}})\,(\text{Spot, now}, l)$
$\quad \wedge\ (-_{\text{location}}$ to be in front of $-, -_{\text{location}}, -_{\text{time}})\,(l, \text{Dick, now}, p)$
$\quad \wedge\ (-_{\text{location}}$ to be to the left of $-, -_{\text{location}}, -_{\text{time}})\,(l, \text{Dick, now}, p)\,)$

Analysis We can use more than one location-orienting predicate to focus on a location or locations.

Example 4 Spot is barking in front of and behind Dick.
$\exists l\, \exists p\, (\, (-$ to bark $-_{\text{time}}, -_{\text{location}})$ (Spot, now, l)
$\land\ (-_{\text{location}}$ to be in front of $-, -_{\text{time}}, -_{\text{location}})\, (l, \text{Dick}, \text{now}, p)$
$\land\ (-_{\text{location}}$ to be behind $-, -_{\text{time}}, -_{\text{location}})\, (l, \text{Dick}, \text{now}, p)\,)$

Analysis We know this can't be true: nothing can be both in front of and behind an object at the same time. Yet it could be, if the time is long enough. If the present is several minutes long, Spot could bark in front of Dick and (then) in back of Dick.

But to say that misconstrues how we use location-orienting predicates. It's not that Spot can be in front of and behind Dick during some time. It's that there is no location that is oriented both as in front of and behind Dick at one time—if Dick presents an orientation during that time. The example can't be true. We formulate a meaning axiom:

$\forall l\, \forall p\, \forall x\, \forall t\, \neg\, (\, (-_{\text{location}}$ to be in front of $-, -_{\text{time}}, -_{\text{location}})\, (l, x, t, p)$
$\land\ (-_{\text{location}}$ to be behind $-, -_{\text{time}}, -_{\text{location}})\, (l, x, t, p)\,)$

Example 5 Zoe: Where's Spot?
Dick: *He's in the house*.

Analysis Dick and Zoe's house presents an orientation of inside. So we can use a location-orienting predicate to formalize what Dick said, taking "DZH" as a name for Dick and Zoe's house:

$\exists l\, \exists p\, (\, (-$ to exist $-_{\text{time}}, -_{\text{location}})$ (Spot, now, l) $\land$
$(-_{\text{location}}$ to be in $-, -_{\text{location}}, -_{\text{time}})\, (l, \text{DZH}, p, \text{now})\,)$

We can relate the predicate "$(-_{\text{location}}$ to be in $-, -_{\text{location}}, -_{\text{time}})$" to the within-location predicate with a meaning axiom:

$\forall l\, \forall p\, \forall x\, \forall t\, (\, (-_{\text{location}}$ to be in $-, -_{\text{location}}, -_{\text{time}})\, (l, x, p, t) \to W_{\text{location}}(l, p)\,)$

Could we replace the location-orienting predicate "to be in" with a formula that uses the within-relation on locations?

$\exists l\, \exists p\, (\, (-$ to exist $-_{\text{time}}, -_{\text{location}})\, (x, t, l)$
$\land\ (-$ to exist $-_{\text{time}}, -_{\text{location}})\, (y, t, p) \land W_{\text{location}}(l, p)\,)$

But Spot's doghouse is in the yard and at the same time Spot could be within the yard and not in his doghouse. Perhaps we could assume that there is a specific location that counts as the inside. But that would be as doubtful as assuming that minimal locations exist.

Example 6 Spot is sitting in front of Dick and Zoe.

Analysis We have two choices for formalizing this. We could say take the example to mean that each of Dick and Zoe presents an orientation, and that Spot is in front of both of those:

$\exists l_1 \exists l_2 \exists l_3$ (((— to sit —$_{time}$, —$_{location}$) (Spot, now, l_1)
$\wedge$ (—$_{location}$ to be in front of —, —$_{time}$, —$_{location}$) (l_1, Dick, now, l_2)
$\wedge$ (—$_{location}$ to be in front of —, —$_{time}$, —$_{location}$) (l_1, Zoe, now, l_3)

But it seems better to view Dick and Zoe together presenting an orientation, for which we can use a conjunction of terms:

$\exists l_1 \exists l_2$ ((— to sit —$_{time}$, —$_{location}$) (Spot, now, l_1)
$\wedge$ (—$_{location}$ to be in front of —, —$_{time}$, —$_{location}$) (l_1, Dick $\wedge$ Zoe, now, l_2)

From this we can conclude that both Dick and Zoe exist at the present time of the example.

Example 7   Spot is sitting next to Dick.

$\exists l$ ((— to sit —$_{time}$, —$_{location}$)/ next to (Dick)) (Spot, now, l)

Analysis   I don't see "to be next to" as a location-orienting predicate, for Spot could be next to Dick regardless which way Dick is facing.

Example 8   Spot is running towards Dick.

$\exists l$ ((— to run —$_{time}$, —$_{location}$)/ towards (Dick)) (Spot, now, l)

Analysis   We can't use "(—$_{location}$ to be towards —, —$_{time}$, —$_{location}$)" as a location-orienting predicate because the example would be true if Dick were running too, without a fixed location so long as Spot kept going in whatever direction Dick is going. The orientation of Dick does not matter in the evaluation of "towards Dick". But it does matter that Dick is there: for the example to be true Spot had to intend to run in that direction because Dick was there. But we can't replace the example with "Spot was running with the intention of arriving at the place where Dick was" because it could be true if Spot was running in order to stop a little before Dick and sit and beg for a treat. The intentionality is completely captured with "towards" because our predications are extensional.

Example 9   Spot is barking between Zoe and Suzy.

Analysis   We're not talking about orientation here, just the relation of locations. It seems that we are using "Spot", "Dick", and "Suzy" to pick out locations. But all we need is that there are some locations in which these exist that are in the right relation. We could do that if we adopt a predicate that relates locations only:

between (—$_{location}$, —$_{location}$, —$_{location}$)

Then we could formalize the example as:

∃l_1 ∃l_2 ∃l_3 (((— to bark —$_{time}$, —$_{location}$) (Spot, now, l_1)
 ∧ (— to exist —$_{time}$, —$_{location}$) (Zoe, now, l_2)
 ∧ (— to exist —$_{time}$, —$_{location}$) (Suzy, now, l_3)
 ∧ between (—$_{location}$, —$_{location}$, —$_{location}$) (l_1, l_2, l_3))

The between-relation is fundamental in formalizations of Euclidean geometry. But those formalizations take space to be composed of indivisible dimensionless points. Here there is no clear interpretation for all models of what "between" should mean. Still, we might agree how to use it in some models and formalize the example as above.[42]

Example 10 *Spot ran from Dick to Zoe.*

∃t ∃l (((— to run —$_{time}$, —$_{location}$) / from (Dick, t) / to (Zoe, t)) (Spot, t, l)
 ∧ ($t <$ now))

Analysis Dick and Zoe are not giving orientations here. Spot could run in any direction starting where Dick is and be said to be running from Dick, and he could run from any direction to Zoe. But the two location restrictors together do give a direction, going from the one location to the other. This suggests viewing "from — to —" as a single restrictor, one that indicates direction. Since the physical objects in it would stand for only the locations where they are, we could add a pure location restrictor:

from–to (—$_{location}$, —$_{location}$)

This along with "between (—$_{location}$, —$_{location}$, —$_{location}$)" would give us a way to formalize not only the relation of places but directions in a geometry of locations. However, this would not allow for the intentional reading of "from" and "to" which we can have with the formalization above.

---

[42] Here is how one linguist adopts the dimensionless-point conception of space in his analysis of language. Leonard Talmy, in "How Languages Structure Space" says:

... the Reference Object [the boulder or boulders] is treated as a single point by *near*:

a. The bike stood near the boulder.;

as a point pair by *between:*

b. The bike stood *between* the boulders (i.e., two of them).

This smacks of too much geometry and too little life. I don't see how this could account for:

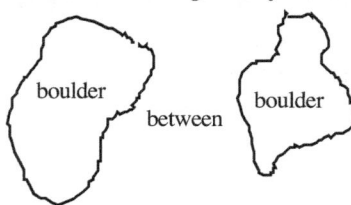

51 Parts of Physical Things

What counts as a part of a thing? Our answer is constrained by what kinds of things we are considering.[43] Whether a hair on the head of my dog Birta is a part of her depends on whether we are paying attention to hairs, which in our formal framework means whether her hairs are in the universe of our model. Can we define a part-whole relation in our temporal and spatial logic?

The right front paw of my dog Birta is a part of her. Wherever she goes, it goes. If a mountain lion were to bite it off, it would no longer be part of her. Anywhere Birta is, her paw is. Or rather, any location that we count as making true that Birta exists is also a location in which her paw exists. We can take that to define what it means for one physical thing to be part of another.

Parts of physical things at times An object b is a *part of* an object o *at time* t iff given any location p in which o exists at time t, b exists at time t in p.

$$\text{Part}\,(-,-,-_{time})\,(x, y, t) \equiv_{Def} \forall l\,(\,(-\text{ to exist }-_{time},-_{location})\,(y, t, l)$$
$$\rightarrow (-\text{ to exist }-_{time},-_{location})\,(x, t, l)\,)$$

This satisfies the minimal conditions for a part-whole relation: The relation is reflexive because Part $(-,-,-_{time})\,(x, x, t)$ is a tautology. It is transitive because the conditional, $\rightarrow$, is transitive. And by the assumption that locations can distinguish physical things, if b is a part of o and o is a part of b, then o = b.

We took as an informal constraint on where physical things are that if o is in location p at time t , then all of o is in that location. So all parts of o are in that location. We can't use this new predicate to enforce that condition since it depends on already being clear about what it means for a physical thing to exist in a location.

We don't need to worry about "accidentally" getting a part by this classification because locations distinguish things. We can't have so few locations that Spot's dish counts as part of Spot.

Is this new predicate a good formalization of the notion of part-whole? Let's consider some examples of how it classifies.

1. The right rear wheel of my car is part of my car now. But when I left it at a tire shop to be repaired and drove away in my car, it was not a part of my car. It wasn't an essential part of my car: my car could exist without the wheel being a part of it.

Essential parts of physical things An object b is an *essential part of* an object o iff given any time t and location p such that o exists at time t within location p, b exists at time t within p.

$$\text{Essential-Part}\,(-,-)\,(x, y) \equiv_{Def} \forall t\,\text{Part}\,(-,-,-_{time})\,(x, y, t)$$

---

[43] See Appendix 6, "Parts of Things" in Volume 1.

If we want to say that the wheel was still part of my car, only a detached part, we are using a notion of "part of" that is about what we consider to be the typical, or correct, or intended constitution of a thing. It is a worthwhile notion to investigate, but not one we have the tools for here.

2. The black king of my chess set is part of my chess set. But is my chess set a thing? To talk of it as a thing and the parts of it as belonging to it is to treat "— is a piece of my chess set" as a predicate and belonging as the application of that predicate. Such a notion of "part of" should be investigated in a second-order predicate logic.

3. I have an apple. I cut it into two halves. Do I still have the apple? Neither half could be considered a detached part of the other; only the two parts together, at least in a collective sense, are the original apple. Yet the definition of a part of a physical thing at a time allows for the two parts together to count as the original apple because there is no requirement that the parts be connected.

I don't see how to require that the physical thing be made up of parts that are spatially connected. There would have to be some region in which both the part and the whole exist that is spatially connected. Yet different parts would not have to be spatially connected: my heart and my liver are parts of my body, but they are not spatially connected except that both are always (I hope) part of my body.

4. Are my heart and liver essential parts of my body? Each could be replaced, so it seems not. But would that still be me? Are they essential parts of me? Consideration of times and locations won't help resolve these questions.

5. Is the hole in a doughnut a part of the doughnut? If you eat all of the doughnut, then you eat every part of it, so if you eat a doughnut, and the hole is part of it, then you eat the hole, which is absurd (are holes easy or hard to digest?).

But perhaps when you take the first bite of the doughnut, the doughnut, as a whole, no longer exists. In particular, the thing that was the hole in the doughnut no longer exists. On that view, to say that you ate all of the doughnut would be false if it meant all of the doughnut that existed: you ate all of it except the hole. But what if you popped the whole doughnut into your mouth and swallowed it? Then you would have eaten the hole. That, too, seems absurd.

Moreover, when you moved the doughnut to your mouth, you moved the whole doughnut. So the hole can't be a region of space, because that's different after you move the doughnut.

For the hole in a doughnut to be a part of the doughnut the hole has to be a thing. That seems to me to take the shape of a physical thing to be a thing itself.[44]

---

[44] Robert Casati and Achille Varzi in *Parts and Places* argue that holes are things and that a hole in a doughnut is a part of the doughnut. They give an analysis of parts and locations that involves topological notions. See also Robert I. Binnick, *Parts and Places*: *The Structure of Spatial Representation*.

6. The second movement of Beethoven's 6th Symphony is part of that symphony. But even if we accept that both of those are things, and that they are in time, they don't seem to be in space. A definition in which times replace locations does not seem appropriate for a part-whole relation for things in time but not space because simultaneity does not seem sufficient for being a part.

To see a symphony, or a poem, or a play, or a book as temporal and not spatial reflects a pre-occupation with things in place of doing. A poem, as much as a computer program, is directions. It is meant to be instantiated in a person by reading it, as a computer program is meant to be instantiated on a computer. The relation to experience/the world is made by a person who connects the reading of it to his/her experience, while for the computer program the connection to "the world" can be made in various ways.

7. I gave my dog Birta a bone last night. She chewed and swallowed a piece of it. That piece of bone was for a time within any location at which Birta existed. So by our definition, for that time it was part of Birta. Is that right?

On one intuition of what "part of" means, the piece of bone isn't a constituent of her. I suspect that the notion of a physical constituent will be hard to make clear. Perhaps a molecule in Birta's brain would be a physical constituent of her. Perhaps a hair on her head would be a physical constituent of her. But is the bile in her a physical constituent of her? And if so, are the contents of her stomach a physical constituent of her? Whatever notion we have of a physical constituent, it doesn't seem that we can explicate it using only the notions of space and time and existence.

We have different notions of what it means to say that one thing is a part of another, notions which in turn depend on what we think is a thing. The part-whole relation defined in terms of space, time, and existence is one way to conceive of parts and wholes of physical things.

Metaphysical Bases of Logics of Time and Space

We are not investigating the nature of time and space here but how our assumptions about the nature of time and space affect what we consider to be true and what we consider to be valid inferences. In order to do so, I have developed formal logics. For each I have set out the metaphysical assumptions that on which I base it. Here I'll summarize those bases, hoping to make clearer how they shape and limit the range of applicability of the formal systems.

Classical propositional logic
We assume that we are among people who talk and reason together, searching for rules to guide us in good reasoning. We assume that from the flow of our talk we can pick out bits, units we call words and sentences. Some of those sentences are meant to be descriptions of the world, however you conceive that. They are true or false, but not both. We call them (representatives of) propositions.

The simplest logic we can devise to formalize any reasoning assumes no more metaphysics than this. We consider how we can reason with propositions as wholes, with conventions for formalizing the ordinary language connectives "and", "or", "not", "if . . . then . . ." as truth-functional. This is classical propositional logic.

Some views of time
The formal systems developed in this text are meant to be compatible with many ways of conceiving of time, the most basic of which I list here.

Time as objective
In his *Principia*, Isaac Newton defines:
> Absolute, true, and mathematical time, in and of itself and of its own nature, without reference to anything external, flows uniformly and by another name is called duration. Relative, apparent, and common time is any sensible and external measure (precise or imprecise) of duration by means of motion; such a measure—for example, an hour, a day, a month, a year—is commonly used instead of true time. p. 54

Chris Sinha et al. say in "When Time is Not Space":
> In our Western cultural and cognitive world, we are accustomed to the notion that 'Time' is an autonomous, abstract conceptual domain. We are not referring here to the phenomenal experience of time as duration, or as a fundamental aspect of events (Bergson, *Time and Free Will*, 1910), but about the way in which time is *thought about* and *talked about*. Our usual cultural presupposition is that time, in this latter sense, constitutes a domain of thought-about, reflective experience, schematized in linear or cyclic terms, that is in some sense independent of the events that occur "in time". This abstract conceptual domain we shall refer to as *Time as Such*.*
>
> * [footnote] . . . we emphasise that the abstract notion of Time as Such is

specifically a consequence of cultural and cognitive practices of its measurement, and that its abstraction from such practices depends upon its symbolic organization and material anchoring.

Time as Such, the idea that time is real, independent of us and our experience, I call *Zeit an Sich* to draw comparison to *Ding an Sich*, the view that a thing is independent of us, real beyond any properties it might have or we ascribe to it.[45]

Time as subjective

Vyvyan Evans says in *The Structure of Time*:

> I argue that our experience of time cannot be equated with an objectively real entity inhering in the world "out there". Nor can it be equated with a second-order concept parasitic on "more basic" kinds of experiences, such as external sensory experience. Rather, I argue that time appears ultimately to derive from perceptual processes, which in fact may enable us to perceive events. As such, temporal experience may be a pre-requisite for abilities such as event perception and comparison, rather than an abstraction based on such phenomena. p. 9

David Park in "The Myth of the Passage of Time" says:

> I think that the so-called passage of time, however one chooses to define it, has such a questionable status in experience, in logic, and in epistemology that its occurrence in any philosophical discussion that does not carefully label it as a subjective impression is a mistake. p. 113

And Benjamin Lee Whorf in "The Relation of Habitual Thought and Behavior to Language" says:

> Our 'length of time' is not regarded [by the Hopi] as a length but as a relation between two events in lateness. Instead of our linguistically promoted objectification of the datum of consciousness we call 'time,' the Hopi language has not laid down any pattern that would cloak the subjective 'becoming later' that is the essence of time. pp. 206–207

Time, in this view, is not independent of us and our experience. I say "experience" not "perceptions" because to take time as based on perception would suggest that we derive temporal relations from *Zeit an Sich*.

Time as mass

I view time as mass: every part of time is time, and there are no smallest parts. We pick out parts of it, "point to" bits of it, as we pick out parts of water or mud. But that can be only an analogy with water and mud, for we cannot touch time as we can put our toe in water or get mud on our shoes. It is an analogy even to say that time is, that time has some distinctness beyond our experience or classification of our descriptions of the world as before or after. We do not pick out times but rather *establish* times using descriptions.

---

[45] Chris Sinha and Vera da Silva Sinha explore the idea of Time as Such in an excellent study of time across cultures in "Time and Events—in language, mind and world".

However, we can view time as both mass and real, saying that descriptions of the world ordered as before or after *pick out* times that are independent of us.

Time as pointillistic
Mathematicians and scientists following Newton have taken time to be "composed" of instants, parts so small that they have no measure, but an infinity of which together can constitute an interval, such as a second, that does have measure.

If this view is meant to describe time as real independent of us, then the instants and intervals are as real as dogs and cats. We pick them out imperfectly, but they are there, ready for us to identify. I find such talk misleading, creating a reality of the (almost) abstract from our process of abstracting. And the method of picking out such real times is not clear to me except as we use descriptions of them. A fully subjective interpretation of time as pointillistic takes instants to be the smallest "parts" of time we wish to pay attention to, parts whose dimension is considered negligible.

A fuller range of views of time is presented in a clear and insightful discussion by James Glieck in "Time Regained!". I'll leave to you to evaluate whether those are compatible with the formal systems developed here.[46]

Temporal propositional logic: relative time
Whether we view time as mass or pointillistic, as real or subjective, we can and do pick out or establish times with descriptions, such as the time when Dick and Zoe's dog Spot choked on a bone, the time when Suzy's cat Puff screeched. We can agree that the propositions "Spot choked on a bone" and "Puff screeched" pick out or establish times—if they are true. To say that a false proposition is about a time assumes that times are there and we give true or false descriptions of them. But "Spot meowed" would be a false description of any time, no use for us in taking account of times as ordered. First we must have the time we are discussing, and for that we can and do use true propositions.

We can relate times as before or after by relating the true propositions that pick out or establish them using formal equivalents of "before". We relate only the simplest such temporal propositions, ones we treat as atomic in classical propositional logic. So "Spot choked on a bone or Puff screeched" does not pick out or establish a time any more than "If Spot choked on a bone, then Puff screeched".

But ordering as before or after is ambiguous. Did Spot finish choking on a bone before Puff screeched, or was there some overlap? To be specific we relate the beginnings and endings of times established by true atomic propositions. Beginnings and endings can be understood as entirely or principally distinctions we impose on our experience. Or beginnings and endings might be real, in the world, distinctions we recognize but do not create. No matter, so long as we can agree that the time a

---
[46] I prefer the characterization of time given by Scott Adams in his *Dilbert*, June 10, 2014:
>Imagine a donut, fired from a cannon at the speed of light while rotating.
>Time is like that, except without the cannon and donut.

true atomic proposition establishes or picks out has or can be viewed as having a beginning and ending, we can adopt four temporal propositional connectives to formalize: ending before beginning; beginning before ending; beginning before beginning; and ending before ending.

In our work in formal logic, the assumption that propositions are (represented by) types is always provisional. In some cases we may decide or note that two equiform sentence tokens are meant as (or are meant to represent) distinct propositions. This is what happens when we reason taking times to be picked out or established with true atomic propositions. In "Spot barked; then Dick yelled; ;then Spot barked" the two equiform occurrences of the words "Spot barked" are meant as (representing) two distinct propositions. Hence, we need to index sentence types to have propositions. We can use numerals and have "(Spot barked)$_2$" and "(Spot barked)$_7$". In this way we arrive at a temporal propositional logic based on classical propositional logic.

Past, present, future

Central to our conception of time as English speakers is the division of past, present, and future. Every sentence in English is tensed via its verbs. So if we want to formalize reasoning from English, we must account for the past-present-future divide. But not every language makes the division, and some have no tenses at all. For example, Benjamin Lee Whorf in "An American Indian Model of the Universe" says:

> [The Hopi speaker] has no general notion or intuition of *time* as a smooth flowing continuum in which everything in the universe proceeds at an equal rate, out of a future, through a present, into a past; or, in which, to reverse the picture, the observer is being carried in the stream of duration continuously away from a past into a future. p. 27

So instead of modifying our logic to account for a past-present-future divide, we introduce conventions on formalizing that allow us to apply the temporal propositional logic of relative time to reasoning in English. We restrict our formalizing to models in which all beginnings and endings of times established or picked out by true atomic propositions can be compared, as if lined up in a row. Then we adopt one or several true atomic propositions as establishing a present, with the past before the time interval those establish and the future after.

Classical predicate logic

In classical propositional logic atomic propositions are unitary: they are simply true or false with no consideration of their internal structure. To formalize more, to parse atomic propositions as having internal structure, some view of how language and the world connect must be adopted. For classical predicate logic we assume that the world is made up, at least in part, of individual things. Then restricting our attention to propositions that are about things, we parse atomic propositions into names and what's left over when the names are taken out of the proposition, what we call (representatives of) "predicates". To connect this structure to the world of our experience, we further assume that each thing we are reasoning about can be picked out and named, perhaps temporarily, perhaps solely in theory or only subjectively. Things can be counted, and we can talk of all things and some things.

No new metaphysics is needed to extend classical predicate logic to take account of the internal structure of atomic predicates: relative adjectives, adverbs, conjunctions of terms, and more, though issues about extensionality have to be resolved.

In classical predicate logic, when a thing exists is not taken into consideration, and reference to a thing in no way involves time. The evaluation of a predication in no way involves time. Things are understood as either outside time or supratemporal, existing through time. In order to take time into account while reasoning about the world as made up of things, we have to resolve how to use an atemporal or supratemporal notion of reference with things that do not exist now, as in the past or future. We agree that the methods we use for referring to objects we can access in our immediate experience can serve as "theoretical" ways to establish reference for objects outside our immediate experience. This does not resolve but leaves a tension throughout our work between the atemporal conception of objects needed for reference and the assumption that the things we are considering are in time.

Times as things
"Sometimes Spot barks." "Puff never barks." "Spot always barks when Dick feeds him." "Spot barked three times." We say how often, we reason about how often, and in doing so we seem to be treating times as things we can count. We might believe that times are "out there", real in the world independent of us, ready to be identified. We might believe that times are our subjective ordering of experience, picked out by true propositions though not necessarily ones of a semi-formal language. We might believe that times are parts of a mass of time, real or noted only by some containing description. No matter, so long as we agree that we can identify times in some way or ways, we can treat them as things we can count.

With this assumption we can extend classical predicate logic to allow for quantification over times along with quantification over individual things. In each model a separate universe of times is given, however those times are specified. We adopt new variables and name symbols to stand for those times.

We assume that there is an ordering of times, not necessarily linear. We talk about the ordering in the formal language with the predicate "$-\ <_{time}\ -$". We assume further that times are intervals, whether those be parts of a mass, or collections of instants, or just a single instant. One interval of time can be within another, as a week is within a month. So we assume a within-relation on times, a part-whole relation that we talk about in the formal language with "$W_{time}(-,-)$". We further assume that every time is determined by its parts: those times, if any, that are within it. We set out how the part-whole relation interacts with the ordering: a part cannot be earlier nor later than its whole in the ordering, nor can overlapping times be compared, and a part is related in the ordering to times outside the whole as the whole is.

Because we can compare times directly, we have no need for tenses or temporal connectives, and hence no need to talk of beginnings and endings of times. To give examples related to English, we replace tensed ordinary predicates with infinitive versions, now with a blank marked for a time term. Then we can have "(— to bark —$_{time}$) (Spot, t)" and "(— to be a cat —$_{time}$) (Puff, b)", where "b" is a time name symbol. These are propositions when we say what t and b stand for. For example, we can let t stand for the time when Puff bit Spot and "b" stand for June 24, 2018. We treat quantification over times in the same manner as we treat quantification over individual things. So to say that at some time Spot barked, we can use "$\exists t$ (— to bark —$_{time}$) (Spot, t)". To say that every time that Spot barks, Dick yells, we can use "$\forall t$ ((— to bark —$_{time}$) (Spot, t) → (— to yell —$_{time}$) (Dick, t))".

In classical predicate logic we take as primitive the truth or falsity of an atomic predication. Here we take as primitive the truth or falsity of an atomic predication at a time. But we put conditions on that beyond what we have in classical predicate logic. We assume that if a predication is true at a time, it is true at all times within that time, and if true at all times within a time, it is true at that time.

To say that Puff is, was, and always will be a cat is not to say that he was a cat at the time when there were dinosaurs. Rather, he is, was, and will be a cat at any time he exists. To formalize talk of when things exist, we add an existence predicate to our logic, "(— to exist —$_{time}$)". This is not a new kind of existence, only a way to take account of when things exist. We certify that and make explicit that we are talking about only things that exist in time by requiring that each thing in the universe exists at some time. We assume, too, that there are no gaps in the times of when a thing exists: things do not come into existence, go out of existence, then come back into existence. We assume further that each atomic predicate is positive for existence: if true of an object at a time, the object exists at that time.

By further allowing for the internal structure of atomic predicates to be parsed in terms of predicate restrictors and negators, conjunctions of terms, and the other forms we usually consider, we have a formal logic with considerable scope for formalizing ordinary reasoning that takes account of time.

Past-present-future with quantifying over times

To formalize propositions and inferences from English, we need to take account of our past-present-future divide.

We cannot assume that all times are lined up in order, for overlapping times are not comparable. Rather, we restrict our attention to quasi-linear models in which every two times that are not overlapping are comparable. We then take a particular time in a model as the present, naming it *now*, whether that be an instant or an interval. A predication is of the past if the time of it is before *now* in the ordering; it is of the future if the time of it is after *now* in the ordering; and it is of the present if the time of it is within *now*. For example, we can formalize "Spot barked" as "(— to bark —$_{time}$) (Spot, t) ∧ ($t <_{time}$ *now*)". This is a convention for formalizing, not part of the logic. Even for a single model we have choices of what we take to be the time of *now*, and with distinct choices a semi-formal wff has distinct interpretations.

Using this way to formalize tensed propositions, we can test how our logic classifies propositions and inferences of English. We can define whether a predicate such as "(— to be a cat —$_{time}$)" is a perpetuating attribute (once true of a thing, always true of that thing thereafter when the thing exists) or an essential attribute (if true of a thing at any time then true of it at all times that the thing exists). We can become clearer about what we mean by "individual thing in time".

Space

There are similarities and differences between how we talk and reason about space and how we talk and reason about time.

For most of us, it seems, space like time is mass: every part of space is space and there is no smallest part of space. Space does not come in identifiable parts, except as we pick out or establish those with descriptions, either verbal or measuring. Space in this sense is subjective. Yet some view space as independent of our experience, a *Raum an Sich* whose parts are there, ready to be picked out. Newton in his *Principia* describes both views:

> Absolute space, of its own nature without reference to anything external, always remains homogeneous and immovable. Relative space is any movable measure or dimension of this absolute space; such a measure or dimension is determined by our senses from the situation of the space with respect to bodies and is popularly used for immovable space ... pp. 54–55

Physicists following him in viewing space as real and objective adopted a pointillistic view of space using variables for indivisible points in space. But we can understand a point to be a smallest "part" of space we choose to pay attention to, like a dot a carpenter makes with a pencil, taking space as subjective and pointillistic.

Despite these similarities, there are many differences between how we talk and reason about time and about space. We use spatial metaphors to describe time: a long time ago, a short time from now, the distant past. We don't use temporal metaphors to describe space. Some language cultures take space and time to be so distinct that they use no metaphors from the one to describe the other.[47]

We use true atomic propositions to establish times. The time when Spot is sitting is picked out with "Spot is sitting". But we cannot use true atomic propositions to establish a place. There is too much uncertainty, too little agreement about the where of "Spot is sitting". At best, we have a range of places, larger or smaller, that we consider to be where Spot is sitting. And we have no ordering of space, no basic relation and connectives we commonly use to establish spatial relations among true atomic propositions. So I have found no way to develop a propositional logic of spatial connectives.

---

[47] See Chris Sinha, et al. "When Time Is Not Space".

Instead, we can turn to how we talk of locations as things. "Suzy stopped at three places on her way home", "My cell phone is somewhere in this house", "Everywhere that Mary went, the lamb was sure to go". As we used talk of times as things in building a temporal predicate logic, we can attempt to integrate talk of locations as things into a predicate logic.

We talk of things in time without talk of locations. But we do not normally talk of things at locations without talk of times. So we build on the temporal logic of quantifying over times to take account of space. We add quantification over locations to our predicate logic for quantifying over times and things by adopting new variables for locations, assuming that each model has a universe of places in addition to a universe of individual things and a universe of times. We restrict our reasoning to things in time and space, what we call *physical things*.

Building on our temporal predicate logic, we continue to use predicates marked for time, such as "— to bark —$_{time}$". For each of those we add a version marked for location, such as "— to bark —$_{time}$, —$_{location}$". That a temporal predicate is true of an object at a time need not entail that it is true of that object at that time in some location. Some temporal predicates are locational, and some might not be, such as "— to be brothers —, —$_{time}$".

We pick out, we refer to times as if they are there. It is less clear how we can pick out or refer to locations. But to formalize any talk of where things are, to take space as well as time into account in a predicate logic, we need to assume that we have some way to pick out or refer to specific locations. The lack of agreement on how we do that can be reflected in how we specify the method of reference for particular models.

We talk of one time being within another. We talk of one location being within another. So we add to our models a within-relation on locations satisfying the same conditions as the within-relation on times, adding to the formal language the predicate "$W_{location}(-,-)$".

True at a time means true at all parts of the time. But for locations, what is true at a location need not be true at a location within that one: "Spot is a dog" is not true of a location too small to enclose Spot. Rather we interpret "—$_{location}$" to mean "within location —". Then if a temporal predicate is true of an object at a time in a location, it is true of that object at that time in any location that contains that one.

Each physical thing determines a time: the time when it exists. But, as we noted, a physical thing determines not a location but a range of locations at a time: all those that enclose the object. We pick those out as we pick out the time of a physical thing by using an existence predicate marked for both time and location, "— to exist —$_{time}$, —$_{location}$". We certify that we're considering only physical things by requiring that every individual thing in a model exists at some time and location. We also assume that predicates are positive for existence in time: if a predication is true of an object at a time and within a place, then the object exists at that time and within that place.

We focus better on what time we are talking about by noting what atomic predications are true of the time. We focus better on what location we are talking about by nothing what atomic predications are true within that location. We focus better on times with talk of the ordering of times. But we have no comparable relation for locations. We can, however, use the orientation of a physical thing in a location to relate locations at a time. There is no orientation of a physical thing in time; it is only a metaphor to say that Suzy faced the future bravely.

With these assumptions added to those of our temporal predicate logic we arrive at a temporal-and-spatial predicate logic compatible with a wide range of views of time and of space. In examining examples of formalizing using this logic, we become clearer about what we mean by "physical thing".

There is much more we could do comparing time and space, testing with formalizing the scope and limits of our temporal-and-spatial logic. But *tempus fugit*. Though space does not fly.

How general are our logics?
I have tried to ensure that the logics of time and space here are compatible with a wide variety of conceptions of time and space and how we understand time and space in our ascriptions of truth and falsity. Perhaps our temporal propositional logic could be seen as so basic as to be universal in formalizing reasoning that takes account of time. But it does not seem that the predicate logic of quantifying over times, much less quantifying over times and places, is universally applicable. It is not just that this or that assumption is debatable, perhaps held by some but denied by others. The greater problem is the tension between our assumption that things exist in a way that transcends any particular time—so that we can have atemporal reference—and that things exist at particular times. And there is the tension between assuming that things exist in a way that transcends any particular location—so that we can have alocational reference—and that things exist in particular locations.

That tension, encountered in many examples here, will lead us in the next volume to consider how to talk and reason about the world as the flow of all with no focus on individual things. To have some idea of that, to see how limiting our assumptions about the world as made up of things in time and space can be, see the essays in *Language and the World*.

APPENDICES

Appendix A Tenses as Propositional Operators

Consider the sentence:

(1)   John loves Mary.

This is not a proposition. It is not true or false. In order to be true or false, references must be supplied for the words "John" and "Mary." But more, we need to know what time (1) is meant to describe. If "John" means the person John Paul Jones, and "Mary" means Mary Magdalene, and the time is meant to be November 16, 2012, then (1) is false. If "John" means the man who married Mary Schwartz Rodrigues of Socorro in 2010, and "Mary" refers to Mary Schwartz Rodrigues, then it is a true proposition about January 16, 2011, but is a false one if it is meant to be about April 13, 2008.

Even with fixed references for the names, we cannot take (1) to be a proposition unless a time is specified. Otherwise (1) could be true at some times and false at others. Yet a proposition is true or false, not both true and false, not true sometimes and false another. When we encounter a sentence that appears to be true at some time and false at another, we know we have an ambiguity. We can resolve that ambiguity by specifying a time that the sentence is meant to describe. Thus, we have to view (1) as a *scheme* awaiting references for the names and a designated time in order to become a proposition. Let's assume in what follows that references have been supplied for "John" and "Mary."[1]

Arthur Prior suggested that we treat "in the past" and "in the future" as adverbs of sentences as wholes.[2] In his approach there are two "temporal operators":

P   meant to be understood as "in the past"
    or "in the past it is true that"

F   meant to be understood as "in the future"
    or "in the future it is true that"

Those readings are ambiguous between "sometime in the past" and "always in the past" and between "sometime in the future" and "always in the future." In the formal systems, "sometime in the past" and "sometime in the future" are normally meant.

Using these operators, we can convert (1) into two distinct propositions:

P (John loves Mary)

F (John loves Mary)

Each is true or false. Taking the default time indication to be the present, such a system treats (1) as being about now.

So let's consider:

(2)   P (John loves Mary)

This is meant to be understood and analyzed as:

---

[1] If propositions are taken to be abstract objects, the issue is whether the sentence under discussion stands for, represents, points to, or somehow indicates a proposition.
[2] See Prior's *Past, Present, and Future*. Steven T. Kuhn and Paul Portner in "Tense and Time" give a survey of work in this area.

(3)   At some time in the past, "John loves Mary" is true.

In this methodology, "P" is said to be an operator on propositions. But the sentence (1) as it appears in (2) cannot be a proposition. If the sentence (1) were a proposition within (2), it would have to be about a time, and prefacing it with "in the past" would make no sense. For example, if the sentence (1) in (2) were taken to be a proposition about now as I'm writing, February 11, 2013, then (3) would mean:

In the past (John loves Mary on February 11, 2013)

That's just nonsense.

Rather, P takes (1) as a scheme and converts that into the proposition (2), where we understand the past as relative to the time I am writing this. That's why I've used quotation marks around (1) in (3).

The sentence (3) gives the conditions under which (2) is true. It says that there is some time in the past that could be assigned to (1) that makes it into a true proposition. The formal semantics build on that. I'll describe those briefly.

First, some conception of time is given. Let's assume for this discussion that time is linear and that it is made up of points, whether those be tiny, such as a nanosecond, or large, like last week. So long as those are ordered linearly, we have a timeline T. Implicit in (3) is that at each time, each atomic scheme, such as (1), is either true or false. That's formalized by assigning to each atomic scheme p a subset of the timeline: those times at which the scheme is true when interpreted as being about that time. Let's notate by $T(A)$ the collection of times at which the scheme A is true. We also need to take some time n as being the now of our timeline. Then in the formal semantics:

P (John loves Mary) iff there is some t in T such that t is before n
and t is in T(John loves Mary)

For "F (John loves Mary)" the condition is the same except "after" replaces "before."

Suppose that "January 6, 1947" is a time in T. At that time, the people to whom "John" and "Mary" refer did not exist. Some would say that "John loves Mary" evaluated at that time is nonsense. But that's not allowed here. On this approach, any atomic sentence that is evaluated at a time at which one or more names lacks reference must be treated as false. Falsity is the default truth-value.

We can extend these semantics to make each time a model not just of the atomic schemes but combinations of those using the formal connectives ∧ for "and" and ¬ for "not", using the classical interpretation of those connectives. Thus, for every time t in the timeline we will have a model M_t of classical propositional logic. The collection of those models constitutes the model M.

This is clear enough. But the operators P and F are allowed to be iterated:

P P (John loves Mary)

P F (John loves Mary)

F P (John loves Mary)

F F (John loves Mary)

P P P (John loves Mary)

P F P (John loves Mary)

244   *Appendix A*

So consider:

(4)   FP(John loves Mary)

Its truth conditions should be:

At some time in the future, "P (John loves Mary)" is true.

But "P (John loves Mary)" is a proposition, evaluated by (3). It is either true or false, and that can't change by prefacing it with F.

The usual presentation of the semantics for such a system obscures this issue. The truth-conditions for (4) are said to be:

(5)   At some time in the future, at some time in the past relative to that time, "John loves Mary" is true.

Using $\vDash$ to stand for "true in the model," condition (5) is:

There is a time $t > n$ such $M_t \vDash$ P (John loves Mary)

which is iff there is a time $t > n$ such that there is a time $t' < t$
   such that $M_{t'} \vDash$ John loves Mary.

The P in (4) is no longer meant to be relative to now as in (3) but to a time in the future.

Here again we see that the operators P and F are not propositional operators. They convert propositional schemes into propositions. But that means we have to view the piece of language "P (John loves Mary)" as a proposition in (2) and as a scheme in (4). Similarly, we have to view the compound "¬ (John loves Mary)" as a proposition, yet as a scheme in "P (¬ (John loves Mary))." There is an endemic ambiguity of scheme vs. proposition in this formal logic.

This arises because a model here is a collection of models of classical propositional logic; to evaluate (4) we have to survey those models. The same problem occurs with formal modal logics, both in syntax and semantics.[3] I suspect that it occurs in the use of any "propositional operator". It does not occur in classical propositional logic, for there though we view "John loves Mary" as a scheme in determining whether an inference is valid, that is outside the formal system. Metalogic and logic are distinguished, not mixed (see pp. 3–4 in this volume).

We could resolve the ambiguity in Prior's approach. We say that we do not form a new proposition "FP(John loves Mary)" from the proposition "P (John loves Mary)." Rather, what we have, and what the formal semantics assume, is that each of P, F, PP, PF, FP, FF, PFP, ... is a distinct "operator" that when prefixed to a proposition-scheme forms a proposition. Thus, each of "John loves Mary," "P (John loves Mary)," "F (John loves Mary)," "PF (John loves Mary)," and so on is an atomic proposition. In the usual formulations of a propositional logic, an atomic proposition is taken to be a unitary whole with no internal structure. In this approach an atomic proposition does have internal structure: a proposition-scheme + tense marker. The axioms of a temporal logic in Prior's tradition can then be understood as relating propositions in terms of the internal structure of not just compound but also atomic propositions.

But why would we want to have such a proliferation of temporal scheme-into-proposition operators? Prior says:

---

[3]   See my "Reflections on Temporal and Modal Logic".

I want to suggest that putting a verb into a past or future tense is exactly the same sort of thing as adding an adverb to the sentence. "I *was* having my breakfast" is related to "I am having my breakfast" in exactly the same way as "I am *allegedly* having my breakfast" is related to it, and it is only an historical accident that we generally form the past tense by modifying the present tense, e.g. by changing "am" to "was", rather than by tacking on an adverb. In a rationalized language with uniform constructions for similar functions we could form the past tense by prefixing to a given sentence the phrase "It was the case that", or "It has been the case that" (depending on what sort of past we meant), and the future tense by prefixing "It will be the case that". For example, instead of "I will be eating my breakfast" we could say

It will be the case that I am eating my breakfast,

and instead of saying "I was eating my breakfast" we could say

It was the case that I am eating my breakfast.

The nearest we get to the latter in ordinary English is "It was the case that I was eating my breakfast", but this is one of the anomalies like emphatic double negation. The construction I am sketching embodies the truth behind Augustine's suggestion of the "secret place" where past and future times "are", and his insistence that wherever they are, they are not there as past or future but as present. The past is not the present but it *is* the past present, and the future is not the present but *is* the future present.

There is also, of course, the past future and the future past. For these adverbial phrases, like other adverbial phrases, can be applied repeatedly—the sentences to which they are attached do not have to be simple ones; it is enough that they be sentences, and they can be sentences which already have tense-adverbs, as we might call them, within them. Hence we can have such a construction as

It will be the case that (it has been the case that (I am taking off my coat)),

or in plain English, "I will have taken off my coat". We can similarly apply repeatedly such *specific* tense-adverbs as "It was the case forty-eight years ago that".[4]

Prior starts with reasoning in English, sees irregularities in it, then devises a formal logic that has little or no resemblance to reasoning in ordinary English. There are some languages that do treat "in the past" and "in the future" as operators on sentences. In Chinese a phrase roughly translated as "before" can precede a sentence, a paragraph, or even a whole story. In American Sign Language words like "yesterday" or "next month" can precede a sentence, as in "Yesterday, John loves Mary," though the more general operators of "in the past" and "in the future," if I understand correctly, attach only to verbs. But I know of no language in which an operator like "in the past" or "in the future" can preface a sentence that already begins with one of those. Iterations are not allowed. Even in English we do not use an adverb to modify a sentence but a predicate. We understand "Spot is barking loudly" not as "Loudly: Spot is barking" but as "Spot is (barking loudly)".

Prior takes the past-present-future divide as basic, the most fundamental metaphysics of time. But speakers of many languages do not parse the world that way.[5] There is no room in his systems for the simpler "before-after" conception of time. Some have tried

---

[4] "Changes in Events and Changes in Things," pp. 40–41.

[5] See, for example, Hopi as described by Benjamin Lee Whorf in "The Relation of Habitual Thought and Behavior to Language" or Joan L Bybee, Revere Perkins, & William Pagliuca, *The Evolution of Grammar: Tense, aspect and modality in the languages of the world*.

to incorporate propositional connectives for before and after into his systems, but to do so exacerbates the scheme vs. proposition confusion: to formalize "Spot barked; then Dick yelled; then Spot barked" would require viewing "P (Spot barked)" as a scheme.

A bigger problem, though, is that for many past-tense propositions the truth conditions Prior gives are just wrong: "Helen had three husbands" is not true iff "Helen has three husbands" is true in the past.[6]

From the very beginning of Prior's work on temporal logic this confusion of scheme and proposition is evident. In "Tense-Logic and an Analogue of S4", Prior begins by taking sentences such as (1) to be schemes of propositions:

> In the logic of tenses, the ordinary statement-variables p, q, r, etc., are used to stand for statements in what is not now the ordinary sense of the term "statement", though it was the ordinary sense in ancient and medieval logic. They are used to stand for "statements" in the sense in which the truth-value of a statement may be different at different times . . . . The statement "It will be the case that Professor Carnap is flying to the moon", as I understand it, is not a statement *about* the statement "Professor Carnap is flying to the moon", but a new statement about Professor Carnap, formed from the simpler one by means of the operator F.   p. 8

But then Prior begins to talk of those sentences as propositions:

> The idea that it is *necessary* to introduce a special present-tense operator would, moreover, have extremely awkward formal consequences. For to say that such an operator is necessary is to say that the expressions to which we attach it would not be propositions, at all events not tensed propositions, without it. This in turn is to say that tense operators do not form propositions out of propositions, at all events out of tensed propositions; rather, they form propositions out of merely juxtaposed nouns and verbs, or they form tensed propositions out of untensed ones. And from this in turn it would follow that tense operators cannot be iterated or attached to propositions to which tense-operators are already attached; that is, we would have to rule out such forms as "It will be the case that it has been the case that p". And to rule this would be practically to destroy tense logic before we have started to build it.   p. 10

I agree. But I take this to mean that tense logic as he does it is wrong, not that we must persevere in the mistake.

---

[6] Timothy Williamson, *Vagueness*, p. 11, notes that Diodorus Cronus already made this observation. The medievals also realized that Prior's truth conditions are wrong: see Example 11 of Chapter 32 here.

Appendix B The Tapestry of Time

Let us suppose:

    Some of what we do we choose freely to do.

Consider, then, the following picture.

At every moment at which a free choice is made, the world branches. So there is a split at the moment at which I chose to sit down to type: in one branch I sit down to type, in another I do not. But that branching is not just a single splitting. Rather, there is a different path according to any of the choices I might have made: go for a walk, play with my dogs, go inspect the sheep corral, stick my head under a faucet to ease my allergies, ...

Similarly, the world branches depending on whether this electron moves to this or that energy level in this atom, if such movements are random. Particular random happenings and particular free choices in any combination lead each to a different branch.

At each "moment" there is a multiplicity of branchings, beyond our ability to comprehend except one branch in comparison to another specific branch. These branchings continue on. What we call "a branch" is just a particular path through all such moments at which a free choice and/or random happening occurs. We do not have any reason to believe that any branch ever stops; nor do we have any reason to think that any branch does not stop. Nor do we have any reason to believe that each point on each branch has a unique path to it: there might have been many ways to arrive at the same point. If we count memory as part of our path, though, then a conscious free choice creates a branching that cannot be reached by any other branch. But a random act of an electron could lead to a point that could be reached by other random physical happenings. This is a way in which the internal world could differ from the external world.

Laws of nature, if there be any, give the substratum of all these branchings. When I choose to sit down and type, I do not have any choice in how the chemical reactions in my blood continue. Given this particular disintegration of the radon atom, the Geiger counter will make a sound.

Each branch is real, as real as any other. These are not "alternate worlds", "alternate possibilities" compared to the world I am in. They are all real, equally real. At a branching, the "I" up to that point continues in a multiplicity of branches, and in each one it is reasonable to say it is the same "I", for they all come from the same branching. Thus, the "I" of this branch is the same "I" as the one in which I chose to go for a walk instead of typing because they can be traced back to that moment at which the "I"s branched apart due to a free choice. If we had the ability to see all these branches, we could say "That is one way I might have been had I done that instead of this".

The world, then, is a tapestry of branchings so multitudinous as to be beyond comprehension in their details: only the general form is conceivable to us. The tapestry is not flowing; there is no movement in the tapestry, only threads that make up the whole. The tapestry is oriented, and that orientation is what we call the arrow from the past to the future. What we call "now" is the consciousness we have of being right here on this branch. Every point on every branch is as real as any other: the unreality of the past and of the future, for me, is that they are not at the point that I call "now" on this branch where I am.

We, each of us, choose which branch we follow, though equally, another person, exactly the same "I" up to a particular branching point, chooses a different branch, and then a different branch, and a different branch again forever. There are some branchings in which the "I" of when I was nineteen chose to be mean to her, and one branch, followed through all its multiple branchings, in which "I" from nineteen on lived a blameless life, good to the point of being saintly.

In this conception, free will is fully compatible with the assumption that there is an omniscient God who knows the future as well as the past. God would be the only intelligence that could comprehend all branches at once. He can see the point on the branch of Jesus' life at which Judas betrayed him and see a branching where Judas did not betray Jesus: in all the branchings that followed the one choice, Judas is damned; in all the branchings that followed the other choice, Judas is saved. Or perhaps not: in some of those branchings he may have repented and been saved; in some of those branchings he could have betrayed Jesus also. Only God knows.

Now all we need is some empirical evidence to support this view.

Appendix C **Events**

> Tom: You look so upset. What's wrong?
> Suzy: Puff scratched Zoe, then the sink got plugged up, and I don't know, just a lot of bad things happened.

Suzy's upset and can't or won't describe what happened. But then she remembers tripping on the rug, getting a splinter in her finger, ... It's enough to make her want to cry.

Some people talk of "what happened" as an event. There was the event of Puff scratching Zoe, the event of the sink getting plugged up. Events, they say, are as real as anything in the world. They can be counted. There are unexpressed events, like what happened during the first three seconds after the Big Bang. We can and should use them in the semantics of predicate logic, particularly in the semantics for time and space.

The only way I know to describe what people call an event is with a true proposition or noun phrase derived from a proposition. We ascribe to an event the structure of the proposition describing it, following the grammar of the description. Other languages describe "what happened" according to very different grammars.[1] But if events are really a category of the world, independent of our language, as we believe individual things are, and we wish to use them in the semantics of predicate logic, we need criteria for distinguishing one event from another. Does "Puff scratched Zoe" describe the same event or a different one from "Suzy's cat extended her claws and dragged them across Zoe's hand"? On what basis do we decide?

Some say an event can be identified with a time and location. The event of Spot sitting can be located in time and space. With time perhaps, using beginnings and endings, we might do that. But with space no, for we have only larger or smaller regions where we can say that Spot was sitting, as we saw in Chapter 40.

I went outside this afternoon. I saw the tree waving in the strong wind. Is that the same event as the tree waving in the wind? Is it the same event as my being outside and watching the tree waving in the wind? I had no sense of an event when I was outside at that moment, only the tree and the wind and the waving and me and . . . whatever I was paying attention to. I did have a strong sense of time, of the now of my being out there, and of the passage of time from when I stepped outdoors until I went back into my office. I had a strong sense of location, of being there—in a place that is vaguely bounded outward. But an event? There's nothing I could point to, physically or intellectually, objectively or subjectively. I am simply at a loss for what was the event.

Perhaps someone can make clear how we can use events in the semantics of predi-cate logic, overcoming these problems.[2] But we have seen that we do not need to talk of events in developing logics that take account of time and space.

---

[1] See *Language and the World: Essays New and Old*.
[2] Appendix 2 of Volume 1 and "Why Event-Talk Is a Problem" for more problems.

Appendix D **Intentions**

    Infinitives used as restrictors 250
    Wffs used as restrictors 252
    Meaningfulness 254
    Chinese 254
    Examples: Open wffs used as restrictors 255
    Examples: Infinitives used as restrictors 256
    Examples: Modifiers of modifiers 257
    Examples: Relations 258
    Examples: Fears, thoughts, knowledge, wishes, feelings 258
    Examples: Intensions 260
    Examples: It doesn't matter that there isn't one 261
    Examples: Qua 263
    Examples: Saying and iterating 263
    A formal logic 264
    Conclusion 265

We understand "Spot wants to play" to be about Spot and what he wants: a desire he has. We understand "Flo thinks that coyotes are dogs" to be about Flo and what she is thinking: a thought she has. We understand "Suzy believes that Spot will bite Puff" to be about Suzy and what she is believing: a belief she has. It seems natural to us to talk of desires, thoughts, and beliefs as things.

There is a different way to understand such sentences that invokes no talk of desires, thoughts, beliefs, but rather ways of wanting, thinking, believing. I'll try to lead you to see this by presenting a method of formalizing such examples in an extension of our logic of quantifying over times.

Infinitives used as restrictors
In English we often use the infinitive form of a verb as a restrictor:

(1)   Spot wants to play.

(2)   Suzy will like to ski.

(3)   Manuel tried to walk.

If (1) is true, so is "Spot wants"; the infinitive "to play" is used to restrict the application of "— wants". If (2) is true, we can conclude "Suzy will like"; the infinitive "to ski" is acting as a restrictor. From (3) we can conclude "Manuel tried"; the infinitive "to walk" is acting as a restrictor.

Yet these seem odd. Can we say that to play is how Spot wants? Can we say that Suzy will like in a skiing-manner? Can we say that Manuel was trying in a walking way? In (1) the predicate "— wants" is restricted to apply to those things that want to play, and there is no adverb-like way of saying that in English. If you like, you could say that "to play", "to ski", "to walk" are adverbs of a kind we haven't seen before. They tell how, but not in a way we can paraphrase with a word that ends in "-ly" or with a phrase that ends with "-like manner".

Reading the infinitives as restrictors seems odd also because we think of "wants", "likes", and "tries" as directed: no one wants except to want something; no one likes except to like something; no one tries except to try something. These are intentional words or words for emotions, and hence there must be an object for them. But emotions need not be directed.[1] It seems incomplete to me to say "I want" without specifying what I want, but it is nonetheless true or false. Compare:

> is mad
> is mad at
> hates

What counts as an intentional predicate is tied to our division of verbs into transitive and intransitive, a division that is not the same in other languages that make that distinction and does not even exist in languages that do not have verbs.[2]

To show that we're treating an infinitive as a restrictor, let's use angle brackets < >. Then we can have an atomic predicate "(— wants $—_{time}$) / <to play>", and we can formalize (1) as:

(4)   ((— to want $—_{time}$) / <to play>) (Spot, now)

We can formalize (2) as:

(5) $\exists t$ (((— to like $—_{time}$) / <to ski>) (Suzy, t) $\wedge$ (now $<_{time} t$))

We can formalize (3) as:

(6) $\exists t$ (((— to try $—_{time}$) / <to walk>) (Manuel, t) $\wedge$ ($t <_{time}$ now))

Since (4), (5), and (6) are atomic, the assignment of a truth-value to each is primitive in a model, relative to our understanding of the words in them and the world. But then how is the use of the infinitive "to play" in (4) related to the predicate "— to play $—_{time}$"? Do we need further semantic conditions beyond those we already have for restrictors to relate an infinitive used as restrictor and the predicate based on that infinitive? We do not have that if Spot wants to play then Spot plays, or that Spot played, or that Spot will play. I don't see any relation we can codify in our system. We can only point in a model to all predications involving the infinitive "to play", whether as the basis for a predicate or used as a restrictor, and say that in total the evaluations of those determine the meaning of "to play" in the model.

Compare how in classical predicate logic we codify in a model the meaning of the predicate "— plays". We assign truth-values to the atomic predications in which it appears based on our understanding of the words in the predications. Some logicians say we should invoke some semantic value beyond the truth-values of the predications, such as the extension of a predicate. But, as I explain in *Predicate Logic*, this is no different from taking the truth-values of the atomic predications as primitive, just clothed in the jargon and doubtful metaphysics of sets, relations, and sequences.

Still, how do we understand these infinitives used as restrictors? Consider:

---

[1] See my paper "The Directedness of Emotions" in *Cause and Effect, Conditionals, Explanations*.
[2] See the essays in *Language and the World*, particularly "Language and the World".

Spot wants to play.
Dick wants to play.
So Dick and Spot want the same thing.

This is to treat Dick and Spot as having some mental condition, an intention, that they share. But an intention is not a thing. Rather, in formalizing (1) as (4), we treat "to play" as characterizing how Spot wants. We need not appeal to Spot's mental life. We infer from Spot's actions that he wants to play, just as we infer from the actions of Dick that Dick wants to play. Intentions here, if we are to use that word, are ways of being in the world, ways of acting if you prefer. But even that is misleading because it, too, suggests that an intention is a thing—a way—and then Spot and Dick could share that way. Rather, Spot and Dick are acting, and we interpret, we characterize that acting, with certain words. Compare:

Spot barked loudly.
Therefore, Spot barked.

The restrictor "loudly" characterizes how Spot was barking; it does not stand for a thing. We don't have to assume that Spot has intentions. Intending is what we and dogs do.

Each of (4), (5), and (6) is atomic. The infinitives "to play", "to ski", and "to walk" are not used in them as bases for predicates, nor as names of things (intentions). How we understand them is part of the larger issue of how we understand any word, whether "dog", "to want", or "brown". I have given my story of that in "Language-Thought-Meaning"; along those lines you could say that these infinitives convey the concepts they are meant to elicit, where a concept is not a fixed thing but a way for a person: meaning is embodied. There is not a mapping of words to meanings but words used in our paths in life.

Wffs used as restrictors

In (1), (2), and (3) it is not said explicitly but we understand that the term that fills the blank in the predicate is also meant to be the "subject" of the infinitive. But consider:

(7) Dick wants Spot to bark.

Here the "subject" of the infinitive is not the subject of the predicate. The phrase "Spot to bark" is used to restrict the application of "wants"; it tells how Dick wants. If (7) is true, then so is "Dick wants", where the time for that is now, and the time he wants for Spot to play is also now. We can formalize (7) if we allow the infinitive to have blanks that are filled, marking such use with bold curly brackets { }.

(8) $((- \text{ to want } -_{\text{time}}) / \{(- \text{ to bark } -_{\text{time}}) (\text{Spot, now})\} (\text{Dick, now}).$

Here the restrictor is:

(9) $(- \text{ to bark } -_{\text{time}}) (\text{Spot, now})$

This is a closed wff. We have always treated closed wffs and only closed wffs as propositions in a model. But (9) is not being used as a proposition in (8); it is used as a restrictor, part of the internal structure of (8). It is (8) that is a closed atomic wff that is a proposition in a model. Its truth-value does not depend on whether (9) is evaluated as true or false.

How then are we to understand (9) used as a restrictor? To evaluate (8) in a model, as with any atomic wff, we use how we understand the words in it. We assign no semantic value to (9) as a restrictor that can be used to build the truth-value of (8) in terms of its parts.

Compositionality—the principle that the semantic value of the whole is determined by the semantic values of its parts—fails here, just as it fail in classical predicate logic with restrictors: we do not assign a semantic value to any restrictor, not even "loudly".

We ensure compositionality in classical predicate logic and in our logic of quantifying over times by abstracting from meaning to consider only truth-values of atomic propositions, references for names, and references for variables relative to assignments of references. We treat those semantic values as things, summing up the parts to the whole. But we know that meaning is not that way in our lives. Meanings are not things but are in the body, ways of being in the world using language. They are not "in our heads". There is nothing we can point to—in the world external to us or in our minds—that is the meaning of "dog" but only how you and I and others use that word to navigate in the world, negotiating meaning.[3] The whole is greater than the sum of its parts.

Similarly, consider

(10) Suzy hopes that Zoe will call.

The phrase "Zoe will call" restricts the application of "hopes"; it tells us in what way or how Suzy is hoping. Here the "subject" of the infinitive is not the same as the subject of the predicate, and the time for the infinitive is not the time of the main predicate, which is now. We can formalize (10) as:

(11) $((- \text{to hope} -_{\text{time}})/\{A\})(\text{Suzy}, \text{now})$

where we take:

(A) $\exists t \, ((- \text{to call} -_{\text{time}})(\text{Zoe}, t) \land (\text{now} <_{\text{time}} t))$

In (11), (A) is a restrictor: it tells us how Suzy is hoping, not what Suzy is hoping. It has the form of a closed wff, though not atomic. The truth-value of (A) in the model is irrelevant to the truth-value of (11), for (10) and hence (11) could be true whether Dick calls or Dick does not call. The sentence "Zoe will call" in (10) is used not as a proposition but as a restrictor, meant to direct us not to a way the world is or could be, but to how Suzy wants, no more a proposition than "to Dick" is in "Spot ran to Dick".

Some say that "Zoe will call" is used as a proposition in (10), it's just that it appears in an *oblique* or *intensional context*. That is, substituting a sentence extensionally equivalent to "Zoe will call" need not preserve the truth value of (10), for (10) could be true while "Suzy hopes that Melinda's niece will call" could be false if Suzy does not know that Zoe is Melinda's niece.

But if "Zoe will call" is being used as a proposition in (10), what if it's false? Then we have to say that Suzy is hoping for something false. I can't make sense of that. (If the future tense bothers you, consider "Zoe hopes that Dick washed the dishes"). What we can have as true in a model is:

(12) $((- \text{to hope} -_{\text{time}})/\{A\})(\text{Suzy}, \text{now}) \land \neg A$

Actually, if "Zoe will call" is being used as a proposition in (11), we should rewrite (10) as:

Suzy hopes that "Zoe will call" is true.

And here insert a resolution of the liar paradox.

---

[3] See my "Language-Thought-Meaning".

With the definition of proposition we started with in Chapter 1, "Zoe will call" is not used as a proposition in (10), nor is its formalization in (11). It is a sentence. And being a sentence, it is meaningful.

But what does it mean to say a sentence is meaningful?

Meaningfulness

If not a proposition, perhaps "Zoe will call" in (10) can be looked at as a scheme of propositions. We would understand it as the contexts in which it could be true or could be false. In formal logic that amounts to surveying the models in which it is true and the models in which it is false. But even in the most extensive formal logics, the meaning of a proposition in a model is severely constrained. This is a tractable but very truncated notion of meaning.[4]

Another way to think of meaning for sentences such as "Zoe will call", as I suggested above, is to survey the wffs in which it appears and is true, and those in which it appears and is false; that division sets out how we (can) understand the sentence.

But all this is abstracting from meaning, not a basis for the work here. What are we abstracting?

The sentence "Zoe will call" is meaningful. I could leave that open to many interpretations, but that is not enough here. I take the notion of meaningful as in my essay "Language-Thought-Meaning". Suzy understands the words. She might have some idea of what contexts would make "Zoe will call" true and which would make it false, though probably she has never considered those beyond thinking of one or two possibilities, as most of us do in trying to plan. Perhaps we could say that she has some concept associated with that sentence, in the sense of a concept not being a fixed thing, not a mental thing nor an object, as I explain in "Language-Thought-Meaning". And that concept gives the way that Suzy hopes; it is part of her disposition. Suzy's whole body is involved in her hoping that way.

The dispositions for meaningfulness for using particular words, such as "dog" or "justice", are constrained by norms of our language community. This is what allows us to extend the idea of disposition for meaningfulness to sentences, where perhaps no one previously used "Zoe is a bipedal mammal" nor will use it again. What we have are norms of use—which include but are not limited to grammar—that constrain how we use a sentence and how we understand the sentence.

I think that most of us understand "Zoe will call" not as a scheme of propositions but in terms of how we understand its parts as put together according to our grammar and our norms of use.[5]

Chinese

Instead of desires, thoughts, and beliefs, we can talk of ways of wanting, thinking, and believing. That is very odd for speakers of English and other Indo-European languages, seemingly too bizarre to be taken seriously. But it is natural for those who wrote and spoke ancient pre-Han Chinese, as I have learned from Chad Hansen. They did not focus on the world as made up of things, they did not talk of propositions, they made no mind-body

---

[4] See my "Possibilities and Valid Inferences".
[5] Another way to explain meaningfulness might be to return to the medieval analysis of a predicate or proposition having signification but not supposition.

distinction, and they did not have verbs, much less intentional verbs. It is our translations of their talk that introduce those.[6] In "Chinese Language, Chinese Philosophy, and 'Truth'" Hansen says:

> I shall argue that classical Chinese philosophy had a different conception of both knowledge and belief. The classical Chinese grammatical structures that we translate as belief expressions were simple two-place predicates—action expressions. I call the expressions "term-belief" contexts. Where Western philosophy of mind dealt with input, procession, and storage of content (data, information), Chinese philosophers portrayed heart-mind as consisting of dispositional attitudes to make distinctions in guiding action. Sentential belief statements represent a relation between a person and a sentence believed. Term-belief, in Chinese, represents a way of responding rather than a propositional attitude.
>
> No single character or conventional string of ancient Chinese corresponds in a straightforward way to "believes that" or "belief that." No string or structure is equivalent to "believe" or "belief" in the formal sense that it takes sentences or propositions as its object. Where English would use a structure such as "King Wen believes that Ch'ang An is beautiful," pre-Han Chinese employed two different structures. The simplest uses the descriptive predicate term as the main verb, "King Wen beautifuls Ch'ang An."
>
> This belief structure of ancient Chinese language signals a different philosophy of mind as well as a different epistemology. It does not generate a picture of some "mental states" with a sentential, propositional, or representational content. Corresponding to King Wen's "belief" is a disposition to discriminate among cities. He discriminates among cities in such a way that Ch'ang An falls on the beautiful side. "Beautiful-ing" a city involves both linguistic and non-linguistic dispositions, for example, King Wen's disposition to classify and distinguish things, to issue orders to his bearers, court artists, and so forth. The most straightforward evidence that he discriminates is his tendency to utter "beautiful" when the dialogue context makes Ch'ang An a topic of discussion. If we think of speech acts rather than beliefs, we will grasp the action-oriented implications of term-belief structure. Students of Chinese learn to talk about the structure as having either a "causative" or "putative" reading. We are taught to translate the sentence discussed above as either "King Wen beautified Ch'ang An" or as "King Wen regards Ch'ang An as beautiful," depending on the context.
>
> ... Deeming ... to be beautiful or "beautiful-ing" are things we do. They are not merely the "having" of some mental "content." The dispositional analysis more naturally reflects the syntax of either term-belief structure than does the mental content analysis.   pp. 500–501

Open wffs as restrictors

Can an open wff serve to describe a way of wanting, thinking, or believing? Consider:

> Spot smells a steak that he wants to eat.

Using the methods we've devised, we would have:

---

[6] And similarly for many others languages, as described in *Language and the World: Essays New and Old*.

256   *Appendix D*

(13) $\exists x(((- \text{ to smell } -_{\text{time}})/\text{obj}(x))(\text{Spot, now})$
  $\land (- \text{ to be a steak } -_{\text{time}})(x, \text{now})$
  $\land (- \text{ to want } -_{\text{time}})/\{((- \text{ to eat } -_{\text{time}})/\text{obj}(x))(\text{Spot, now})\})$

The restrictor is an open wff:

(14) $((- \text{ to eat } -_{\text{time}})/\text{obj}(x))(\text{Spot, now})$

The quantifier that governs the variable in it and that closes the entire wff is outside the restrictor. Can we understand (14) as saying how, in what way, Spot wants when there is no value for the variable x? In evaluating (13) in a model, we do assign a value to x:

(13) is true iff there is some assignment of references σ such that:

$\vee_\sigma(((- \text{ to smell } -_{\text{time}})/\text{obj}(x))(\text{Spot, now})) = T$ and
$\vee_\sigma((- \text{ to be a steak } -_{\text{time}})(x, \text{now})) = T$ and
$\vee_\sigma((- \text{ to want } -_{\text{time}})/\{((- \text{ to eat } -_{\text{time}})/\text{obj}(x))(\text{Spot, now})\}) = T$

When a value is assigned to the variable, (14) is used as a restrictor just as a closed wff is.

Examples: Infinitives used as restrictors

Example 1   Dick wants to play.

Analysis We can formalize this as an infinitive restricting a predicate:

$((- \text{ to want } -_{\text{time}})/\langle \text{to play} \rangle)(\text{Dick, now})$

Or we can make the subject and time of the infinitive explicit:

$((- \text{ to want } -_{\text{time}})/\{(- \text{ to play } -_{\text{time}})(\text{Dick, now})\})(\text{Dick, now})$

But don't we understand the example as meaning that Dick wants to play sometime after when he wants, not at the same time as he wants?

$((- \text{ to want } -_{\text{time}})/\{\exists t((- \text{ to play } -_{\text{time}})(\text{Dick}, t) \land (\text{now} <_{\text{time}} t))\})(\text{Dick, now})$

Example 2   Wanda wants to be a veterinarian.

Analysis We can formalize this as an infinitive used as a restrictor:

$((- \text{ to want } -_{\text{time}})/\langle \text{to be a veterinarian} \rangle)(\text{Wanda, now}))$

But making the subject and time of the infinitive explicit, we understand the example to mean that Wanda wants now to be a veterinarian in the future, since if she were a veterinarian now she wouldn't want to be one.

$((- \text{ to want} -_{\text{time}})/\{B\})(\text{Wanda, now}))$

(B) $\exists t((- \text{ to be a veterinarian} -_{\text{time}})(\text{Wanda}, t) \land (\text{now} <_{\text{time}} t))$

Note that a classification predicate is used in the restrictor.

Example 3   Manuel tried to walk.

Analysis We can formalize this as an infinitive restricting a predicate:

$\exists t(((- \text{ to try } -_{\text{time}})/\langle \text{to walk} \rangle)(\text{Manuel}, t) \land (t <_{\text{time}} \text{now}))$

Since Manuel is paralyzed and he has been taking physical therapy, there seems no choice but to take a wff as restrictor to be for the same time as the trying:

$\exists t\,(\,((-\text{ to try }-_{\text{time}})\,/\,\{(-\text{ to walk }-_{\text{time}})\,(\text{Manuel}, t)\})\,(\text{Manuel}, t)\,)\,\wedge\,(t <_{\text{time}} \text{now})\,)$

Example 4 *Manuel wants to dance reggae.*

Analysis If we formalize the example as an infinitive used as a restrictor, we have:

$((-\text{ to want }-_{\text{time}})\,/\,(<\text{to dance}>)\,/\,\text{reggae})\,(\text{Manuel}, \text{now})$

But the example is ambiguous. Should it be understood as Manuel wants to dance reggae now — which he can't because he is confined to a wheelchair?

$((-\text{ to want }-_{\text{time}})\,/\,\{((-\text{ to dance }-_{\text{time}})\,/\,\text{reggae})\,(\text{Manuel}, \text{now})\})\,(\text{Manuel}, \text{now})$

Or should we understand that he wants to dance reggae sometime in the future?

$((-\text{ to want }-_{\text{time}})\,/\,\{C\})\,(\text{Manuel}, \text{now})$

(C) $\exists t\,(\,((-\text{ to dance }-_{\text{time}})\,/\,\text{reggae})\,(\text{Manuel}, t) \wedge (\text{now} <_{\text{time}} t)\,)$

Example 5 *Suzy expects Zoe to call.*
Dick expects Zoe to call.

Analysis We can formalize these using infinitives as restrictors:

$((-\text{ to expect}-_{\text{time}})\,/\,(<\text{to call}>))\,(\text{Suzy}, \text{now})$

$((-\text{ to expect}-_{\text{time}})\,/\,(<\text{to call}>))\,(\text{Dick}, \text{now})$

But does Suzy expect Zoe to call her or Dick? Does Dick expect Zoe to call Suzy or him? Or neither?

The use of an infinitive is ambiguous as to its implicit time or subject or object. I see no rule that can guide us beyond our understanding of the words used in context or asking the person who makes the claim. Should we, then, exclude the use of infinitives as restrictors because they are ambiguous?

It is not that the use of an infinitive is ambiguous, it is just incomplete. But then everything we say is incomplete, less than a full description. Compare formalizing "Spot barked". We have to infer from the context or ask the speaker whether it's meant as a description completed in the past or continuing into the present.

The ambiguity of the use of infinitives is due to the ambiguity of our ordinary speech, which our formalizing can show but not cure.

Examples: Modifiers of modifiers

Example 6 *Spot wants to run quickly.*

Analysis We formalize this with a restrictor within a wff used as a restrictor:

$((-\text{ to want }-_{\text{time}})\,/\,\{((-\text{ to run }-_{\text{time}})\,/\,\text{fast})\,(\text{Spot}, \text{now})\})\,(\text{Spot}, \text{now})$

Example 7 *Zoe almost wanted to dance.*

Analysis We formalize this with a restrictor of a negated predicate.

$\exists t\,(\,(((-\text{ to want }-_{\text{time}})\,/\,\text{almost})\,/\,\{(-\text{ to dance }-_{\text{time}})\,(\text{Zoe}, t)\})\,(\text{Zoe}, t)$
$\wedge\,(t <_{\text{time}} \text{now})\,)$

Example 8   Zoe wanted to almost dance.

Analysis We formalize this with a negator modifying a predicate in the wff used as a restrictor:

$\exists t\,((($— to want $-_{\text{time}})/\{((($— to dance $-_{\text{time}})/$almost$)$ (Zoe, t)$\})$ (Zoe, t) $\land$ ($t <_{\text{time}}$ now))

Example 9   On May 27th, Dick planned to play tennis with Suzy the next day.

Analysis We can formalize this:

$\exists t_1\,(($ (— to plan $-_{\text{time}})/\{D\}$) (Dick, t) $\land$ $W_{\text{time}}(t_1,$ May 27 2017))

(D) $\exists t_2\,(((($— to play $-_{\text{time}})/$tennis$)/$with (Suzy)) (Dick, t_2) $\land$ $W_{\text{time}}(t_2,$ May 28 2017))

With enough context we can formalize a lot about making plans.

Examples: Relations

Example 10   Zoe wanted to be taller than Dick some day.

Analysis This was true when Zoe and Dick were children. We can formalize it as:

$\exists t\,(($ (— to want $-_{\text{time}})/\{E\}$) (Zoe, t) $\land$ ($t <_{\text{time}}$ now))

(E) $\exists w\,(($ (— to be taller than —, $-_{\text{time}}$) (Zoe, Dick, w) $\land$ ($t <_{\text{time}} w$))

A binary predicate is used in a restrictor here, and one of the variables in (E) is bound within (E), and the other is bound outside (E).

Example 11   Dick is throwing a ball for Spot to fetch.

Analysis The examples we've seen have been formalized as a unary predicate restricted by an infinitive or wff. Here we have a predicate "to throw" which we classify as binary (transitive) in English. However, we can use the variable restrictor "obj ( )" for direct objects and so eliminate the need to talk of transitive versus intransitive verbs, formalizing the example as:

$\exists x\,(($ — to be a ball $-_{\text{time}}$) (x, now) $\land$ ((— to throw $-_{\text{time}})/$obj$(x))/\{F\}$) (Dick, now)

(F) ((— to fetch $-_{\text{time}})/$obj(x)) (Spot, now)

Example 12   Zeke and Isaiah want to be friends.

Analysis We can formalize this using conjunctions of terms rather than taking "to be a friend of" as binary:

((— to want $-_{\text{time}})/\{G\}$) (Zeke $\land$ Isaiah, now)

(G) $\exists t\,(($ — to be a friend $-_{\text{time}}$) (Zeke $\land$ Isaiah, t) $\land$ (now $<_{\text{time}} t$))

I don't have an example of an infinitive or wff used as a restrictor of a relation. We don't understand relations as involving intentions, as far as I know. Nor do I have an example of a wff used as a negator.

Examples: Fears, thoughts, knowledge, wishes, feelings

Example 13   Suzy is afraid that Spot will bite Puff.

Analysis We can formalize this as:

((− to be afraid −$_{time}$) / {H}) (Suzy, now)

(H) $\exists t$ (((− to bite −$_{time}$) / obj (Puff)) (Spot, t) ∧ (now <$_{time}$ t))

What is the fear that Suzy has? We describe it with a sentence, "Spot will bite Puff", formalized as (H). Is that the same fear she has that Dick's dog will bite Puff? Is that the same fear she has that Puff will be bitten? Is that the same fear she has that Puff will be run over by a car? We can describe these "fears" only by invoking specific wffs. Then we ask whether the resulting ways of fearing used as restrictors yield equivalent propositions.

Example 14 *Dick and Zoe both saw something that they thought was a coyote.*

Analysis Some would conclude from the example:

Dick and Zoe had the same thought.

But what is the thought that both Zoe and Dick had? Is it some state of their brains? How do we draw equivalences of brain states? Is the thought conscious? Is it part of the world outside their mental lives?

These questions seem intractable. But all we need is that how they are thinking is the same, using the same semi-formal wff as restrictor.

$\exists x \exists t$ (((− to see −$_{time}$) obj(x)) (Dick, t) ∧ (t <$_{time}$ now)

∧ ((− to see −$_{time}$) obj(x)) (Zoe, t) ∧ (t <$_{time}$ now)

∧ ((− to think −$_{time}$) / {(− to be a coyote −$_{time}$) (x, t)}) (Dick, t)

∧ ((− to think −$_{time}$) / {(− to be a coyote −$_{time}$) (x, t)}) (Zoe, t)))

Example 15 *Flo knows that Spot is a dog.*
Therefore, *Flo knows something.*

Analysis What is the thing that Flo knows? To formalize this example, we would need to allow for quantifying over closed wffs used as restrictors of " (− to know −$_{time}$)".

Example 16 *Flo knows how to whistle.*

Analysis We can formalize this as:

((− to know −$_{time}$) / <to whistle>) (Flo, now)

In the Western tradition we distinguish between knowing how and knowing that. But here both are seen as ways of knowing, distinguished only by whether the restrictor of "to know" is an infinitive or a closed wff.[7]

Example 17 *Spot barked that he was hungry.*

Analysis We don't have to invoke Spot's thoughts or mental states to evaluate this. Just as we normally do, we say he was barking in a way we interpret as showing he was hungry.

$\exists t$ (((− to bark −$_{time}$) / { (− to be hungry −$_{time}$) (Spot, t)}) (Spot, t) ∧ (t <$_{time}$ now))

---

[7] If I understand correctly from Chad Hansen's *A Daoist Theory of Chinese Thought*, pre-Han Chinese philosophers made no distinction between knowing how and knowing that. Indeed, the latter was not even considered, for they used no notion of proposition.

Example 18 Dick wishes he were rich.

Analysis Rewriting the example as "Dick wishes that he were rich", we can formalize it:

$$((- \text{ to wish } -_{\text{time}})/\{(- \text{ to be rich } -_{\text{time}})(\text{Dick, now})\})(\text{Dick, now})$$

But can "Dick wishes" be true absent any wish?[8]

Example 19 Zoe: How do you feel?
  Dick: *I've got a pain in my shoulder.*
  Zoe: Me, too.

Analysis Is the pain that Dick has the same as the pain Zoe has? Can I feel the pain you have in your foot? Can anyone feel my pain? We are led to ask these questions by treating pains as things.

But Zoe is asking how Dick feels. He could say that he feels bad, or he feels happy, or he feels hungry. Or he feels pain-in-my-shoulder. That's to treat pain-in-my-shoulder as a way of feeling. We can parse the example as:

$$(- \text{ to feel } -_{\text{time}})/(\text{pain}/\text{in}(\text{shoulder})/\text{of}(\text{Dick}))(\text{Dick, now})$$

Whether you can feel my pain becomes a question of whether you can feel in the same way I do. And for that we look to behavior rather than inaccessible mental states.

Hopes, fears, desires, thoughts, wishes, feelings—we invoke these to find a thing that is the cause of how Suzy or Dick or Spot is acting. But all we need in evaluating such examples, and all we can use, is our observation of how they act, including how they talk. To say that "Spot wants to play" is true because Spot has a desire to play is no more than rewriting "Spot wants to play".

Examples: Intensions

Example 20 (a) *Walter believes that Marilyn Monroe was an actress.*
      (b) *Marilyn Monroe is Norma Jean Mortensen.*
  Therefore (c) *Walter believes that Norma Jean Mortensen was an actress.*

Analysis Is the inference valid? Walter doesn't know that Norma Jean Mortensen was Marilyn Monroe, so how could he believe that Norma Jean Mortensen was a great actress?

Some would say that the terms "Marilyn Monroe" and "Norma Jean Mortensen" are used here *intensionally*. But all that means is that the truth-values of (a) and (c) depend on some semantic value of "Marilyn Monroe" and "Norma Jean Mortensen" other than their reference(s). In this case, it seems, the evaluations depend on what those terms mean to Walter. But we have no way to grasp that semantic value except to track how the atomic sentences that involve referring to Walter that include the terms "Marilyn Monroe" and "Norma Jean Mortensen" are evaluated. Or at least I see none. Talk of intensions as opposed to extensions does not clarify here.[9]

We can formalize the example:

---

[8] See my "The Twenty-First or "Lost" Sophism on Self-Reference of John Buridan" at www.AdvancedReasoningForum.org.

[9] See Chapter 13 of Volume 1, *The Internal Structure of Predicates and Names* for a fuller discussion of intensions and extensions.

(a′)   (((− believes −$_{time}$)/{I}) (Walter, now)
(b′)   Marilyn Monroe ≡ Norma Jean Mortensen
Therefore, (c′) (((− believes −$_{time}$)/{J}) (Walter, now)

where (I) and (J) are:

(I) $\exists t$ ((− to be an actress −$_{time}$) (Marilyn Monroe, t) ∧ ($t <_{time}$ now))

(J) $\exists t$ ((− to be an actress −$_{time}$) (Norma Jean Mortensen, t) ∧ ($t <_{time}$ now))

The inference is not valid because (I) and (J) are different restrictors. The reference(s) of the names in them is not part of the evaluation of (a′) and (c′), for there is no semantic value of (I) or (J) used as restrictors that is built up from the parts of (I) and (J). Believing is something we do, and there are ways we do it.

Some would say that the conclusion follows with the additional premise "Walter believes that Marilyn Monroe is Norma Jean Mortensen ". We can formalize that as:

((− believes −$_{time}$)/{ Marilyn Monroe ≡ Norma Jean Mortensen}) (Walter, now)

Examples: It doesn't matter that there isn't one

Example 21   *Isabel thinks that Bertha is a witch.*

Analysis   We can formalize this as:

((− to think −$_{time}$)/{K}) (Isabel, now)

(K)   ((− to be a witch$_{habitual}$ −$_{time}$)) (Bertha, now)

It doesn't matter to the truth-value of the formalization that (in a model) there are no witches.

Example 22   *Flo wants a dog.*

Analysis   Following the grammar, it seem we should formalize this as:

(a) $\exists x$ (((− to want −$_{time}$)/obj (x)) (Flo, now) ∧ (− to be a dog) (x, now))

But this is wrong. For (a) to be true, there has to be at least one specific dog that Flo wants. Yet Flo wants to have a dog without her wanting being directed to any one dog: Flo has a disposition, not a mental condition (thought) directed at some object. It is not correct to take "a dog" as a direct object of the verb "wants".

So what is the role of "a dog" in (a)? If (a) is true, then so is "Flo wants". That is, "a dog" is acting as a predicate restrictor. Perhaps we could use the infinitive "to be a dog" to formalize the example as:

((− wants) /<− to be a dog>) (Flo, now)

But this formalizes "Flo wants to be a dog". Yet we know that the example is short for:

(b)   Flo wants to have a dog

This we can formalize:

(− to want −$_{time}$)/{L} (Flo, now)

(L) $\exists x$ (((− to have$_p$ −$_{time}$)/obj (x)) ∧ (− to be a dog) (x, now))

The existential quantifier in the restrictor does not entail picking out any dog for the formal wff to be true. Rewriting the example as (b) is an unavoidable analysis before formalization.

Example 23   (a) *Suzy hopes someone will call.*
              (b) *There is someone Suzy hopes will call.*

Analysis Are these equivalent?

For (a) to be true, there need not be any particular person Suzy is hoping will call. We formalize it as:

(a′) $(\,(\,-\text{ to hope }-_{\text{time}})\,/\,\{M\}\,)\,(\text{Suzy, now})$

     (M) $\exists t\,\exists x\,(\,(-\text{ to call }-_{\text{time}})\,(x, t)\,\wedge\,(-\text{ to be a person }-)\,(x, t)\,\wedge\,(\text{now} <_{\text{time}} t)\,)$

For (b) to be true, there must be a some person that Suzy hopes will call. So the existential quantifier is outside the restrictor:

(b′) $\exists x\,(\,(-\text{ to want }-_{\text{time}})\,/\,\{N\}\,)\,(\text{Suzy, now})\,)$

     (N) $\exists t\,(\,(-\text{ to call }-_{\text{time}})\,/\,\text{obj}(\text{Suzy})\,(x, t)\,\wedge\,(\text{now} <_{\text{time}} t)\,)$

The semi-formal wffs (a′) and (b′) are not equivalent.

Example 24   (a) *Dick ate a steak.*
              (b) *Dick wanted a steak.*

Analysis From (a) "There is a steak" follows. So we formalize (a) as:

    $\exists t\,\exists x\,(\,\text{Max}((-\text{ to eat }-_{\text{time}})\,/\,\text{obj}(x))\,(\text{Dick}, t)$
        $\wedge\,(-\text{ to be a steak }-_{\text{time}})\,(x, t)\,\wedge\,(t <_{\text{time}} \text{now})\,)$

In the grammar of English, (b) has the same form as (a). But we can't formalize it in parity with (a). The phrase "a steak" in (b) does not refer to anything. We can't conclude "There was a steak" from (b), for (b) could be true even if there had been no steaks in the world. We don't want to populate our universe with Dick's imaginary steaks or possible steaks. I think we understand (b) as:

    Dick wanted to eat a steak.

We can formalize this as:

(c) $(\,-\text{ to want }-_{\text{time}})\,/\,\{O\}\,)\,(\text{Dick, now})$

     (O) $\exists t\,\exists x\,(\,(-\text{ to be a steak }-_{\text{time}})\,(x, t)\,\wedge$
          $(-\text{ to eat }-_{\text{time}})\,/\,\text{obj}(x))\,(\text{Dick, now})\,\wedge\,(t <_{\text{time}} \text{now})\,)$

There is no object referred to in (c).

Example 25   (a) *Jim wrote a book.*
              (b) *Jim was writing a book when he died.*

Analysis From (a) we can conclude "There is a book that Jim wrote". From (b) we cannot conclude that there is or was a book that Jim was writing: the continuous tense indicates that he did not complete it. But that's the problem: there is no "it", no reference of any kind for "a book", not even a future one.

The lack of an object is clearer if we write out more fully what we mean by (b):

    Jim wrote with the intention that his writing would be a book.

Jim's action was completed. Indeed, at any moment that he stopped you could say his action was completed: it was the action that started when it could first be characterized as writing and it stopped when he stopped. What is not completed is what he wanted to do, or what he intended to do, or perhaps what we imagine him to be intending to do. We formalize (b):

$\exists t\,((\,(-\text{ to write }-_{\text{time}})\,/\{P\}\,)\,(\text{Jim},t)\,\wedge\,(-\text{ to die }-_{\text{time}})\,(\text{Jim},t)\,\wedge\,(t<_{\text{time}}\text{now})\,)$

(P) $\exists w\,\exists x\,(\,(-\text{ to be a book }-_{\text{time}})\,(x)\,\wedge\,(t<_{\text{time}}\text{now})\,)$

Examples: Qua

Example 26 Dean Furtz respects Dr. E as a teacher but not as a scholar.

Analysis Here we have two propositions we can formalize as joined by $\wedge$:

(a) Dean Furtz respects Dr. E as a teacher.
(b) Dean Furtz does not respect Dr. E as a scholar.

We understand (a) to mean that Dean Furtz respects Dr. E as someone who teaches regularly, habitually, and similarly for (b). Using the analysis of the habitual tense from Chapter 35, we can formalize these:

(a´) $(-\text{ to respect }-_{\text{time}})\,/\,\text{obj}(\text{Dr. E})\,/\,\{Q\}\,(\text{Dean Furtz, now})$
 (Q) $(-\text{ to be a teacher}_{\text{habitual}}-_{\text{time}})\,(\text{Dr. E, now})\,)$

(b´) $(-\text{ to respect }-_{\text{time}})\,/\,\text{obj}(\text{Dr. E})\,/\,\{R\}\,(\text{Dean Furtz, now})$
 (R) $\exists t\,(\,(-\text{ to be a scholar}_{\text{habitual}}-_{\text{time}})\,(\text{Dr. E, now})\,)$

Since (Q) and (R) are different, the truth-values of the atomic propositions (a´) and (b´) are independent; neither follows from the other.

Example 27 As a horse trainer, Gerald has a lot of patience.

Analysis We couldn't formalize this in Volume 1 (Example 11 of Chapter 33), though we could eliminate the mass term "patience" by rewriting the example:

(a) As a horse trainer, Gerald is very patient.

We can't take "as" to introduce a modifier here: it isn't that Gerald is patient, now or habitually, in the manner of a horse trainer. Rather, he's patient *when* he's training a horse: no matter that a cat scratches a child who screams and the horse nips him, he doesn't lose his temper. We can formalize (a) as:

$\forall t\,(\,(-\text{ to be a horse trainer }-_{\text{time}})\,(\text{Gerald},t)\,\rightarrow$
 $(\,(-\text{ to be patient }-_{\text{time}})\,/\,\text{very})\,(\text{Gerald},t)\,)$

Patience here is not construed as something one has but as how one acts.

Examples: Saying and iterating

Example 28 Dick said that Zeke is a criminal.

Analysis There is no use-mention confusion here: "Zeke is a criminal" is a sentence but not used as a proposition or as a quoted part of speech. It is used as a restrictor, telling how Dick said. We can formalize the example as:

$\exists t\,(\,((-\text{ to say }-_{\text{time}})\,/\{S\}\,)\,(\text{Dick},t)\,\wedge\,(t<_{\text{time}}\text{now})\,)$

(S) $(-\text{ to be a criminal }-_{\text{time}})\,(\text{Zeke},t)$

Example 29 *Tom wants Suzy to think that he is sleeping.*

Analysis We can formalize this as:

((— to want —$_{time}$) / {T}) (Tom, now)

(T) ((— to think —$_{time}$) / {U}) (Suzy, now)

(U) (— to sleep —$_{time}$) (Tom, now)

A wff is used as a restrictor in a wff that is used as a restrictor. So for a formal system, we will need to build the formal language in stages, at each stage allowing any wff of the previous stage to be used as a restrictor.

A formal logic

Syntax

We define the formal language L in stages.

L_0
This is the formal language of **QT+interior** (Chapter 23).

L_1
We add to the definition of *formal modifier* of L_0 the following clauses:

If I is an infinitive symbol, then <I> is a restrictor.
It is an *infinitive used as restrictor*.

If A is a wff of L_0, then {A} is a formal restrictor of degree 1.
It is a *wff used as restrictor*.

We add to the definition of *formal ordinary atomic predicate* the following clauses:

If Q(—, —$_{time}$) is unary formal atomic predicate of degree *n*, then (Q/<I>) (—, —$_{time}$) is a unary formal atomic predicate of degree *n* + 1.

If Q(—, —$_{time}$) is unary formal atomic predicate of degree *n*, and {A} is a formal wff used as a restrictor, then (Q/{A}) (—, —$_{time}$) is a unary formal atomic predicate of degree *n* + 1.

L_{i+1}
This is defined as is L_1 except replacing L_0 in that definition with L_i.

L
A concatenation of symbols is a *wff* iff it is a wff of stage *i* for some *i*.

I'll let you prove that there is one and only one way to parse each wff of L.

Semantics

A model is defined as it was for **QT+interior**.

To formalize reasoning in English, we can add the semantic conditions for quasi-linear time (Chapter 24, p. 124).

Conclusion

We've seen how it is possible to understand talk of intentions without invoking mental objects, or different notions of propositions, or intensional contexts. There are ways of hoping, wanting, trying, knowing, rather than objects of hoping, wanting, trying, and knowing.

This is not meant to replace how we talk ordinarily. But we often do talk this way when we talk about what dogs want or know. We infer from behavior for them, as we do with each other. We look at how they act, not what they think. But behavior is not enough. It's true now that I wish I were a millionaire, but if I had not told you that, there was nothing in my behavior which would have clued you to it.

This is not meant to replace the current Western metaphysics of hopes and desires, thoughts and intentions. It is an alternative metaphysics, minimal in that it invokes human capabilities but not mental objects. It is how the ancient Chinese philosophers understood their talk. Its great advantage is that we can now formalize much talk of intentions that has been outside the scope of formal methods. Now we can compare and analyze in a clearer way. Perhaps this will stand as a challenge to those who base their understanding of intentions on mental objects to give clearer analyses.

Appendix E Are Things Needed to Pick Out Locations?

There is a view that locations are determined by objects and relations among those—or at least we can orient ourselves spatially only by reference to things, physical things. As A. Irving Hallowell says in "Cultural Factors in Spatial Orientation":

> Long ago, Poincaré pointed out that the notion of space must be understood as a function of objects and their relations. There is no such thing as space independent of objects. Relations among objects and the movements of objects are a necessary condition of space perception. . . .
> While it remains an open question how far the purely psycho-physical dimensions of perception may be influenced by culturally constituted experiential factors, schematic perception, involving the meaningful aspects of experience, can hardly be understood without reference to an articulated world of objects whose relations and attributes become meaningful for the individual, not simply through the innate psychological potentialities he brings to experience but, above all, through the significance for experience that the development, patterning, transmission, and accumulation of past experience, in the form of a cultural heritage, have come to imply.
> pp. 184–185

Hallowell assumes that not only is the world made up of things, but we must perceive the world through our experience of things. But many people speak mass-process languages in which the concept of an individual thing is quite secondary, as you can read in *Language and the World*.

But even for us speakers of English, Hallowell is wrong. I run along a dock and at the end do a somersault into a lake, where for the sheer joy of it I turn over and over again in the water. I become disoriented, not sure which way is up and which is down. I quickly orient myself by swimming toward the light, knowing that the light is above the surface of the water, while the dark is below, towards the bottom of the lake. The light above is not a thing. I go walking in the desert where I live and orient myself towards home by walking towards the ridge of mountains in the distance. The ridge of mountains is not a thing. I go for a walk in a forest on the mountain near where I live and I become disoriented, so I walk uphill to get a view of the land. Upwards is not a thing. Camping in the forest I go to look for more wood for the fire and I find my way back to camp by the smell of smoke. The smell is not a thing. Or I return to camp by going in the direction of the sound of voices. Sounds are not things.

Though we can use nouns in English to describe light, mountain ridges, up, smells, and sounds, that does not make them physical objects of the sort that Hallowell is telling us we need for orientation. They are masses, like mud and water, except for upwards, which is neither mass nor thing. Though we speakers of English (and other languages that use words for things in this way) often orient ourselves in relation to things, that is not the only way we orient ourselves. There is no necessity in our orienting ourselves in relation to things. And if there is no necessity, a propositional logic of spatial connectives that does not take account of the internal structure of atomic propositions is not ruled out.

Appendix F Descriptive Names in the Logic of Quantifying over Times

Our logic **QTL** for taking account of time is an extension of classical predicate logic, a 2-sorted version. So the method of introducing descriptive names built on classical predicate logic with non-referring names taken as nil introduced in Volume 1 can be used here, though it will require some work to set it out carefully. I'm going to assume we can do so and look at some examples of formalizing in such a theory.

Example 1 *The wife of Humbert at May 1, 1826 is not the same as the wife of Humbert at June 3, 1831.*

$\quad$ the x ((((— is a wife —$_{time}$) / of (Humbert)) (x, May 1 1826))
$\quad\quad \not\equiv$ the x ((((— is a wife —$_{time}$) / of (Humbert)) (x, June 3 1831))

Analysis We can formalize that an informal descriptive name does not refer to the same thing at different times.

Example 2 *The first cat that scratched Zoe is a female.*

$\quad$ (— to be a female —$_{time}$) (c, now)

where c is: the x ($\exists t$ (((— to be a cat —$_{time}$) (x, t) $\land$
$\quad\quad$ ((— to scratch —$_{time}$) / obj(Zoe)) (x, t)) $\land$ ($t <_{time}$ now)
$\quad\quad \land\ \forall y\ \forall w\ ((x \not\equiv y) \land$ (— to be a cat) (y, w) $\land$ ($w <_{time}$ now)
$\quad\quad \land$ ((— to scratch) / obj(Zoe)) (y, w)) $\rightarrow$ ($t <_{time} w$))))

Analysis The time used in the formalization of "to be a cat" is that of the description in which it is embedded, in this case some time in the past. For this formalization to be true, the thing that was a cat and scratched Zoe must exist now, that is, in the present of the model, since the atomic predicate "(— to be a female —$_{time}$)" is positive for existence. But it need not still be a cat. If we think that being a cat is an essential or perpetuating attribute in time, we should add that as a meaning axiom. In the unlikely event that no cat ever scratched Zoe, the formalization is false since the descriptive name is nil. I'll let you formulate a general method for describing the first thing in time that satisfies an open wff, as well as the second, and the third.

$\quad$ We have a choice of what to take as the time for formalizing a common noun that appears in a descriptive name: the time of the main predication or the time of the description in the name. I suggest that we adopt the following convention.

> The formalization of a common noun in a descriptive name takes the tense of the description of which it is a part, not the tense of the predicate that is applied to that description.

That the common noun describes the thing at the one time will not ensure that it describes the thing at the other time unless we adopt a meaning axiom to ensure that.

Example 3 *The first cat that scratched Zoe is small.*
Analysis The word "small" is a relative adjective, so it has to be used to modify a

268   *Appendix F*

predicate. But if all we know about the thing is that it exists now, we have no predicate we can use to formalize the example. If, though, we take that being a cat is an essential attribute or perpetuating attribute in time and have a meaning axiom for that, we can formalize the example as "((— to be a cat $-_{time}$) / small) (c, now)", where c is as in the last example.

Example 4   Dick yelled when Spot barked.

Analysis If we read the example as "Dick yelled at the time that Spot barked" we can use a descriptive name in the formalization for the time when Spot barked:

the t ((— to bark $-_{time}$) (Spot, t) ∧ ($t <_{time}$ now))

But Spot barks a lot, so this does not refer. Generally, we need an extensive formula to pick out a specific time.

Example 5   Dick yelled when Spot ran to Flo's house.

∃t (Max (— to yell $-_{time}$) (Dick, t) ∧ W(t, d))

where d is: the t (Max ((— to run $-_{time}$) / to (x)) (Spot, t)
      ∧ ((— to be a house $-_{time}$) / of (Flo)) (x, t))

Analysis There was only one time that Spot ran to Flo's house, so we can use a descriptive name for that.

Example 6   3 p.m. September 11th, 1998 is the first time that Spot barked.

(— to bark $-_{time}$) (Spot, 3 p.m. September 11 1998)

∧ ∀t ((— to bark $-_{time}$) (Spot, t) ∧ ¬ W$_{time}$(t, 3 p.m. September 11 1998)

→ (3 p.m. September 11 1998 ≤ t) ∨ O$_<$(3 p.m. September 11 1998, t))

Analysis There is no reason to use a descriptive name to formalize the example because we already have a name for the time.

Example 7   The time when Birta existed overlaps the time when Chocolate exists.

O$_<$(b, c) ∧ (b $<_{time}$ now) ∧ O$_<$(c, now)

where b is: the t (Max (— to exist $-_{time}$) (Birta, t))
and c is: the t (Max (— to exist $-_{time}$) (Chocolate, t))

Analysis The use of "the time" here is justified because maximal intervals of existence are times. Since the past tense is used for when Birta exists, and the present is used for when Chocolate exists, we know that the time b overlaps d from before.

Example 8

In 1898, the eldest son of Martha Roosevelt personally led the charge
    up San Juan Hill.
The eldest son of Martha Roosevelt is identical with the 26th U.S. President.

In 1898, the 26th U.S. President personally led the charge up San Juan Hill.

Analysis The person referred to by the descriptive names is usually called "Theodore Roosevelt".[1]

---

[1] The example comes from Nicholas Rescher and Alasdair Urquhart, *Temporal Logic*, p. 139.

This example seems to be invalid. After all, Theodore Roosevelt wasn't President when he led the charge. But that is to read more into the example than is said. Yes, the two descriptions do pick out the same man, so by extensionality whatever predicate is true of one is true of the other. What is odd is that we understand the conclusion to mean that the person who led the charge up San Juan Hill was the 26th President at that time. If we include that in a formalization (which I'll leave to you), the example will be invalid.

Example 9 *The moment when Spot stopped barking at Zeke, Flo started to cry.*
$\exists w \, (\text{Max} \, (— \text{ to cry } —_{time}) \, (\text{Flo}, w) \wedge W_{time}(b, w) \wedge \forall t \, (W_{time}(t, w) \rightarrow (b \leq_{time} t)))$
where b is:
<u>the</u> $t \, (\, \exists t_1 \, \text{Max}((— \text{ to be bark } —_{time})/\text{at} \, (\text{Zeke})) \, (\text{Spot}, t_1)$
$\wedge \, (t_1 <_{time} \text{now}) \wedge W_{time}(t, t_1) \wedge \forall t_2 \, (W_{time}(t_2, t_1) \rightarrow (t_2 \leq_{time} t)))$

Analysis By saying "the moment" it's assumed that there is a distinct time at which Spot stopped barking and Flo started to cry. Since there was only one time when Spot encountered Zeke, we can use a descriptive name to pick out the time when he stopped barking. We do not have to say that the time when Spot stopped barking is an instant, for if it had parts it would not be less than those. We do not need to use a descriptive name for when Flo began to cry since that's the same as when Spot stopped barking. That's good, since there were many times when Flo began to cry.

Example 10 *Hera was the wife of Zeus before Athena was born.*

Analysis To formalize this we need a logic for reasoning with non-referring names that are not taken as nil and which also allows for reasoning about time. That requires somehow reconciling fictional times and real times. Is "Sherlock Holmes was a detective before Fred Kroon was born" true? I'll leave that development to those who have more experience in reasoning about fiction.

BIBLIOGRAPHY

- Only works cited in the text or elsewhere in the bibliography are listed.
- Citations are to the most recent English reference listed, unless otherwise noted.
- Works cited in the text without attribution are by the author of this book, Richard L. Epstein.

ALLEN, James F.
    1983   Maintaining Knowledge about Temporal Intervals
              Communications of the ACM, vol. 26, pp. 832–843.

AZAR, Betty Schrampfer
    1989   *Understanding and Using English Grammar*
              Second edition, Prentice Hall Regents.

BINNICK, Robert I.
    1991   *Time and the Verb: A Guide to Tense and Aspect*
              Oxford.
    1999   *Parts and Places*: *The Structure of Spatial Representation*
              MIT Press.

BYBEE, Joan L., Revere PERKINS, and William PAGLIUCA.
    1994.   *The Evolution of Grammar: Tense, aspect and modality in the languages of the world.*
              University of Chicago Press.

CASATI, Roberto and Achille VARZI
    1999   *Parts and Places*: *The Structure of Spatial Representation*
              MIT Press.

EPSTEIN, Richard L.
    1990   *Propositional Logics* (*The Semantic Foundations of Logic*)
              Kluwer. Second edition, Oxford University Press, 1995. Second edition with corrections, Wadsworth, 2000.
    1992   A Theory of Truth Based on a Medieval Solution to the Liar Paradox
              History and Philosophy of Logic, vol. 13, pp. 149-177.
    1994   *Predicate Logic* (*The Semantic Foundations of Logic*)
              Oxford University Press. Reprinted, Wadsworth, 2000. Reprinted Advanced Reasoning Forum, 20 12.
    1998   *Critical Thinking*
              Wadsworth. Fifth edition with Michael Rooney, Advanced Reasoning Forum, 2017.
    2006   *Classical Mathematical Logic* (*The Semantic Foundations of Logic*)
              Princeton University Press.
    2011   *Cause and Effect, Conditionals, Explanations*
              Advanced Reasoning Forum.
    2012A   *Reasoning in Science and Mathematics*
              Advanced Reasoning Forum.
    2012B   Models and Theories
              In EPSTEIN, 2012A, pp. 19–51.
    2013A   Mathematics as the Art of Abstraction
              In *The Argument of Mathematics*, eds. Andrew Aberdein and Ian Dove, Springer-Verlag, 2013, pp. 257–289. Reprinted in EPSTEIN, *Reasoning in Science and Mathematics*, Advanced Reasoning Forum, 2012.

    2013B   Truth and Reasoning
                In EPSTEIN, 2013C, pp. 101–127. Reprinted in EPSTEIN, 2015, pp. 57–83.
    2013C   *Prescriptive Reasoning*
                Advanced Reasoning Forum.
    2013D   *The Fundamentals of Argument Analysis*
                Advanced Reasoning Forum.
    2014   Reflections of Temporal and Modal Logic
                Logic and Logical Philosophy, vol. 24, no. 1, pp. 111–139.
                DOI 10.12775/LLP.2014.020. Reprinted in EPSTEIN, 2015, pp. 98–126.
    2015   *Reasoning and Formal Logic*
                Advanced Reasoning Forum.
    2015A   Possibilities and Valid Inferences
                In EPSTEIN, 2015, pp. 1–30.
    2021   *Language and the World: Essays New and Old*
                Advanced Reasoning Forum.
    2021A   Language-Thought-Meaning
                In EPSTEIN, 2021, pp. 60–85.
    2021B   Why Event-Talk Is a Problem
                In EPSTEIN, 2021, pp. 86–89.
EPSTEIN, Richard L. and Walter A. CARNIELLI
    1989   *Computability: Computable Functions, Logic, and*
                the Foundations of Mathematics
                Wadsworth. Third edition, Advanced Reasoning Forum, 2008.
EVANS, Vyvyan
    2003   *The Structure of Time: Language, Meaning and Temporal Cognition*
                John Benjamins Publishing Company.
FRASER, J. T., F. C. HABER, and G. H. MÜLLER, eds.
    1972   *The Study of Time*
                Springer-Verlag.
GALE, Richard M.
    1967   *The Philosophy of Time*
                Doubleday Anchor.
GLEICK, James
    2013   Time Regained! Review of Lee Smolin, *Time Reborn*
                In *New York Review of Books*, June 6, 2013, pp. 46–49.
GORANKO, Valentin, Angelo MONTANARI, and Guido SCIAVICCO
    2004   A Road Map of Interval Temporal Logics and Duration Calculi
                Journal of Applied Non-Classical Logics, vol. 14, p. 9–54.
HALLOWELL, A. Irving
    1967   Cultural Factors in Spatial Orientation
                In Hallowell, *Culture and Experience*, University of Pennsylvania Press, pp. 184–202.
HAMBLIN, C. H.
                Instants and Intervals
                In FRASER, HABER, and MÜLLER, pp. 324–331.
HANSEN, Chad
    1985   Chinese Language, Chinese Philosophy, and "Truth"
                Journal of Asian Studies, Vol. XLIV, no. 3, pp. 491–519.

1992 *A Daoist Theory of Chinese Thought: A Philosophical Interpretation*
Oxford University Press

JESPERSEN, Otto
1924 *The Philosophy of Grammar*
Henry Holt and Company.

KRETZMANN, Norman
1982 Continuity, Contrariety, Contradiction, and Change
in *Infinity and Continuity in Ancient and Medieval Thought*, ed. N. Kretzmann, Cambridge University Press, pp. 270–296.

KUHN, Steven T. and Paul PORTNER
2002 Tense and Time
In *Handbook of Philosophical Logic*, Second edition, vol. 7., eds. D. Gabbay and F. Guenthner, Kluwer, 2002.

MYRO, George
1986 Identity and time
In *The Philosophical Grounds of Rationality*, eds. Richard E. Grandy and Richard Warner, Oxford University Press. Reprinted in *Material Constitution* ed. Michael Rea, Rowman & Littlefield, 1997, pp. 148–172.

NEWTON, Isaac
1686 *Philosophiae Naturalis Principia Mathematica*
Translated by I. Bernard Cohen and Anne Whitman, assisted by Julia Budenz, as *The Principia*: *Mathematical Principles of Natural Philosophy*, University of California Press, 1999.

ØHRSTRØM, Peter and Per F.V. HASLE
1995 *Temporal Logic*: *From Ancient Ideas to Artificial Intelligence*
Kluwer.

PARK, David
1972 The Myth of the Passage of Time
In FRASER, HABER, and MÜLLER, pp. 110–121.

PRIOR, Arthur
1956 Tense-Logic and an Analogue of S4
In Arthur Prior, *Time and Modality*, Oxford University Press, pp. 8–17.

1962 Changes in Events and Changes in Things
Preprint. Reprinted in *The Philosophy of Time*, eds. Le Poidevin and MacBeath, Oxford University Press, 1993, pp. 35-46. Also in PRIOR, 1968 (2003), pp. 7–19.

1965 Time, Existence, and Identity
Proceedings of the Aristotelian Society, vol. 66, pp. 183–192. Reprinted in PRIOR, 1968, pp. 93–101.

1967 *Past, Present and Future*
Oxford University Press.

1968 *Papers on Time and Tense*
Oxford University Press. 2nd edition, 2003.

QUINE, Willard van Orman
1939 Designation and Existence
The Journal of Philosophy, vol. 36, no. 26, pp. 701–709.

1953 On What There Is
In Quine, *From a Logical Point of View*, Harvard University Press, Second ed., 1961, pp. 1–19.

RENNIE, M. K.
- 1974 *Some Uses of Type Theory in the Analysis of Language*
  Monograph Series, No. 1, Department of Philosophy, Research School of Sciences, Australian National University.

RESCHER, Nicholas and Alasdair URQUHART
- 1971 *Temporal Logic*
  Springer-Verlag, 1971.

RIESEN, John J.
- 1983 The Generation and Early Development of Spatial Inferences
  In *Spatial Orientation: Theory, Research, and Application*, Plenum Press.

SINHA, Chris and Vera da Silva SINHA
- 2021 Time and Events—in language, mind and world
  Preprint at psyarxiv.org, accessed April, 2022.

SINHA, Chris, Vera da Silva SINHA, Jörg ZINKEN, and Wany SAMPAIO
- 2011 When Time Is Not Space: The social and linguistic construction of time intervals in an Amazonian culture
  Language and Cognition, Vol. 3, no. 1, pp. 137–169.

SINHA, Vera da Silva
- 2018 *Linguistic and Cultural Conceptualisations of Time in Huni Kui, Awety´, and Kamaiurá Communities in Brazil*
  Ph.D. thesis, University of East Anglia.

TALMY, Leonard
- 1983 How Languages Structure Space
  In *Spatial Orientation: Theory, Research, and Application*, Plenum Press.

TAYLOR, Richard
- 1955 Spatial and Temporal Analogies and the Concept of Identity
  Journal of Philosophy, vol. LII. Reprinted in *Problems of Space and Time*, ed. J.C.C. Smart, Macmillan Co., 1964.
- 1957 The Problem of Future Contingencies
  Philosophical Review, vol. 66, pp. 1–28.

VARZI, Achille. *See* CASATI and VARZI.

WAISMANN, Friedrich
- 1968 The Decline and Fall of Causality
  In *How I See Philosophy*, Macmillan, pp. 208–256.
- 1968 How I See Philosophy
  In Waismann, *How I See Philosophy*, Macmillan, pp. 1–38.

WHORF, Benjamin Lee
- 1936? An American Indian Model of the Universe
  From the unpublished writings of Whorf, edited by E. A. Kennard and G. L. Trager, *International Journal of American Linguistics*, vol. 16, no. 2, pp. 67–72, 1950.
  Reprinted in *ETC.: A Review of General Semantics*, vol VIII, no. 1, pp. 27–33, 1950, from which the quotation in this text is drawn.
- 1941a The Relation of Habitual Thought and Behavior to Language"
  In *Language, Culture, and Personality: Essays in Memory of Edward Sapir*, Sapir Memorial Publication Fund, 1941, pp. 75–93.
  Reprinted in EPSTEIN, 2021, pp. 200–226.

    1941b   Languages and Logic
                Technology Review vol. 43, 1941 pp. 250–252, 266, 268, 272
                Reprinted in EPSTEIN, 2021, pp. 225–236.
    1945   Grammatical Categories
                Language, vol. 21, pp. 1–11 (written in 1937).
                Reprinted in EPSTEIN, 2021, pp. 174–187.
WILLIAMSON, Timothy
    1994   *Vagueness*
                Routledge.

Index of Examples

underlined page numbers indicate an example in the example-analysis format
In the alphabetization, a blank and punctuation come before all letters.
Examples that begin with a symbol or numeral are at the end.

A little boy will become a famous man. <u>153</u>
All dodos were birds. 142
All dogs are mammals. 142, 162
All dogs bark. Spot is a dog. Therefore Spot barks. 70
All dogs will be greyhounds. 142
All men are mortal. Socrates is a man. Therefore, Socrates is mortal. 70
All police officers are men 142
A sausage falls from the grill. Spot barks. Dick yells. Spot eats it. He wags his tail. It was the best of times; it was the wurst of times. 162
An electron changed energy level in a radon atom. 63
An object can't be both P and not-P. <u>140</u>
Antichrist will be an orator. <u>153</u>
Anubis is a wild big dog. 6
As a horse trainer, Gerald has a lot of patience. <u>263</u>
Augustine will convert. 128

Barking is not an attribute of a thing. <u>219</u>
Bill is a bachelor. <u>146</u>
Birta hates every cat. <u>144</u>
Birta is a dog. <u>138</u>, <u>218</u>
Birta is a dog, but she will be a princess. <u>138</u>
Birta is a dog if and only if she is a domestic canine. <u>146</u>
Birta is a puppy. <u>146</u>
Bob and Dick are brothers. 183
Bruno doesn't exist anymore. <u>136</u>

Caesar was assassinated. <u>44</u>, <u>65</u>, <u>146</u>
(Complements of locations are locations. <u>214</u>

Dean Furtz respects Dr. E as a teacher but not as a scholar. <u>263</u>

December 21st or December 22nd is the shortest day of the year. <u>158</u>
Dick always yelled after Spot barked. <u>64</u>
Dick and Zoe both saw something that they thought was a coyote. <u>259</u>
Dick ate. 160, 161
Dick ate a steak. Dick wanted a steak. <u>262</u>
Dick ate dinner. 48
Dick ate when Zoe arrived. 161
Dick drinks. 163
Dick expects Zoe to call. <u>257</u>
Dick had been eating when Zoe arrived. 161
Dick had eaten. 160
Dick had eaten when Zoe arrived. <u>53</u>, 160
Dick has been studying. 162
Dick is a student. All students study hard. Therefore, Dick studies hard. 3
Dick is sleeping ↔ ¬ (Spot is barking) <u>42</u>
Dick is throwing a ball for Spot to fetch. <u>258</u>
Dick is yelling in front of where Spot is barking. 222
Dick likes hamburger, while Wanda is a vegetarian. <u>39</u>
Dick said that Zeke is a criminal. <u>263</u>
Dick spent more time training Spot than Suzy spent training Puff. <u>158</u>
Dick talked only when Spot wasn't barking. <u>44</u>, <u>64</u>
Dick wants Spot to bark. 252
Dick wants to play. <u>256</u>
Dick was eating. 161
Dick was eating when Zoe arrived. 161
Dick was quickly irritated and chased Spot. <u>151</u>
Dick will begin cooking no later than when Zoe arrives. <u>39</u>
Dick will cook dinner when Zoe arrives. <u>39</u>, 162
Dick will not sing and dance. <u>152</u>
Dick will sing and dance. <u>152</u>
Dick will sing and not dance. <u>152</u>
Dick wishes he were rich. <u>260</u>
Dick yelled after Spot barked. <u>35</u>
Dick yelled at the same time as Spot barked. <u>36</u>
Dick yelled at 3 p.m. 11
Dick yelled during the time that Spot barked. <u>38</u>
Dick yelled, then Spot began barking, and then Spot finished barking before Dick stopped yelling. <u>36</u>

276   Index of Examples

Dick yelled when Spot barked. <u>39</u>, <u>268</u>
Dick yelled when Spot ran to Flo's house. <u>268</u>
Dick yelled while Spot was barking. <u>38</u>
Dick yelled within the time that Spot was barking. <u>37</u>
Do you see that bird? Yes, I see it. 162

Electrons have spin. 18, <u>43</u>, <u>140</u>
Every instant of time has the same length. <u>159</u>
Every location contains a point of space. <u>213</u>
Every location is contained in some other location. <u>213</u>
Every region of space is a location. <u>214</u>
Every thing happens at once. <u>113</u>
Every thing is somewhere. <u>216</u>
Every time contains an instant of time. <u>109</u>
Every time Dick feeds Spot, Spot barks. <u>147</u>
Every time Dick yells near Spot, a little time after Spot jumps. <u>158</u>
Every time Dick yells near Spot, Spot jumps. <u>148</u>
Every time is contained in some other time. <u>112</u>
Every time one billiard ball hits another, the second one moves. <u>147</u>
Every time Spot barked Dick yelled. 123
Every time that Spot barks, Dick yells afterwards. 56
Every two locations are within some location. <u>213</u>
Every two times are within some time. <u>113</u>
Everywhere that Mary went the lamb was sure to go. 171, <u>217</u>

Flo knows how to whistle. <u>259</u>
Flo knows that Spot is a dog.
   Therefore, Flo knows something. <u>259</u>
Flo wants a dog. <u>261</u>

Helen had three husbands. 246
Hera was the wife of Zeus before Athena was born. <u>268</u>
I never hit an old man.
   My brother is an old man.
   Therefore, I never hit my brother. 4, <u>152</u>
I will return before there are three full moons. <u>156</u>
If Birta is a dog, then Birta is a mammal. <u>138</u>, <u>219</u>

If Birta is barking squealingly, then Birta is chasing a rabbit. <u>218</u>
If Dick yelled, then Spot barked. Dick yelled. So Spot barked. <u>131</u>
If nothing happens, there is no time. If nothing exists, there is no space. <u>215</u>
If something existed, then something exists, and something will exist. <u>134</u>
If Spot barks, then Dick will yell. Spot barked. So Dick yelled. <u>53</u>, <u>131</u>
If Spot barks, then he barks loudly. <u>148</u>
If Suzy went to bed, then she took off her clothes. <u>37</u>
If Suzy went to bed, then she took off her clothes. Suzy went to bed.
   Therefore, Suzy took off her clothes. <u>37</u>
If Tom and Suzy have a baby, it will be born in a hospital. If a baby is born in a hospital, the parents will have to pay a doctor.
   Therefore, if Tom and Suzy have a baby, they will have to pay a doctor. 75
I'm in front of the grocery store. 221
In 1898, the eldest son of Martha Roosevelt personally led the charge up San Juan Hill. The eldest son of Martha Roosevelt is identical with the 26th U.S. President. Therefore, in 1898, the 26th U.S. President personally led the charge up San Juan Hill. <u>268</u>
It will be the case that I am eating my breakfast. 245
Isabel thinks that Bertha is a witch. <u>261</u>

Jim wrote a book. Jim was writing a book when he died. <u>262</u>
John F. Kennedy was President of the U.S.A. in 1963. 183
John loves Mary. 242–244
Juney was a dog that no longer exists. 93
Just once Dick yelled before Spot barked. <u>64</u>

King Wen believes that Ch'ang An is beautiful. 255

Lala will be away less time than when Ugo was sick. <u>156</u>

Manuel tried to walk. 250, <u>256</u>
Manuel wants to dance reggae. <u>257</u>
Maria used to be a student. <u>136</u>

Marilyn Monroe ≡ Norma Jeane Mortensen 70, 89, 261
Most of the time Spot is happy. <u>149</u>
Much of the time Spot is happy. <u>149</u>

Nearly every time that Dick feeds Spot, Spot barks. <u>148</u>
No one ever trained Puff. <u>148</u>
No two things can be in the same place at the same time. 188
Nothing can be in two places at the same time. 188
Nothing existed. <u>134</u>
Nothing exists. <u>134</u>
Nothing exists everywhere. <u>215</u>

On May 27th, Dick planned to play tennis with Suzy the next day. <u>258</u>
Once a dog, always a dog. <u>139</u>

Picking out the location of my ranch. <u>216</u>
Predications are extensional. <u>217</u>
Puff didn't bark. 123
Puff is nearly coughing. So Puff exists. <u>136</u>
Puff never barks. <u>148</u>
Puff scratched Zoe. 11, 48, 249
Puff scratched Zoe and Tom sprained his ankle before Dick ate dinner and after Suzy visited Dick. 110
Puff scratched Zoe while Suzy was at the grocery store. 11

Ralph barks. 4
Ralph barks. All dogs bark.
   Therefore, Ralph barks. 3
Ralph barks. All dogs bark.
   Therefore, Ralph is a dog. 3
Ralph does not exist. <u>133</u>
Rex was the father of Bruno.
   Therefore, Bruno existed. <u>135</u>
Richard L. Epstein will die in 2029. 75
Rolando drives. <u>163</u>
Rudolfo nearly existed on February 18, 1931 at 2:46 p.m. <u>136</u>
Slowly, Tom escorted Zoe and Dick pushed Manuel. <u>151</u>
Socrates did not exist when Julius Caesar did. <u>134</u>

Socrates is dead. 96, <u>135</u>
Socrates is famous. <u>136</u>
Socrates was a philosopher. 96, 128
Socrates was shorter than Julius Caesar. <u>137</u>
Some day Flo will be a woman. <u>149</u>
Some dogs are terriers. 143
Some dogs were terriers. 143
Some dogs will be terriers. 143
Sometime after Dick stopped eating, Spot began to bark. <u>36</u>
Sometime between when Zoe arrived and Dick started cooking, Spot ran away. <u>40</u>
Sometimes when Dick feeds Spot, Spot barks. <u>147</u>
Space is dense. <u>215</u>
Space is three-dimensional. <u>214</u>
Spot almost barked.
   Therefore, Spot did not bark. <u>154</u>
Spot almost barked at 4 p.m. June 2, 2002.
   Therefore, Spot did not bark at 4 p.m. June 2, 2002. <u>154</u>
Spot barked. 12, 56, 58, 63. 122, 123, 126, <u>219</u>
(Spot barked)$_1$. Then Dick yelled.
   Then (Spot barked)$_2$. 12
Spot barked and Dick yelled before Dick fell, and then Zoe laughed. <u>41</u>
Spot barked and Dick yelled before Dick fell or Zoe laughed. <u>42</u>
Spot barked and Puff ran. 124
Spot barked around 7:12 am May 5th, 2005. <u>36</u>
Spot barked at 7:12 am May 5th, 2005. <u>36</u>, 115
Spot barked at 12:19 a.m. August 18, 2017. 86
Spot barked at almost 4 p.m. June 2, 2002. <u>154</u>
Spot barked at exactly 7:12 am May 5th, 2005. <u>36</u>
Spot barked before Dick yelled. 10, 14, <u>35</u>, <u>45</u>, <u>64</u>
Spot barked just once. <u>147</u>
Spot barked just twice. <u>43</u>, <u>63</u>
Spot barked longer than Dick yelled. <u>156</u>
Spot barked loudly at 1 p.m. May 3, 1998. 114, <u>150</u>
Spot barked loudly at 1 p.m. May 3, 1998 in comparison to how Spot barked at all times. 151
Spot barked loudly at 1 p.m. May 3, 1998 in comparison to those things that were barking then. 151
Spot barked loudly at 1 p.m. May 3, 1998 in Kroon Park. <u>219</u>
Spot barked loudly. Therefore, Spot barked. <u>150</u>, 252
Spot barked loudly at 1 p.m. May 3, 1998.
   Therefore, Spot barked at 1 p.m. May 3, 1998. 94
Spot barked shortly before Dick yelled. <u>158</u>

Index of Examples

Spot barked that he was hungry. 259
Spot barked. Then Dick yelled. And then
　Spot barked. 11, 36, 234
Spot barked, then Dick yelled and Zoe got
　upset. 40
Spot barked (said by Dick on May 3, 2002). 130
Spot barked (said by Zoe on April 7, 2004). 130
Spot barked twice. 43, 63
Spot barked when either Dick was jumping
　or Zoe was laughing. 41
Spot barked yesterday. 38
Spot breathed. Spot is breathing. Spot will
　breathe. 126
Spot chased Puff 114
Spot didn't bark at 7:12 am May 5th, 2005. 115
Spot didn't bark between the time that Suzy
　opened the gate and Dick yelled. 42
Spot didn't bark when Caesar was
　assassinated. 44, 64
Spot exists. 124
Spot hardly ever barks at Tom. 149
Spot has been barking since Suzy opened
　the gate. 39
Spot is a dog. 4
Spot is barking. 127
Spot is barking between Zoe and Suzy. 225
Spot is barking in front of and behind Dick.
　224
Spot is barking in front of and to the left of
　Dick. 223
Spot is barking in front of Dick. 221, 223
Spot is barking in front of where Dick is
　yelling. 222
Spot is barking in front of where the band is
　playing. 223
Spot is barking not loudly at 7:12 am May 5,
　2005. 115
Spot is biting Puff. So Puff exists. 135
Spot is in front of where Dick was at 9 a.m.
　June 1, 2002. 221
Spot is running towards Dick. 225
Spot is sitting. 177, 179–181, 183, 189, 237
Spot is sitting in front of Dick and Zoe. 224
Spot is sitting next to Dick. 225
Spot smells a steak that he wants to eat. 255
Spot wants to play. 250
Spot wants to play. Dick wants to play.
　So Dick and Spot want the same thing. 252

Spot wants to run quickly. 257
Spot was a dog. 126
Spot was a puppy. 126
Spot was barking at 3 p.m. June 3 2003 in front
　of where Dick was yelling at 2:58 p.m.
　June 3, 2003. 222
Spot will bark. 127
Spot will bark and then Dick yelled. 51, 53
Suzy arrived shortly after Tom arrived. 157
Suzy expects Zoe to call. 257
Suzy hopes that Dick will call. 253
Suzy hopes someone will call.
　There is someone Suzy hopes will call. 262
Suzy is a cheerleader.
　All cheerleaders have a liver.
　Therefore, Suzy has a liver. 3
Suzy is afraid that Spot will bite Puff. 258
Suzy is going to the grocery store after I go to
　her apartment to take care of Puff. 11
Suzy took off her clothes and went to bed. 37
Suzy was a student throughout 1998. 130
Suzy will jump. 129
Suzy will like to ski. 250

The first cat that scratched Zoe is a female. 267
The first cat that scratched Zoe is small. 267
The last time Spot barked Dick didn't yell. 123
The moment when Spot stopped barking at Zeke,
　Flo started to cry. 269
The ordering of times is dense. 113
The time when Birta existed overlaps the time
　when Chocolate exists. 268
The time when Lala was away was longer than
　the time between when Ugo was sick and Mizza
　killed a bear. 156
The train whistled. 123
The wife of Humbert at May 1, 1826 is not the
　same as the wife of Humbert at June 3, 1831. 267
There are largest units of space. 213
There are largest units of time. 112
There are no smallest units of space. 212
There are no smallest units of time. 110
There are only instants of time. 109
There are only points in space. 212
There are parallel timelines. 110
There are smallest units of time, and every
　interval of time is a collection of those
　and nothing more. 110

There are smallest units of space, and every location is a collection of those and nothing more. 213
There are times other than those named by fixed time markers. 113
There are unicorns. 133
There is a place which contains no smallest unit of space. 216
There is a smallest unit of space. 212
There is a smallest unit of time. 109
There is a time when nothing exists. 134
There is a time which contains no smallest unit of time. 110
There is more than one location. 212
There is more than one time. 109
There is no location that contains all other locations. 213
There is no time that contains all other times. 112
There is something that always existed, exists, and always will exist. 134
There is something that was a dog. 133
There is something now that was a dog. 133
There is something that will exist. 134
There never was a unicorn. 148
There was a time when both Dick yelled and Spot barked. 38, 64
There was a time when Spot barked after Suzy opened the gate. 56
There was something that was a dog. 133
There was something that was not a dog. 136
There was some time between when Zoe arrived and Dick started cooking. 44
There were unicorns. 133
Time flows. 113
Time has no beginning and no ending. 112
Tom and Dick sang. So Tom sang and Dick sang. 152
Tom ate lunch yesterday. 131
Tom had a dog. 52
Tom had a dog that will bark. Therefore, there will be a dog that will bark. 138
Tom has a dog. 53
Tom played football. 48
Tom spoke carefully not clearly October 9, 2021. 116
Tom wants Suzy to think that he is sleeping. 264
Tom will have a dog. 52
Tom will telephone in a day. 154
Tom will telephone in a week. 157
Tom will telephone in three minutes. 157

Walter believes that Marilyn Monroe was an actress.
   Marilyn Monroe is Norma Jean Mortensen.
   Therefore, Walter believes that Norma Jean Mortensen was an actress. 260
Wanda remembers Bruno. 135
Wanda wants to be a veterinarian. 256
When did Suzy jump? When Spot barked and Puff screeched. 131
When Spot barks, he barks loudly. 148
When Zoe cooks, Dick washes up. 43, 65
Where is the band going to play tomorrow? Over there where Dick is. 188
Where is the lawn chair? Over there where Dick is. 188
Where's Spot? He's in the house. 224
Where was Spot barking? 216

Zeke and Isaiah want to be friends. 258
Zoe almost wanted to dance. 257
Zoe: How do you feel? Dick: I've got a pain in my shoulder. Zoe: Me, too. 260
Zoe is a woman. 85
Zoe is tall. 115
Zoe sang at 4:15 p.m. November 4, 2003. 85
Zoe wanted to almost dance. 258
Zoe wanted to be taller than Dick some day. 258
Zoe will call. 253–254

$2 + 2 = 4$ 43, 70
2:49.000000001 p.m. May 3, 2004 is an instant of time. 109
7 is a prime number. 137
3 p.m. September 11th, 1998 is the first time that Spot barked. 268

$\forall i$ (Spot barked)$_i$ 63
$\forall i \, \exists j \, (\, (\text{Spot barked})_i \, \wedge_{eb} (\text{Dick yelled})_j \,)$ 63
$\exists i \, (\, (\text{Spot barked})_i \, \wedge_{eb} (\text{Dick yelled})_i \,)$ 63
$\exists x \, (x \equiv \text{Socrates}) \, \leftrightarrow \, \exists t \, (- \text{ to exist } -_{time}) \, (\text{Socrates}, t)$ 93
$(- \text{ to bark } -_{time}) \, (\text{Spot}, 3\!:\!00 \text{ p.m. May 3 2002})$ 130

Index of Symbols

"iff" abbreviates "if and only if"

¬, →, ∧, ∨   14, 20, 98, 190

$\wedge_{bb}, \wedge_{eb}, \wedge_{be}, \wedge_{ee}$ 14–15, 20

<   17, 81, 101, 198

≤   17, 81

T   18

⟨T, ◇⟩   18, 21

b_T, e_T 19, 21

$((p)_1)\ ((p)_2)\ ((p)_3)\ \ldots$ 20

$p, q, r, s, q_1, q_2, \ldots, r_1, r_2, \ldots, s_1, s_2, \ldots$ 20

$p_0, p_1, \ldots$ 20

$A, B, C, A_0, A_1, B_0, B_1, C_0, C_1, D_0, D_1, \ldots$ 20, 100, 195

$L(\neg, \rightarrow, \wedge, \vee, \wedge_{bb}, \wedge_{ee}, \wedge_{be}, \wedge_{eb}, p_0, p_1, \ldots)$ 20

Γ, Δ, Σ   20

b_p, e_p 21

ν, ν(p)   21

t, t(p)   21

T, F   21, 103, 200

$\approx_{bb}, \approx_{ee}, \approx_{be}, \approx_{eb}, \approx_T$ 22, 26, 49

M   22, 104, 205

⟨ν, ⟨T, ◇⟩, t⟩   22

M ⊨ B   22, 104

M ⊭ B   22, 104

M ⊨ Γ   22

M_R 24

M_A 24

⊢   26

⊬   26

Γ ⊢ A   26

Γ ⊬ A   26

[z]   31

$\wedge_w$ 37

$\wedge_O$ 38

∧ since   39

B(p, q, r)   40

$\underline{B}$(p, q, r)   40

$p \wedge_{eb} A$ 40

⋀   40

$A \wedge_{eb} B$ 41, 42

W   42

$⋀_i$ 56

$W_{i,j}$ 56

W(r, p, q)   42

Omni   43

Atemp   43

Present (p)   52

Past (p)   52

Future (p)   52

A(i)   57

$i, j, k, j_0, j_1, \ldots, k_0, k_1, \ldots$ 57

$n, m, n_0, n_1, \ldots, m_0, m_1, \ldots$ 57

$x_0, x_1, x_2, \ldots$ 80, 190

$t_0, t_1, t_2, \ldots$ 80, 190

$b_0, b_1, b_2, \ldots$ 80, 190

$c_0, c_1, c_2, \ldots$ 80, 190

$t, t', w, w', w_1, w_2, \ldots$ 80, 130

∀   57, 98, 190

∃   57, 98, 190

T   17, 80, 101, 173, 198

U   80, 101, 173, 198

⟶   81

$w_1 < w_2 < w_3$ 81

$- <_{time} -$ 81

$W_{time}(-, -)$ 82

$O_<(w_1, w_2)$ 83, 102

$O(w_1, w_2)$ 83, 102, 198

W (also W_{time})   82, 101, 176, 198

Index of Symbols

$-_{time}$ 85, 98, 190
$- \equiv -$ 91, 190
$- \equiv_{time} -$ 91, 190
$(-\text{ to exist }-_{time})$ 93, 190
$(-\text{ to exist }-_{time}, -_{location})$ 186, 190
I_j^n 98, 190
$I(-, \ldots, -, -_{time})$ 98
$A(u/x)$ 100
$u \equiv v$ 100, 195
$t \equiv_{time} w$ 100, 195
$u \not\equiv v$ 100, 195
$t \not\equiv_{time} w$ 100, 195
$t <_{time} w$ 100, 195
$W_{time}(t, w)$ 100, 190, 195
$t \leq_{time} w$ 100, 195
$A(w/t)$ 100, 195
$w_1 <_{time} w_2 <_{time} w_3$ 100, 196
$O_{time}(w_1, w_2)$ 100
$O_<(w_1, w_2)$ 100, 196
c 102
σ 102
$\sigma(x)$ 102
$\sigma(t)$ 102
$\tau \sim_t \sigma$ 102
ν_σ 103, 200
$\nu_\sigma \vDash A$ 103, 104, 200
$M \vDash B$ 104
$\vDash A$ 104
$\Gamma \vDash A$ 104
M_t 105
instant (t) 109
obj$(-)$ 114, 117, 190
$\vec{d}$ 114
$R_0, R_1, \ldots$ 117, 190

R_n^i 117, 190
$R(d_1, \ldots, d_k)$ 191
$N_0, N_1, \ldots$ 117, 190
$\sim$ 117, 191
$\wedge$ 117
$+$ 117
$\&$ 117
$\cup$ 117
now 122
Max C(t) 123
$\exists!$ 129
x, y, z 130
to have$_p$ 138, 163
$(-_{time}$ is shorter than $-_{time})$ 156
$l_0, l_1, \ldots$ 173, 190
$e_0, e_1, e_2, \ldots$ 173, 190
$l, l', p, p', p_0, p_1, p_2, \ldots$ 173
P 173
p, p', p'', p$_1$, p$_2$, $\ldots$ 173
$W_{location}$ 176, 198
$W_{location}$ 176, 190, 195
$-_{location}$ 180, 186
$\equiv_{location}$ 190
$\not\equiv_{location}$ 196
$A(x_1, \ldots, x_n, t)$ 182
$A(x_1, \ldots, x_n, t, l)$ 182
locational (A) 185
omnilocational (A) 185
alocational (A) 185
$A(p/l)$ 195
point (l) 212
$P(-_{location}, -, -_{location}, -_{time})$ 223
$<>$ 251
$\{\}$ 252

Index

Italicized page numbers indicate a definition, statement of principle, or a quotation.

n indicates a footnote

The summaries are not indexed.

abbreviations, 20, 100, 185, 195–196
"above", 222–223
abstract objects, 1, 2n, 43, 242n
abstracting,
    for meaning, 253–254
    from reasoning, 5, 160
    in science and mathematics, 189, 214
    instants are a result of —, 79, 233
    numbers are the result of —, 137
    things are result of —, 78, 172
    to get a universe of things, 172
Adams, Scott, *233n*
adjectives,
    order of, 6
    relative, 115
    See also modifier, predicate.
adverbs,
    as modifier, 250–251
    time indications as —, 87–88, 245
    See also modifier, predicate.
African languages, 92
"after", 35
agreements, 4, 143, 179, 181
"all" 142–143
Allen, James, 19
alocational predicate, *184*
ambiguity,
    of "before", 14
    resolved by specifying time, 242
    scheme vs. proposition —, 244
    use of infinitive as restrictor, 257
American Indian languages, 80, 92, 125
American Sign Language, 245
"and then", 37
anti-reflexivity, *17*, *81*, 102
anti-symmetry, *17*, *81*, 101
apple, 228
"anymore", 136
arguments, 2

Aristotelian logic, 85
arithmetic, 62. *See also* mathematics.
arity of a predicate, *98*
aspect of a verb, 160n
assignment(s) of references, *102*, *199*
associativity of ∧, *203*, *209*
associativity of +, *203*, *210*
associativity of &, *204*, *210*
associativity of ∪, *204*, *210*
"at the same time", 36
atemporal
    infinitive predicates are —, 86
    meaning axioms are —, 146
    numbers are —, 137
    predication in classical predicate logic is, 71
    proposition, *43*
    reference is, 89
    See also supratemporal, things are.
atomic predicate, formal, *118*
atomic wff,
    temporal and spatial predicate logic, *194*
    temporal predicate logic, *99*
    temporal propositional logic, *20*, *21*
    temporal propositional logic with index quantification, *57*
attention, paying, 78, 113, 125
attribute(s), 138–141, 218–219
    essential, 71–72, 218–219
    essential in time, *139*, 237
    in space, 181
    perpetuating in time, *139*, 219, 237
Augustine, 245
axiom system,
    for classical propositional logic, *26*, *105*
    for **QT**, *105–108*
    for **QTS**, *207–211*
    for **TPC**, *50*
    for **TPC-linear**, *26–28*
Azar, Betty Schrampfer, 160n

"before", 11, 14, 35
before and after. *See* establishing relative times with before and after.
"began", 14, 36
beginnings and endings,
    billiard ball example, 147
    butterfly example, 15
    compared to boundary of locations, 179
    of intervals of time, 15–*16*
    of time, 112
    of time line, 18–*19*, *21*
    real or by agreement?, 14–15, 17, 18–19, 54, 77, 233–234
    tenses and —, 160
beliefs, 250
Bergson, Henri, 231
between relation in space, 177, 225–226
"between when", 40, 42
billiard ball example, 147
binding of a variable in temporal and spatial predicate logic, *194*
binding of a variable in temporal predicate logic, *99*
binding of an index variable, *57*
Binnick, Robert I., 160n, 228n
branching time, 50, 51, 110–*112*, 140, 113, 214, 247–248
bounded maximal intervals of true predications are times, *123*
butterfly example, 15, 15
Bybee, Joan L., 245n

Carnielli, Walter, 47n
Casati, Robert, 82n, 228n
categorematic vocabulary, *101*.
    See also syncategorematic vocabulary.
caterpillar, 15
cause and effect, 2, 147
change, 73, 89–91, 139, 174, 188–189
chess set parts, 228
Chinese, 5, 245, 254–255, 265
circular time, 17, 231
classical predicate logic, 4, 5
    metaphysical basis of, 71–72, 234–235
    timelessness of —, 70–72
    with internal structure of atomic predicates, 114
    quantifying times and locations in—, *206*

classical propositional logic, 4, 5, 14
    axioms for —, 26, 105
    metaphysical basis of —, 231
    truth in —, 12
    with temporal connectives. *See* **TPC**; **TPC-linear**.
classification predicate, 85, 126–127, 181
clock time, 11, 156
closed interval, *18*
closed wff, *57*, *99*, *194*
    is a proposition, 104, 205, 252–253
closure of a wff. *See* universal closure of a wff.
coherence theory of truth, 1n
collections. *See also* extensionality; set theory.
commas, 86
commutativity of ∀ in temporal propositional logic with index quantification, *60*
commutativity of ∧, *202*, *209*
commutativity of +, *203*, *209*
commutativity of &, *203*, *210*
commutativity of ∪, *204*, *210*
compactness theorem, 61
comparatives, 137
complement of a location, *214*
complete collection of assignments of references, *102*, *199*
complete collection of wffs, *28*
complete substitution instance, *58*
completed action, 53, 127–128, 160n, 263
completeness theorem,
    for **QiTCL**, 60–61
    for **QT**, 108
    for **QT+internal**, 120
    for **QTS**, 211
    for **TPC-linear**, 33
    for **TPC**, 50
compositionality, 253
compound proposition/wff,
    comparing times of—, 40–42
    in temporal and spatial predicate logic, *194*
    in temporal predicate logic, *99*
    in temporal propositional logic, *20*, *21*
    in temporal propositional logic with index quantification, *57*.
    picks out a time?, 16
    See also atomic wff.
computable functions, 47
concepts, 252, 254

284   Index

conditional, does not pick out a time, 16
conjunction does not pick out a time, 16
conjunction, internal,
    of predicates, *118*
    of restrictors, 115, 118, 120
    of terms, 114, 117, 258
        duplicated reference in —, *203*
    See also disjunction of predicates.
connected regions, 177–178, 187, 228
connectives,
    classical —, *20, 190*
    external —, *190*
    internal —, *190*
    temporal —, *20*
consciousness, , 125, 232, 247. *See also*
    time, as experienced; subjectivity.
consequence relation. *See* semantic
    consequence; syntactic consequence.
consistent collection of wffs, *28*
constituent, physical, 229
context of utterance, 130–132
continuity of existence in time, 94–*95, 103,*
    107, 123–124, 140, 186, *201*
continuous (progressive) tense,
    future, 161
    past, 161
    present, 127–128, 162
    See also tenses.
conventions on formalizing,
    "after" as the reverse of "before", 35
    "not", 136
    past-present-future, 122, 129
    relative uses of "all", 143
    relative uses of "some", 144
    tenses, 127–129
    See also abbreviations.
co-ordinate system for space, 171, 216

death, 96–97
decidability of **TPC-linear**, 25
deduction theorem, *29*
default time indication, *242*
default truth-value, falsity is, 243.
    See also nonsense.
degree of conjunction of terms, *117*
degree of formal modifier, *117–118*
degree comparisons, 137

deleting parentheses,
    in temporal predicate logic, 100
    in temporal propositional logic, 20
dense ordering, *46*–47, 113
density of space, 215
dentist's office, 158
descriptions,
    for parts of space, 171–173
    for parts of time, 78
    of the world, propositions are, 1, 10–12
descriptive names, 267–269
designating. *See* reference.
desires, 250
determinism, 75–77
Diodorus Cronus, 246n
direct object, 251, 258, 261. *See also* obj (—).
discrete ordering, *46–47*
disjunction does not pick out a time, 16
disjunction of predicates, 114, 118–120
disposition, 71, 163
    for meaning, 254, 261
    to discriminate, 255
distant in time compared to distant in space,
    168–170
distribution of $\forall$ in temporal propositional logic
    with index quantification, *60*
downward and upward closure of existence in time,
    94, 103
    in a location, *186, 201, 208*
downward and upward closure of truth in time,
    87, 103, 200
downward and upward closure of truth in time in
    a location, *182, 201, 208*
duplicated reference in conjoined terms, *203, 209*
"during", 131

electrons, 18, 43, 140, 247. *See also* physics; science.
emotions, 251
endings. *See* beginnings and endings.
endpoints of an interval, 18, 123
equality predicate, *91, 98, 190*
    axioms for —, *107*
    See also index of notation ≡; identity.
equality predicate for locations, *190*
    axioms for —, *207*
equality predicate for times, *91, 190*
    axioms for —, *107*
    characterizing, 112

equiform word or sentence, 1–2.
　　See also sentence type.
equivalence class, 31
equivalence of ∪ and ∨, *204*
equivalence relation, *30*
Eskimo language, 92
essential part of physical thing, *227*
establishing relative times with before and
　　after, 10-13, 233–234
Evans, Vyvyan, 232
events, 85n, 128n, 231–232, 249
event-based time, 125
evidence used for referring, 74, 169
excluded middle, 76, 140–141
existence, 133–137
　　different kinds?, 93
　　downward and upward closure of — in
　　　　time, 94, *103*
　　downward and upward closure of — in
　　　　time at a location, *186*
　　in classical predicate logic, 70–72, 235
　　in space and time, 186–187, 238
　　in time, 79, 89–91, 93–96, 123–124, 236
　　outward closure of — in space, *186*
　　supratemporal, 73, 91
　　See also predicate is positive for existence.
existence at a time and location, then
　　existence at that time, *186, 208*
existence at a time, then existence at some
　　location at that time, *186, 208*
existence predicate,
　　modifiers and, 115
　　temporal —, *93*, 236
　　temporal and spatial —, *186*
existential quantifier. See quantifier(s).
experience,
　　abstracting and —, 189, 214
　　beginnings and endings and —, 233–234
　　computer programs, poems, and —, 229
　　identity and —, 89–90
　　reference and —, 235
　　reversibility of time and —, 46
　　stability imposed on —, 174
　　space independent of —?, 237
　　static vs. process and —, 85
　　things and —, 234–235
　　things needed for —?, 266
　　time as given by —, 231–232

experience (continued)
　　time as linear and —, 111
　　time as mass and —, 78–79, 165, 232–233
　　See also consciousness; subjectivity; time,
　　　　as experienced; time, objective.
extension of a predicate, *58*, 235
extensionality of predicates, 90
　　axioms for, *106*
　　in temporal and spatial predicate logic, 175,
　　　　200, 207, 217
　　in temporal predicate logic, 91, 96, *103*
external logical symbols, *117*

fairy tales, 139
false proposition picks out no time, 11
falsity is the default truth-value, 243.
　　See also nonsense as false.
fears, 259–260
fiction, 74–75
filling a blank with a term, *98, 151, 220*
"finished", 36
finite model, *25*, 46–47
fixed time marker(s), *80*, 84, 112, 113, 156
form of a proposition, 3–4.
formal atomic predicate, *118*
formal conjunction of terms, *117*
formal language,
　　of classical predicate logic with quantifying
　　　　over times, 98–99
　　　　with internal structure of atomic
　　　　　　predicates, *117–119*
　　of classical predicate logic with quantifying
　　　　over times and locations, *190–194*
　　of classical propositional logic with temporal
　　　　connectives, 20
　　of classical propositional logic with temporal
　　　　connectives and quantifying over relative
　　　　times via indices, 57
formal logic, *3*
formal modifier(s), *117–118*
formalizing. See conventions on formalizing;
　　meaning axiom.
free choice, 247–248
free index variable, 57
free individual variable, 99
free location variable, *194–195*
free time variable, 99
"from — to —" 226

286   Index

future. *See* future tense; past-present-future divide.
future tense, 51–53
   continuous —, 161
   convention on formalizing —, *128*
   perfect —, 161–162
   simple —, 127–128
   See also determinism; past-present-future divide; tenses.

Gale, Richard M., 91–*92*
geometry, 177, 214–215, 226.
   See also mathematics; science.
Glieck, James, 233
God, 248
Goranko, Valentin, 19
grammar, 6, 249, 254

habitual tense, 160n, 163–164.
   See also tenses.
half-open, half-closed interval, *18*
Hallowell, A. Irving, 266
Hamblin, C. L., 19
Hansen, Chad, 254–*255*, 259n
"hardly ever", 149
"has". *See* "possesses".
Hasle, Per F. V., *153–154*
Hausa, 92
holes, 228
Hopi, 80, 232, 234, 245n

ideas, 168
identity of individuals,
   across possible worlds, 140n
   criteria of —, 89–91, 94,–95 140n
   equality predicate and —, 89–92
   locations can't be used to define —, 189
   predicate logic criterion of —, 189n
   timeless, 70
   See also equality predicate; equality predicate for locations; equality predicate for times.
"iff" is short for "if and only if"
"in front of" 221–222
index-all rule, *60–62*
index for a sentence type, 12–13, 57
   colors instead of numerals, 37
index variable, *57–58*

individual. *See* individual thing.
individual things, 5, 70, 78–80, 165, 172
   See also existence, predicate is positive for; identity of individuals; physical thing.
individual thing in time, 79, 95, 237
Indo-European languages, 92
inference, 2
infinite model, *25, 46–47*
infinitive form of a predicate, 85–86
   simple, *101, 196*
infinitive symbol, 98
infinitives used as restrictors, 250–252, 256–257, 259, 264
   are ambiguous, 257
instant of time, *21*, 78, 109–110, 122–123, 125, 140, 159, 161, 235
intensional context, 253, 260–261
intentional word, 251, 255
intentions, 2n, 252, 258, 265
internal logical symbol, *117*
internally conjoined atomic predicates, *118, 192*
internally disjoined atomic predicates, *118, 192*
intersubjectivity. *See* subjectivity.
interval of an ordering, *18*, 48–49
   closed, *18*
   endpoints of, *18*
   half-open, half-closed, *18*
   open, *18*
interval(s) of time, 83–*84*
   beginning and ending of an —, 15–*16*, 48–49
   established by a proposition, 15
   not assumed to be a time, 84
   times are —, *84, 102*
inward closure of falsity for locations, *182*

Jespersen, Otto, *92*

knowing, 259
Kretzmann, Norman, 15n
Kroon, Fred, 169
Kuhn, Steven T., 242n

lamb, Mary's, 217
language, ordinary, *passim*.
laws of nature, 247
Lee, Dorothy, 51n
"left of" 222–223

length of time comparison, *156*
length of wff,
  in temporal and spatial predicate logic, *193*
  in temporal predicate logic, *99*
  in temporal predicate logic with internal structure, *119*
  in temporal propositional logic, *20*
  in temporal propositional logic with index quantification, *57*
liar paradox, 132
linear ordering for time, *17*, 48, *122*, 243. *See also* non-linear time; quasi-linear ordering.
local logic, *25*
location(s),
  as things, 175, 238
  created by picking out, 171–173
  distinguish physical things, *189*
  given only by relations of things?, 266
  minimal —, 180, 188–189
  universe of —, 172
location markers, 168, 182
location name symbols, *173, 190*
location-orienting predicates, 221–226, *223*, 239
location terms, *173*
location variables, *173, 190*
locational connective?, 237
locational predicate, *184, 201*, 238
  is positive for existence in time and location, *187, 201, 208*
logic. *See kind of logic*, e.g., classical propositional logic.
logic and reasoning. *See* reasoning.
logical vocabulary. *See* syncategorematic vocabulary.
Łukasiewicz, Jan, 76

mass nature of space, 171–173, 176, 212, 237
mass nature of time, 78, 109, 110, 149, 232–233, 235 *See also* dense ordering.
masses, 78, 109, 168, 172, 266.
mathematics, 62, 70, 137n, 171, 177, 213
  conception of time in —, 233.
  See also science.
maximal interval of a true predication, *123*
meaning, 251–252, 253–254
  dispositions for —, 254

meaning axiom, 71–72, 97, 135–136, 146, 184, 219, 224
measuring time,
  in temporal predicate logic, 148, 155, 156–159
  in temporal propositional logic, 11, 45
  See also clock time; fixed time marker.
medieval logic, 15n, 153–154, 254n
memory, 12, 169, 247
Mende, 92
mental states (objects), 252, 254, 255, 259–260, 261, 265
metalogic not distinguished from logic, 244
metaphors, time for space, space for time, 237
metaphysics, *passim*
  as basis of formal logics, 231–239
  minimal, 231
metavariables. *See the Index of Notation.*
mind-body distinction, 254–255
minimal location, 180, 188–189
minute as measure of time, 157
model(s), 4
  finite —, *25*
  infinite —, *25*
  of pure language of time, *105*
  of true at a time, *105*
  reduced model, *24*
  temporal and spatial predicate logic, *205*
  temporal predicate logic —, *104*
  temporal predicate logic with internal structure, *119*
  temporal propositional logic —, *22*
  See also sufficiency of collection of models.
modifier, predicate, 114–120, *118*
  formal, *117–118, 191*
  need not be omnitemporal, 151
  need not be omnilocational, 220
  used omnitemporally, 151, 220
modus ponens, *28*, 60
Montanari, Angelo, 19
moon, 156
"most", 136, 149
mountain lion, 97
"much", 149
mud, 78, 165, 168, 171, 232
Muybridge, Eadweard, *181*
Myro, George, 92n

name(s),
    continuity of existence in time and —, 95, 140
    descriptive —, 267–269
    of locations, 171
    of times, 79–80
    of individuals reflect belief in supratemporal things, 73, 91
    fictional —, 269
    refer, 133
    non-referring —, 243, 269
    simple —, *101, 196*
name symbol
    location —, *173*
    time —, *80*
naming,
    individuals, 73–75
    times, 79–80
"nearly every", 149
Neg, 202, 208
NEG-time 115
negation, does not pick out a time, 16, 42
negator of predicate, 114–120, 190–191, 257–258
    pure —. *See* pure negator.
negators of restrictors and pure negated restrictors, *202, 204, 209*
"never", 148
Newton, Isaac, *231*, 233, *237*
"next to", 225
NN′, 202, 209
non-linear time, 48–50. *See also* branching time; circular time.
nonsense, 77
    as false, 184, 243
"not", convention on formalizing, *136*
now, 122, 129–130, 131, 236–237, 247
numbers, 43, 79, 168
numerals
    as adjectives, 137
    for indices, 37, 63

obj (—), *114, 117, 190*
objectification, 125
objectivity. *See* subjectivity.
oblique context, 253
Øhrstrøm, Peter, *153–154*
omnilocational predicate, *184*

omniscience, 248
omnitemporal modifier, 151
omnitemporal proposition, *43*, 137, 138, 140, 142–143, 146
open interval(s), *18*
    in semantics of **TPC-linear**, 33–34
open wff, *57, 99, 194*
    as restrictor, 255–256
ordering(s), *17*, 81
    anti-reflexive ,*17, 81*
    anti-symmetric, *17, 81*
    transitive, *17, 81*
    linear, *17*
    temporal, *81*
    with endpoints, 18–19
ordinary atomic predicate,
    formal —, 98, *118–119, 193*
    simple —, *86, 196*
    See also existence, predicate is positive for.
orientation, spatial, 221–226, 266
outward closure of existence in space, *186, 201, 208*
outward closure of truth for locations, *182, 201, 208*
overlapping locations, 176–178
overlapping times, 38, 82–83
    of maximal intervals is a time, *124*
    not related in temporal ordering, *83, 102,* 198

Pagliuca, William, 245n
pains, 260
parallel timelines, 110, 214
parentheses, deleting,
    in temporal and spatial predicate logic, *196*
    in temporal predicate logic, *100*
    in temporal propositional logic, *20*
Park, David, *232*
parts of physical things, 227–229. *See also Appendix 6, "Parts of Things", in Volume 1.*
part(s) of time, 10, 78–79
    are related to other times as the whole is related, *83, 102*
    descriptions of —, 78
parts determine locations, *176, 199, 207*
parts determine times, *82, 101, 107, 198*
part-whole relation,
    for physical things, 227–229
    for space, *176, 199*
    for times, *82, 198*
past, 51–52, 73

past tense,
   continuous, 161
   convention on formalizing —, *127–128*
   perfect —, 53, 160
   simple, 127
   See also tenses.
past-present-future divide, 73–75, 75–77, 92, 125, 234, 236–237, 242–246
   conventions on formalizing, 122, 129
paying attention, 78–79, 113, 125
PC, *26, 105*
perception, 231. *See also* experience.
Perkins, Revere, 246n
physical thing(s), *168, 175, 238*–239
   distinguished by locations, *189, 202, 208*
   notion of — is basic, 216
   parts of —, *227*–229
Physical Things, the World, and Propositions, *175*
physics, 140, 147–148, 174. *See also* science.
picking out. *See* pointing; reference.
plans, 258
platonist (platonism), 172
poems, 229
Poincaré, Henri, 266
point in space, 78–79, 171, 177, 212–213, 215
pointing,
   to a location, 179–182
   to a time, 10
   to an individual thing, 73–75
   See also reference.
"possesses" ("has"), 138, 163
possibility, 139–140, 247–248.
Portner, Paul, 242n
predecessor, *46*
predicate(s), *234*
   applies to an object or objects in time, 87, 90, *103*
   atomic —, 72, *99, 192*
      modified —, *193*
      See also wff, atomic.
   conjunction of —, *118*
   extensionality of —,. *See* extensionality of predicates.
   formal atomic, *118*
   infinitive —, 85–86
   internally conjoined —, *118, 192*
   internally disjoined—, *118, 192–193*

predicate(s) (continued)
   modified atomic —, *193*
   ordinary —. *See* ordinary atomic predicate.
   simple —, *86, 101, 192*
predicate conjoiner, disjoiner, *117*
predicate is positive for existence, 96–98, 103, 133, 135–137, *200*, 238
   axioms for —, *108, 208*
   conjunction of terms, 152
   habitual predicates, 164
   location-orienting predicates, *223*
   locational predicates, *187, 201*
predicate negator. *See* negator of predicate.
predicate restrictor(s), 114–120
   meaning of, 251–252
   variable —, *117–118*
predication is timeless in classical predicate logic, 70–72, 235
preference, 163
present tense, 51–53, 122, 128–129
   continuous (progressive) —, 127–128, 162
   convention on formalizing —, *128*
   perfect continuous —, 162
   habitual —, 163
   simple —, 127, 162, 163
   used for future, 39, 131, 162
   used for past, 162
   See also now; tenses.
Prior, Arthur, 19, 52n, *92*, 242, *245, 246*
process predicate, 85, 126–127, 181
progressive tense. *See* continuous tense.
proposition(s), *1, 231*
   are linguistic, 1, 10
   are types, 1–*2*, 234
   as description of the world, 1, 10, 12
   as description of how we do or should use words, 1n
   as indexed sentence type, 12–13
   as tokens, 132, 234
   becomes true?, 11, 242
   closed wff used as restrictor is not a —, 252–254
   form of —, 3–4
   scheme of —, 3–*4*, 12, 242–246
propositional operators, 19, 87, 92n, 242–246
puppy, 71
pure language of space, *197*
pure language of space and time, *197*

pure language of time, *104*–105
pure logic of time, 104–*105*
pure negated predicate, *118*
pure negated restrictors, *202, 209*
pure negator, *116*, 120
   and negation, *120, 204–205, 211*

Q*i*TCL, *59*
QT, *104*
QT+internal, *120*
QTL, *124*
QTS, *206*
"qua", 263
quantifier(s),
   binds a variable in predicate logic, *99*
   binds an index variable, *57*
   over indices, 55–*57*
   relativizing —, 142–145
   scope of —, *99*
   timeless, 70, 91, 93
   unique existential —, *129*
quasi-linear ordering, *122*, 124
quasi-trichotomy, *122*
Quine, W. V. O., 70n

Raum an Sich, 237. *See also* space;
   Time as Such.
realization,
   temporal and spatial predicate logic,
      196–197
   temporal predicate logic, *101*
   temporal predicate logic with internal
      structure, *119*
   temporal propositional logic, *21*
   temporal propositional logic with index
      quantification, *59*
reasoning, 1
   abstracting from —, 5
   classical propositional logic and —, 231
   — together, 10
reduced model, *24*, 47
reference (designating, picking out),
   evidence used for —, 74, 169
   existence and —, 70
   extensionality of predications and —, 217
   names refer, 132
   pure —, 79, 172n
   supralocational, 174–175, 238

reference (continued)
   supratemporal, 73, 174–175, 238
   taking account of time for —, 73–75
   timeless, 70–72, 89, 93, 96, 235
   to a time, 79,
      independent of other times and individual
         things, 79
      independent of other times, individual
         things, and locations, 174–175
   to a location, 174–175, 266
   to things distant in space, 168–170
   to things in the past or future, 73–76,
      168–170, 235
   See also descriptive names; pointing.
reflexive relation, *82*
relation,
   anti-reflexive ,*17, 81*
   anti-symmetric, *17, 81*
   reflexive, *82*
   transitive, *17, 81*
relative adjective, 115
relative time, 10–13, 233–234
relativizing quantifiers, 142–145
Rennie, M. K., 96n
RES, *202, 208*
RES-time 114
Rescher, Nicholas, 50n, *75n*, 268n
restrictor(s) of a predicate,
   and +, *203, 210*
   and ∪, *204, 211*
   conjunction of —, 115, 118, 120, 191
   infinitives as —, *250–252, 256–257, 259, 264*
   negator of —, *202, 204, 209*
   simple —, *196*
   time indication as —?, 87–88
   wffs used as —, 252–254, 255–264
reversibility of time, 46
Riesen, John J., 173
RN, 202, 209

SAE (Standard Average European) language, 125
satisfaction of a wff,
   in temporal and spatial predicate logic,
      200–205
   in temporal predicate logic, *103–104*
saying, 130–132, 263–264
scheme of propositions, 3–*4*, 12, 242–246
Sciavicco, Guido, 19

science, 46, 70, 140, 147–148, 189.
   See also electron; physics.
scope of a quantifier
   in temporal and spatial predicate logic, 194
   in temporal predicate logic, *99*
   in temporal propositional logic, *57*
scope of an index quantifier, *57*
second-order logic, 84
semantic consequence,
   temporal propositional logic, *22*
   temporal predicate logic, 104
semi-formal language,
   temporal and spatial predicate logic,
      196–197
   temporal predicate logic, *101*
   temporal propositional logic, *21*
sentence type, 1, 11–13, 20
   times of —s distinct or equal, 16
sentence-type symbol, *57*
set theory notation, 18
set theory, 18, 81, 251
sidereal time, 159
sign language, 5
simple infinitive, *101, 119*
simple modifier, *119*
simple name, *101, 119*
simple ordinary temporal atomic predicate,
   86, 118
simple predicate, *101, 119*
simple present tense. *See* present tense,
   simple.
"since", 39
Sinha, Chris (et al.), *231–232*, 237n
Sinha, Vera da Silva, 80, 125n
social construct, 183
"some", 142–143
space
   as absolute, 237
   as mass, 171–173, 176, 237
   as objective, 237
   as pointillistic, 237
   as subjective, 237
spatial propositional connectives?, 237, 266
stability vs. flux, 166, 174, 239
standard clock, 11, 156
strong inference, 2
subjectivity, 12, 46, 125, 157–158, 162,
   231–235

substitution for a variable in temporal and spatial
   predicate logic, *195*
substitution for a variable in temporal predicate
   logic, *100*
substitution instance, *58*
substitution of equivalent propositions, 33
successor, *46*
sufficiency of collection of models,
   temporal and spatial predicate logic, *205*
   temporal predicate logic, *104*
   temporal predicate logic with internal
      structure, *120*
   temporal propositional logic, *22*
supratemporal, things are, 73, 89–92, 165, 174,
   235, 239. *See also* atemporal.
symphonies, 168, 216, 229
syncategorematic vocabulary, 93, *101, 197*
   existence predicate is —, 93
   time-ordering predicate is —, 81
   within-time predicate is —, 82
   See also categorematic vocabulary.
syntactic consequence
   temporal predicate logic, *104*
   temporal propositional logic, 26
syntactic deduction theorem for temporal
   propositional logic with index quantification, 60

Talmy, Leonard, *226*
tautology,
   temporal and spatial predicate logic, *205*
   temporal predicate logic, *104*
   temporal propositional logic, *22*
Taylor, Richard, 75n, 168n
TCL, *59*
temporal proposition, *12*
   false — does not establishes a time, 12
   only atomic wffs are —, 16
   true — establishes a time, 11, 12
   truth-value taken as primitive, 12–13
tenses, 5, 51–53, 122, 125, 160–162, 234, 237
   are indexicals, 130
   as part of atomic predicate, 129
   convention on formalizing —, *127–128*
   logic of —?, 126
   See also particular tenses, e.g., future
      tense.
term(s),
   conjunction of —, 114, 117, 191

term(s) (continued)
   fills a blank, 91, *98, 119, 193*
   free for a variable, *100, 195*
   individual —, *98, 190*
   location —, *173, 190*
   time —, *98, 190*
theory, 28
thing (individual thing),
   at-a-time, 90–92
   exists at a time and location, then exists at that time, *187, 201*
   exists at a time, then exists at some location at that time, *187, 201*
   exist in time, *103, 107, 200*
   events are?, 249
   in time, 79, 95, 237
   is supratemporal, 73, 89–92, 165, 174, 235, 239.
   names and —, 91
   parts of —. *See* parts of physical things.
   physical —. *See* physical things.
   world made up of, 70, 79, 234, 249
   See also identity of individuals; universe of individual things.
Things Exist in Time, *93, 103*
Things in Time, *93*
Things in Time, the World, and Propositions, *79*
thing-at-a-time, 90–92
thoughts, 1, 250, 259–260
"throughout", 130
time,
   absolute, 231. *See also* time, objective; space, as absolute.
   as experienced. *See* subjectivity. *See also* experience.
   branching. *See* branching time
   independent of us. *See* time, objective?
   mass nature of —, 78, 109, 110, 149, 232–233, 235
   mathematical conception of —, 233
   measures of, 11
   objective?. *See* subjectivity.
   part of —, 10, 78–79
   real?. *See* subjectivity.
   reversibility of —, 46
   subjective?. *See* subjectivity.
Time as Such, 232–233

time assignment, *21*
timeline in temporal propositional logic, *21*
times,
   are intervals, 15–*16*, 84, *102, 199*
   as pointillistic, 233
   as things, 79, 175, 235–236
   descriptions of —, 78–79
   established by true propositions, 10–15
   fictional —, 269
   indexed sentence type, time of, *16*, 234
   name symbols for —, *80, 98*
   named —, 79. *See also* fixed time marker.
   pointing to —, 10, 233–234
   referring to —, 79, 174–175
   relative —, 10, 231
   See also beginnings and endings; terms.
tokens, propositions as, 132
topology, 177
"towards", 225
TPC, *49*–50
TPC-linear, *22*
   is decidable, 25
   is local, 25
   with open intervals, *33*–34
transitive relation, *17, 81*, 101, 102
transitive verb, 251
trichotomy, *17*
   quasi-trichotomy, *122*
true at a time and location, then true at that time, *182*
truth, 1
truth in a location, 179–182
truth in time, 12–13, 86–87, 236
   downward and upward closure of —, *87*
truth-value, 1
   as a primitive, 12–13, 236, 251
"twice", 43
type, *1*
   proposition is a —, 1–2, 12
   sentence —, 1, 11–13, 20
   word is a —, 1–2

unique existential quantifier, *129*
universal closure of a wff,
   temporal and spatial predicate logic, *195*
   temporal predicate logic, *101, 195*
   temporal propositional logic with index quantification, *58*

universal instantiation in temporal propositional logic with index quantification, *60*
universal quantifier. *See* quantifier(s).
universe of individual things, 101, 198
    does not take account of when things exist, 70–71, 165
    not a collection of things, 172
universe of things in time, 74–75, 89
universe of things-at-a-time, 90–91
universe of locations, 173, 198
universe of times, 80, 101, 198
Urquhart, Alasdair, 50n, *75n*, 268n
"used to", 136
utterance, 1

vagueness, 15
valid inference, *2*
    formalized as a conditional, 150, 152, 154
    temporal and spatial predicate logic, *205*
    temporal predicate logic, *104*
    temporal propositional logic, *22*
valid wff,
    temporal and spatial predicate logic, *205*
    temporal predicate logic, *103–104*
    temporal propositional logic, *22*
valuation,
    temporal predicate logic, *103–104*
    temporal predicate logic with internal structure, *119*
    in temporal propositional logic, *21*
variable(s),
    bound —, *99*
    fills a blank, *98*
    free —, *99*, 95
    individual —, *98, 190*
    location —, *173*
    term is free for a —, *100*
    time —, *80, 98*
Varzi, Achille, 82n, 228n
verbs, transitive vs. intransitive, 251

Waismann, Friedrich, *77–79, 140*
water, 10, 78, 165, 168, 171–172, 232
well-formed formula (wff),
    applies to object(s), *103*
    compound, *99*
    false in a model, *104*
    temporal and spatial predicate logic, *193–194*
    temporal predicate logic, *98–99*
    temporal predicate logic with internal structure, *119*
    temporal propositional logic, *20*
    temporal propositional logic with index quantification, *57*
    true in a model, *104*
    true of object(s), *103*
    used as restrictor, 252–254, 255–264
    See also atomic wff; universal closure of a wff.
wff. *See* well-formed formula.
"while" 38–39
Whorf, Benjamin Lee, *5–6*, 51n, 80, *125, 232, 234*, 245n
Williamson, TImothy, 245n
wishes, 260
within-relation,
    for locations, *176*
    for times, *82*
"within the time", 37
world, 1, 4
    is made up of things, 70, 79, 234, 249
    propositions are descriptions of —, 1, 10, 12

Zeit an Sich, 232. *See also* Raum an Sich.

ω-rule, 62

& is equivalent to time-indexed $\wedge$, *115, 120, 204*

www.ingramcontent.com/pod-product-compliance
Lightning Source LLC
Chambersburg PA
CBHW081801300426
44116CB00014B/2201